GEOMETRY
ANCIENT AND MODERN

MATHEMATICS TEXTS FROM OXFORD UNIVERSITY PRESS

Norman L. Biggs: *Discrete Mathematics*
A. W. Chatters and C. R. Hajarnavis: *An Introductory Course in Commutative Algebra*
René Cori and Daniel Lascar: *Mathematical Logic: A Course with Exercises, Parts 1 and 2*
Geoffrey Grimmett and Dominic Welsh: *Probability: An Introduction*
G. R. Grimmett and D. R. Stirzaker: *Probability and Random Processes*
G. H. Hardy and E. M. Wright: *An Introduction to the Theory of Numbers*
John Heilbron: *Geometry Civilized*
Raymond Hill: *A First Course in Coding Theory*
D. W. Jordon and P. Smith: *Non-Linear Ordinary Differential Equations*
Jiří Matoušek and Jaroslav Nešetřil: *Invitation to Discrete Mathematics*
Tristan Needham: *Visual Complex Analysis*
H. A. Priestley: *Introduction to Complex Analysis*
H. A. Priestley: *Introduction to Integration*
Ian Stewart and David Hall: *The Foundations of Mathematics*
W. A. Sutherland: *Introduction to Metric and Topological Spaces*
Dominic Welsh: *Codes and Cryptography*

For further information on these and other titles
Email: science.books@oup.co.uk
Or Fax: +44(0)1865 267782

Geometry
Ancient and Modern

J. R. Silvester

Department of Mathematics
King's College London

OXFORD
UNIVERSITY PRESS

OXFORD

UNIVERSITY PRESS

Great Clarendon Street, Oxford OX2 6DP

Oxford University Press is a department of the University of Oxford.
It furthers the University's objective of excellence in research, scholarship,
and education by publishing worldwide in

Oxford New York

Athens Auckland Bangkok Bogotá Buenos Aires Calcutta
Cape Town Chennai Dar es Salaam Delhi Florence Hong Kong Istanbul
Karachi Kuala Lumpur Madrid Melbourne Mexico City Mumbai
Nairobi Paris São Paulo Shanghai Singapore Taipei Tokyo Toronto Warsaw

with associated companies in Berlin Ibadan

Oxford is a registered trade mark of Oxford University Press
in the UK and in certain other countries

Published in the United States
by Oxford University Press Inc., New York

A catalogue record for this book is available from the British Library

Library of Congress Cataloging in Publication Data

Silvester, John R.
Geometry: ancient and modern / J.R. Silvester.
p. cm.
Includes bibliographical references and index.
1. Geometry. I. Title.
QA445 .S545 2001   516—dc21   00-067606

ISBN 0 19 850758 5 (Hbk)
     0 19 850825 5 (Pbk)

Typeset using the author's LaTeX files by
Newgen Imaging Systems (P) Ltd., Chennai, India
Printed in Great Britain
on acid-free paper by
T.J. International Ltd, Padstow

This book is dedicated to the memory of
John Alfred Tyrrell BSc PhD AKC
1932–1992

Preface

This book is addressed to those who did little or no geometry at school, or who did some geometry a long time ago and would like to recall it or take it further. The only background knowledge required at the start is sixth-form mathematics (A-level), though as you go along you will need increasing amounts of other mathematics. The book does *not* attempt to be self-contained, or to be particularly systematic, and it borrows ideas and techniques from analysis (limits, continuity, convergence, calculus) and from algebra (group theory, linear algebra) which must be part of any undergraduate mathematics programme, and which ideally would be being studied alongside this material. What this book *does* do is to try to get you going, with pencil and paper, at drawing, reasoning and calculating. If you can get a buzz from solving some of the problems here, or from the satisfaction of understanding how things work or why they are true, then I have succeeded in my task. If you then go on and explore geometry further, or see new ways of thinking about other parts of mathematics, then I am delighted.

The book is based on a course I have taught for a number of years at King's College London. The course is taught to first year mathematics students, and covers most of Chapters 1–5 and parts of Chapters 6–8. Chapters 9–10 are a piece of self-indulgence on my part, and simply extract some of the juicier bits from further geometry courses I have occasionally taught, while avoiding most of the hard work. My excuse for including this is that it might entice you to want to know more, and give you the courage to tackle the standard literature on the subject. The approach throughout is unashamedly old-fashioned, because that is the only way of starting geometry without a horrifying shopping-list of technical prerequisites; and also it is the route into the subject which most present-day geometers actually followed themselves.

There are fashions in mathematics, as in anything else. Euclid, the mainstay of English mathematical education until only a generation or two ago, fell out of fashion, and by the time I had finished my secondary education, geometry itself was being replaced by set theory and group theory and other 'new mathematics'. (It was set *notation* and group *examples* in practice: there was precious little *theory*.) This in its turn has now all but disappeared from school syllabuses, and mathematics itself has largely been replaced—I am being cynical now—by something called data handling, and statistics, which are thought by politicians and other know-alls to be more important for industry and 'wealth creation'.

Today's budding mathematicians (Casio's children) seem to some of us old fogies to have been short-changed. Mathematics is not just about decimals—merely a way, in Tony Gardiner's words, of making all numbers appear equally boring—but about shape and space and abstract patterns of overwhelming beauty and complexity: a mental universe (Tony Gardiner, again), waiting to be explored. It is also—in fact, principally—about deduction and reasoning, and understanding not just *what*, but *why*. It is a lot more fun than data-handling, and a lot more challenging. Geometry provides a door into this magical world. Its problems, at first appealing in their apparent simplicity and inevitability, turn out on deeper investigation to lead us on the one hand to question the nature of the real world, and on the other hand to invent abstract apparatus of the most exotic kind (but mostly beyond the scope of this book, dear reader) to gain a proper understanding of what is going on.

An author has many people to thank, and not just for help or tolerance while immersed in his task. I should start by thanking my parents, and especially my mother, who made me want to study, and whose extra lessons at home got me ahead of the game in my early years. Then East Sussex County Council, who bought me (and every other 11-year-old in the county) a geometry set; what a pity this splendid practice has died out. My teachers in the sixth form at Hove Grammar School for Boys, Messrs. Tabrett ('Tabby') and Baxter, taught mathematics with an energy and enthusiasm that should be the model for teachers everywhere; and because I was preparing for Cambridge, they introduced me to the delights of inversive and projective geometry. At Cambridge I was converted into an algebraist largely by the inspirational lectures of Philip Hall, and I continued in this vein as a PhD student with Paul Cohn in London. Paul's way of doing things still seems to me to be absolutely the right way, and his three volumes [6], [7], [8] are never far from my elbow, and are referred to constantly in this text.

When I arrived at King's College, in 1969, they said 'Ah,—an algebraist!—you can teach complex variables'. It was years before they let me teach what I thought I knew about (algebra), but in the meantime I had become friends with John Tyrrell, who re-awakened my interest in geometry. John had an encylopaedic knowledge of mathematics in general and geometry in particular, and he was a patient and generous colleague and teacher. His lectures were the most popular in the department, with his easy, laid-back style, interspersed with quotations and witticisms, and his clarity of exposition (and handwriting). When I came, later, to teach some of the courses he had taught, I found I could never complete more than half the syllabus, though the students complained that I went twice as fast as he did. (They also complained about my handwriting.) We tried to get John to write a book based on his courses, but he steadfastly refused. This book, I suppose, is the one we tried to get him to write, and he would have made a better job of it.

I have been helped throughout by my colleagues at King's, among whom I mention especially Tony Barnard and Bill Harvey, with whom I have had many stimulating discussions about the best way to do this or that piece of mathematics; and Alan Pears, who carefully read an early draft and made meticulous and invaluable comments. Sometimes I have followed these people's advice, and sometimes I have been headstrong and gone my own way. I am grateful to my students for their comments, and have taken encouragement from those who said they liked my lectures; though modesty forces me to admit that this is probably due more to the attractiveness of the subject than to any particular skill on my part. I also wish to thank my colleague David Lavis for much help and advice on the computing aspects of this project, and especially for sharing his expertise on LaTeX and PSTricks.

Finally, my wife Jill has been helpful and patient throughout. One of the most helpful things she did was to preserve from her student days at King's a complete set of notes of John Tyrrell's courses, and I have raided these shamelessly. As to the patience, I hereby undertake to finish painting the verandah on the very next fine weekend.

King's College London   J. R. S.
July 2000

Contents

1

History and philosophy

Of all branches of mathematics, geometry has the most chequered and complicated history and offers the most bewildering choice of where to start and how to approach the subject. This chapter offers the briefest sketch of some of the more important stages in that history, together with some sort of justification for the particular approach to geometry in this book.

1.1 BEFORE EUCLID

Geometry had its beginnings in ancient Egypt, where tax was levied on land, and the constant washing away of boundaries by floods made it necessary to recalculate areas by measurement. The Egyptians developed practical methods for measuring areas such as rectangles and triangles, but gave no proof of any of their rules. Thales of Miletus (*c*.624–*c*.547 BC) went to Egypt and brought these rules back to Greece, and it was the Greeks who first set out to support such rules with proofs.

No one before them had thought of proving such a thing as that the two base angles of an isosceles triangle are equal; the idea was an inspiration unique in the history of the world, and the fruit of it was the creation of mathematics as a science. [17]

The first proof of the above theorem (Fig. 1.1) is usually attributed to Thales, as is the discovery that the angle in a semicircle is a right angle (Fig. 1.2), but at this stage the

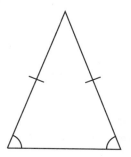

Fig. 1.1   An isosceles triangle

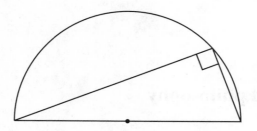

Fig. 1.2   The angle in a semicircle

main reason for developing geometry was still very practical, and Thales seems to have spent some time using the properties of triangles to measure the distances of ships from the shore. The study of geometry solely for its own sake began with Pythagoras, about fifty years after Thales.

1.2   EUCLID

Little is known about Euclid. He probably grew up in Athens, and was invited by Ptolemy to go to Alexandria in about 300 BC to teach in what was in effect the first university. Here he wrote the *Elements*, an account in thirteen books of the mathematics of the day. The earlier books of the *Elements* expound the geometry of Pythagoras, Hippocrates, and Eudoxus; later volumes are mostly concerned with arithmetic, and the work finishes with the proof of the formula πr^2 for the area of a circle, and the construction of the five Platonic (regular) solids. It is more or less obligatory at this point to repeat the much-quoted episode when Euclid was asked by a pupil what he would gain by learning geometry, and Euclid supposedly turned to an assistant and said, 'Give him threepence, since he must gain by what he learns'.

In order to demonstrate the truth of his theorems, Euclid starts with a list of *definitions* and *axioms* (some of which he calls *postulates*, but we shall not distinguish these from axioms). The idea here is that everything in the initial list should be incontrovertible, self-evident, and that everything that follows should be deduced from what has gone before. Thus he starts:

I. *A point is that which has no parts, or which has no magnitude.*
II. *A line is length without breadth.*
III. *The extremities of a line are points.*
IV. *A straight line is that which lies evenly between its extreme points.*

The tendency nowadays, when first confronted by this, is to start to poke fun. What are these *parts*, whose absence apparently will enable us to recognize a point? What is this *length* whose presence is necessary for us to recognize a line? And it is no good trying to patch it up by mentioning distance, for instance, because that has not been defined either, and a glance at IV. shows us that the line in II. is not necessarily straight! Even if we make allowance for when the *Elements* were written, and adopt a we-know-what-he-meant attitude, there are still difficulties, and the very first proposition in Book I contains a flaw. Here (Fig. 1.3) the construction of an equilateral triangle on a given base AB is achieved by drawing two circles, one through A, with centre B, and the other through B, with centre A. These circles meet in two points, and if one of them is chosen as C, then the triangle ABC is equilateral. But as Coxeter [9] and others before him have pointed out, although Euclid's axioms ensure the existence of the two circles, and (once C is found) the equality of the three lengths AB, BC, CA, there is nothing *in the axioms* to say that the two circles will meet at all.

1.3   PARALLELS AND NON-EUCLIDEAN GEOMETRY

The statement in Euclid that caused most concern, however, was the *parallel axiom*. The preceding axioms were mild affairs, asserting that any two points determine a straight line, that all right angles are equal to each other, and so on. The parallel axiom was much more complicated: in modern terms, it says that if two lines AC, BD lie in a plane, with C and D on the same side of the line AB, and if the angles CAB and ABD add up to less than two right angles ($\angle CAB + \angle ABD < 180°$), then the lines AC, BD produced (that is, extended) in the direction implied by the notation, will eventually meet (Fig. 1.4). It is easy to deduce from this that if $\angle CAB + \angle ABD > 180°$, then CA,

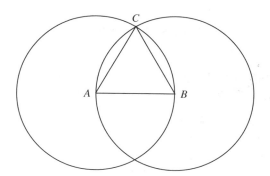

Fig. 1.3   Euclid's construction of an equilateral triangle

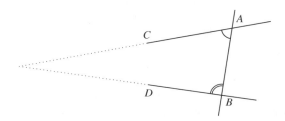

Fig. 1.4 $\angle CAB + \angle ABD < 180°$

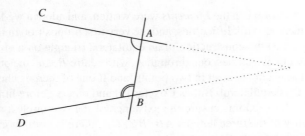

Fig. 1.5 $\angle CAB + \angle ABD > 180°$

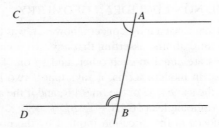

Fig. 1.6 $\angle CAB + \angle ABD = 180°$

DB produced—note the change of direction—will eventually meet (Fig. 1.5), and that if $\angle CAB + \angle ABD = 180°$ then *AC*, *BD*, produced in either direction by as much as you please, never meet, and are therefore parallel (Fig. 1.6). So a consequence of the parallel axiom is that, through a given point *B*, there is one and only one line parallel to a given line *AC*: just choose *D* so that *C*, *D* are on the same side of *AB*, and $\angle ABD = 180° - \angle CAB$. The parallel axiom seems more like a theorem than an axiom; Euclid does not use it until his 29th proposition, and then only to prove the converse of the preceding result. Many unsuccessful attempts were made to *prove* the parallel axiom from the other axioms, and not until the nineteenth century were these attempts shown to be futile by the demonstration that a structure obeying all of Euclid's axioms except the parallel axiom, and for which the latter was actually false, could be built *within* Euclidean geometry. As a consequence, any 'proof' of the parallel axiom from the others could be repeated within this structure, and so the parallel axiom would be both true and false within that structure, and Euclidean geometry itself would therefore be inconsistent. To put it another way, if you believe in Euclidean geometry, you are forced to conclude that you must believe in non-Euclidean geometry also. (There are two sorts of non-Euclidean geometry, *elliptic* and *hyperbolic*, according as you negate the parallel axiom by insisting that there are respectively *no* or *many* parallels to a given line through a given point. We shall have a look at them in Chapter 9.)

This discovery of non-Euclidean geometry was made independently by Bolyai, Gauss, and Lobachevsky, and was the beginning of modern mathematics. Other mathematicians started to investigate whether it might be possible to construct yet more 'geometries' by denying others of Euclid's axioms, which were now no longer regarded as *true* in any absolute sense. Indeed, pure mathematics could now be seen as a game in which we write down any set of definitions and axioms that we find interesting and see what we can deduce from them, and so, in Bertrand Russell's famous words,

mathematics may be defined as the subject in which we never know what we are talking about, nor whether what we are saying is true. [27]

If you are of a practical turn of mind you may protest that all this is quite absurd: we know how to draw lines and measure things, thank you very much, and we don't want to waste our time playing meaningless games. It is true that if you just want to put up a shelf in the garage, or navigate your ship across an ocean, then Euclidean geometry will serve you well. But when very large (astronomical) or very small (sub-atomic) distances are involved, modern science gives plenty of reasons to conclude that whatever is going on is not best explained in terms of Euclidean geometry.

In practical terms, geometry *is* absurd; the geometer is supposed to be able to tell that two lines are not parallel when they meet a hundred or a million times further away than the edge of the known universe, and that two points are distinct when their distance apart is a hundredth or a millionth of the size of the molecules in the lead of a pencil. Much emphasis is placed in school these days on estimation and measurement of length, as a practical exercise, and on the handling of the estimation of errors involved. These activities, which are undoubtedly vital, are *not mathematics*, in spite of the fact that they appear in the National Curriculum for mathematics. Not only does their presence there take valuable time away from mathematics, but the undue emphasis they receive gives the unfortunate impression to the pupils (and, dare one say it, to some of their teachers) that mathematics is not an exact subject, and that when we say (as mathematicians) that two things are equal, we just mean that they are near enough the same for present purposes, or that we cannot distinguish them with the particular measuring devices available. Nothing could be further from the truth!

1.4 HILBERT AND BIRKHOFF

The advent of non-Euclidean geometry led to a careful re-examination of Euclid's axioms, and in 1899 David Hilbert [18] produced a set of axioms for Euclidean geometry that left nothing to chance. The terms point, line, plane, on, between and congruent are *undefined*, and the axioms include the insistence that our geometry is not empty: any two points are on one and only one line, and there are at least three points not on the same line. The curious feature of this approach, to the newcomer, is that we do not start by assuming the existence of 'space', but instead introduce points, lines, planes, often just one at a time, by invoking the axioms (or propositions already proved from the same axioms). To quote [27] again, this gives the subject 'an air of almost wilful pedantry', though Russell clearly approves of this.

An alternative approach was put forward by Birkhoff in 1932 [2]. His undefined terms are *point, line* (lines being certain sets of points), *distance* $d(A, B)$ between points A, B, and *angle* $\angle ABC$ formed by three ordered points A, B, C. In contrast with Hilbert's

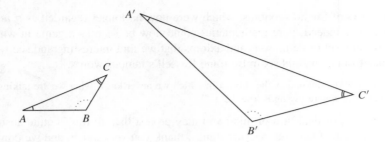

Fig. 1.7 Similar triangles

synthetic approach, Birkhoff's first axiom immediately gives us infinitely many points, as it states that the points $A, B, \ldots$ of any line can be put in $1 : 1$ correspondence with the real numbers by $A \mapsto x_A$, $B \mapsto x_B, \ldots$, say, such that for all A, B we have $|x_A - x_B| = d(A, B)$. With a few more definitions (*between, triangle, side, vertex, ...*), Birkhoff is able to construct plane geometry from only four axioms, the last of which asserts that if two triangles have corresponding sides in proportion (*similar* triangles), then their corresponding angles are equal (Fig. 1.7).

1.5 COORDINATE GEOMETRY

In 1637, René Descartes published his *Géométrie* [31], and showed how to represent points, lines and curves by cartesian coordinates and algebraic equations. This became known as coordinate or analytic geometry, which we now describe.

Imagine a plane placed vertically, like a blackboard fixed to a wall. Choose and mark a point O in the plane, the *origin*. Through O draw two lines in the plane, one horizontal (the *x-axis*) and one vertical (the *y-axis*). On each axis, mark a scale of measurement, as on a ruler; in modern parlance, regard each axis as a 'number-line'. Choose the sense of measurement so that, on the x-axis, positive numbers are to the right and negative numbers are to the left of O (with zero at O); and on the y-axis, so that positive numbers are above and negative numbers are below O (with zero at O, again).

Now take any point P in the plane. Through P, draw a vertical line, which meets the x-axis at a point representing a certain number x, which we call the *first coordinate* or *x-coordinate* of P. Through P, draw also a horizontal line, which meets the y-axis at a point representing a certain number y, which we call the *second coordinate* or *y-coordinate* of P. So P gives rise to an ordered pair of real numbers (x, y), its coordinates. Conversely, given any two numbers (x, y), we can draw a vertical line through the point representing the number x on the x-axis, and a horizontal line through the point representing the number y on the y-axis, and these lines will meet at a point P whose coordinates are (x, y) (Fig. 1.8).

This device enables geometrical problems to be turned into algebra, or trigonometry. In fact, rather than using coordinates as a way of labelling or keeping track of points, we shall start our geometry by turning the device on its head: we shall *define* the plane

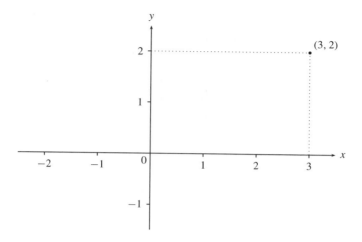

Fig. 1.8 Coordinates

in which we do our geometry to be no more and no less than the set of all ordered pairs (x, y) of real numbers. This neatly gets around all sorts of difficulties about the proper axiomatic basis for geometry, and pushes possible difficulties over from geometry into algebra, or analysis.

1.6 PYTHAGORAS: THEOREM OR AXIOM?

For reasons we shall explain shortly, Pythagoras' theorem is going to be the starting point for our development of geometry; it is going to assume the status of an axiom. For this reason, we shall now give outlines of a couple of proofs, from a naïve point of view, of the theorem of Pythagoras, that in a right triangle the square on the hypotenuse is the sum of the squares on the other two sides. In more modern terms, if a right-angled triangle has sides of length x, y, z, with hypotenuse z (the longest side, or side opposite the right angle), then $x^2 + y^2 = z^2$. The methods are rather different, the first having a cunning diagram and an easy proof, and the second having an easy diagram and a cunning proof, which serves to illustrate that in mathematics, as in most things, there is no free lunch.

Given the right-angled triangle then, as above, first draw *two* equal squares of side $x + y$, and subdivide them as in Fig. 1.9. Each of the two squares contains four copies of the original triangle, and various squares of side x, y, and z, and if we write Δ for the area of this triangle, and equate the areas of the two squares we immediately obtain

$$x^2 + y^2 + 4\Delta = z^2 + 4\Delta$$

from which the result follows on subtracting 4Δ from each side.

The second method requires the addition of only one line to the given triangle ABC, namely the perpendicular AD from A onto the hypotenuse BC (Fig. 1.10). It is easily

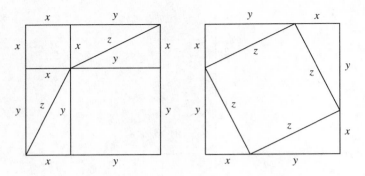

Fig. 1.9   Pythagoras' theorem, first proof

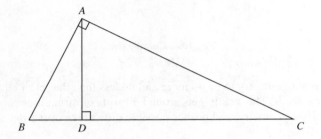

Fig. 1.10   Pythagoras' theorem, second proof

seen that the three triangles *DBA*, *DAC*, and *ABC* are *similar*, and since their three hypotenuses are of length x, y, z respectively, their *areas* are in the ratio $x^2 : y^2 : z^2$. But

$$\text{area}(DBA) + \text{area}(DAC) = \text{area}(ABC)$$

from which it follows that $x^2 + y^2 = z^2$.

In Euclid, the concept of distance was undefined: he wrote about lines (that is, line segments) being *equal*, meaning *of equal length*, apparently without worrying about how one was supposed to decide if this was the case in a particular instance. On the other hand, Birkhoff, as we have seen in Section 1.4, went to the trouble of building measurement of distance into his axioms, with rulers attached to every line in the plane. The particular feature of Descartes' geometry which we are going to exploit is that, once we have decided on a scale of measurement in just two directions, along the axes, Pythagoras' theorem gives us a way of measuring distance in any other direction. In particular, the distance from the origin $(0, 0)$ to the point with coordinates (x, y) is z, where $x^2 + y^2 = z^2$; that is, it is $\sqrt{x^2 + y^2}$. By a slight complication of this, one arrives at the formula

$$\sqrt{(x_1 - x_2)^2 + (y_1 - y_2)^2} \tag{1.1}$$

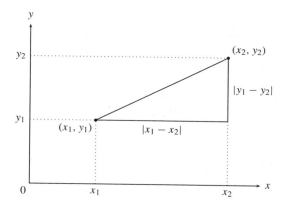

Fig. 1.11 Distance from (x_1, y_1) to (x_2, y_2)

for the distance from the point (x_1, y_1) to the point (x_2, y_2) (Fig. 1.11). We are going to turn this on its head and *define* the distance from (x_1, y_1) to (x_2, y_2) to be the quantity (1.1). Once this is done, it is no longer honest to offer a *proof* of Pythagoras' theorem, as it is assumed in the very structure of our geometry: it has become an axiom.

1.7 PHILOSOPHY OF THIS BOOK

Most of this book is about two-dimensional or plane geometry, and we are going to regard the plane as being

$$\mathbb{R}^2 = \{ (x, y) \colon x, y \in \mathbb{R} \},$$

where $\mathbb{R}$ stands for the set of real numbers. So our points have coordinates (x, y), and every such ordered pair of real numbers (x, y) corresponds to a point. If we wanted to start with one-dimensional geometry, which might be rather dull, then we should have to define the object of study as the line $\mathbb{R}^1$, or $\mathbb{R}$ for short, and a typical point as (x), or x for short, where $x \in \mathbb{R}$. The distance between the points x_1, x_2 we would want to be $|x_1 - x_2|$, and it is comforting to note that this is the same as

$$\sqrt{(x_1 - x_2)^2}, \tag{1.2}$$

cf. (1.1). (Of course, an unadorned square root will always mean the non-negative value: $\sqrt{4}$ means 2, not ± 2 and *certainly* not -2.) We shall make some excursions into three-dimensional geometry, and then we shall be studying three-space $\mathbb{R}^3 = \{ (x, y, z) \colon x, y, z \in \mathbb{R} \}$; points now need three real coordinates, and the distance from (x_1, y_1, z_1) to (x_2, y_2, z_2) is, naturally,

$$\sqrt{(x_1 - x_2)^2 + (y_1 - y_2)^2 + (z_1 - z_2)^2}. \tag{1.3}$$

The pattern is clear, and if we should ever need to venture into four-dimensional space, whatever that might be, all we need to do algebraically is add a fourth coordinate, and extend the distance formulae (1.2), (1.1), (1.3) in the obvious way.

The advantage of basing an approach to geometry on coordinates and algebra is that many of the difficulties are transferred to the algebra, and if we assume (as we shall) that we know most of what we need about real numbers, then many of these difficulties vanish. The danger of this way of doing geometry is that we never learn to use our eyes or think geometrically, being content merely to turn any geometrical problem into an algebraic one; books written like this tend to have their pages densely covered with algebraic equations, with rather few diagrams. We shall try to avoid this trap and have the best of both worlds, by getting away from the algebra as soon as and as often as possible; this may sometimes involve cheating a little, by consigning the duller algebraic calculations to the exercises, or by use of a well-worn escape clause such as *it is easy to show that* ... or maybe ... *details are left to the reader*.

1.7.1   Advice to readers

The exercises are a vital part of this (or any) mathematics text, and you will learn little if you do not attempt at least a good selection of them. They range from the insultingly easy to the challenging. Many of them are part of the flow of exposition, and we shall not hesitate to use them later on in the text as if we had proved them: ... *by Exercise 42 it follows that* ... , etc. From Chapter 3 onwards, solutions are provided to all the exercises; in Chapter 2 the exercises are of a different kind, and solutions are provided only where necessary.

You will need a good desk or table, a geometry set, sharp pencils, eraser, lots of paper, and perseverance. Don't be satisfied with any statement until you have tried it, tested it, and proved it. Intuition, and first impressions, can get you into trouble. If I ask you to come round and feed the cats while I am on holiday, and you turn up with some milk and a tin of Kitty-Chunks, you might be a bit upset if you find that actually I run a zoo. Moral: take nothing on trust, and read the small print, which in mathematics usually means study the definitions very, very carefully and don't just hope they mean what you think they ought to mean. When something is claimed to be true, don't just purse your lips, nod wisely, and pass on, but check that the claimed result follows inevitably from what has gone before. The best attitude is to assume the author is a lying toad, and see how many mistakes, ambiguities and misleading statements you can find in this book.[1] (Some of those you find will be unintentional, so please write and tell me about them and we'll try to avoid them in the next edition, if any.)

If you follow the above advice too literally, of course, you will read the whole book with your nose an inch from the page, and you will check the accuracy of hundreds of statements without ever having the faintest idea what is going on. To understand the flow of an argument, where it is heading, and why it is constructed the way it is, you also have to do the other sort of reading, where you try to keep going and don't stop to worry about every last technical detail. If you get stuck, by all means double back and read the last bit again, but also try reading ahead to see what is about to happen, and often you will find that this extra information helps you sort out your difficulty.

---

[1] Apart from anything else, this will help you develop what educationalists like to call a *transferable skill*. You can use it when listening to party political broadcasts, for instance.

With practice, you will not need to read every passage in both the slow painful and the fast what's-this-all-about-then way, because you will start to recognize familiar types of argument and you will know you *could* supply the technical details yourself, if you had to. But you should still stop and test yourself from time to time, to see if you really can.

That's more than enough of the avuncular advice. Let's begin!

2

Drawings and constructions

In this chapter, you are encouraged to draw for yourself lots of diagrams in order to get a feel for what geometry is about, to develop your visual skills, and to build up a catalogue of geometrical facts and experiences which will help illustrate and motivate later parts of the book. There are few theorems and no formal proofs in this chapter—they come later—but just exercises, interspersed with comments and hints. There are also no diagrams, as *you* are going to draw those!

Most of the exercises in this chapter simply ask you to follow a set of instructions and observe what happens, or verify some statement. Solutions to such exercises are *not* provided: the exercise *tells* you exactly what to do. Often you might wish to repeat this type of exercise with a different configuration of initial points, lines, etc., to see if the same thing happens again. Making it happen twice, or twenty times, or even twenty million times, is no proof that it will always happen, of course, and in any case there is precision to worry about: how good is your eye, how steady is your hand, and how sharp is your pencil?—so did what you thought you saw happen, actually happen at all? Many of these exercises also contain suggestions as to how you might construct a proof, so have a go before reading on to see if a solution is provided or (if not) if a proof is offered elsewhere in the book.

Some exercises call for ingenuity and cunning, and might ask you to construct certain points, lines, etc., that depend on the initial configuration, using your geometrical instruments or a specified subset of them, such as ruler and compasses, or ruler alone. Exercise 2.2 is a relatively straightforward example; there are more in later chapters, some quite a lot harder. Solutions to this sort of exercise *are* provided, but please don't cheat by looking before you have made an honest attempt of your own. Some exercises you are unlikely to solve at a glance, or at the first attempt, so persevere: the satisfaction of solving them yourself is well worth the effort.

2.1 EQUIPMENT

You will need a geometry set: ruler, compasses, protractor, and set squares. Then lead pencils (kept very sharp), eraser, and a supply of paper, preferably unlined and not too flimsy. If possible, use a drawing board to which you can attach your paper with pins or clips; but in any case make sure you have a hard level surface which no one minds having attacked by the point of your compasses. Using a polished table or desk is *not* a good idea!

There are many computer programs which will allow you to construct geometrical diagrams. The drawing programs that come free with computers usually do not allow you to specify coordinates of points or equations of lines: you just click with the mouse, and drag things around the screen until they are near enough where you want them. This is not really a great deal of use for mathematics.

Much better are programs such as Maple, Mathematica, and Derive, which handle algebraic symbols and can work arithmetically to arbitrary precision. To draw diagrams with these, you can specify coordinates of named points, and then use the program to find the equations of whatever lines, circles, etc., you need, in terms of these points. You then plot; and if you don't like the result, or if you just want to experiment, you can go back and alter the coordinates of the original points and run the calculations and plots again.

There are programs, such as *Cabri* and *The Geometer's Sketchpad*, which have been written specifically for exploring geometry, and combine features of both the above types of program. They allow you to choose arbitrary points, lines and circles by clicking with the mouse. They will then allow you to construct such things as the line through two points, the meets of lines and/or circles (see Exercise 2.8 and Definition 3.5), parallel lines, mid-points, bisectors, and perpendiculars. All of these are done by choosing from menus, and clicking with the mouse. Best of all, when you have finished you can use the mouse to drag the original points, etc., to new locations, and all the constructions that depend on these points will move in consequence. This will not give you a proof of anything, but it does give a good feel for what is going on, while keeping all the algebra hidden.

You are strongly advised to *start* with paper and pencil, nonetheless. You can always use a computer later, if you get bored or run out of paper or just want to save a forest or two.

2.2   RULER AND COMPASSES

Rulers as bought in the shops have scales marked on them, usually in centimetres and millimetres, or inches and fractions of an inch. We shall often use our rulers to measure length, but this is not what Euclid had in mind when describing constructions. The ruler is to be thought of simply as a straight edge, a device for drawing straight lines: it can be used for drawing an arbitrarily chosen line, an arbitrary line through a previously chosen or constructed point, or the unique line through two such points (which is called their **join**), and nothing else. This may seem rather silly, so please regard it as just a game, like chess. In chess, it might be fun to move your rook diagonally, but the game you would then be playing would not be chess, and your opponent would rightly have you denounced as a cheat. Likewise, in the game called ruler-and-compasses, measuring or comparing lengths (or other tricks, such as using both sides of the ruler to draw parallel lines) is cheating.

We label our points A, B, ...; and given two points A, B we write AB for their join. Sometimes this will mean the whole of the line through A and B, which we refer to as the *line AB*, and sometimes it will mean only that portion of the line which is between A and B, which we refer to as the *line segment AB*. The distance from A to B is denoted $|AB|$. The instruction *join AB* means you must add the line (or line segment) AB to your diagram. The points A, B separate the line AB into three parts. The middle part (between A and B) is the line segment AB. The part beyond B (on the opposite side from A) is called *AB produced*. So if we say C lies on AB produced, this means C is on the line AB, with B between A and C; and the instruction *produce AB to C* means you already have the line segment AB drawn, and now you need to draw a bit more of the line AB, and to choose a point C on it, with B between A and C.

- Ex.2.1:   *To construct an equilateral triangle. Given a line segment AB, place the compasses on A as centre and draw the circle through B; and similarly place the compasses on B as centre and draw the circle through A. These circles meet at two points; choose one, call it C, and join CA and CB. Is it clear why $|AB| = |BC| = |CA|$? Check with your protractor that $\angle ABC = \angle BCA = \angle CAB = 60°$. (Read 'angle ABC =', etc.)*

There is a similar restriction on the use of the compasses. These can be used to draw a circle, or arc of a circle, with an arbitrarily chosen centre or a previously constructed centre, and passing through an arbitrary point or a previously constructed point. But when the circle or arc is drawn, and the compasses are lifted from the page, Euclid assumed that they would spring shut, or go floppy, so that the same radius could not be transferred elsewhere on the page. These rules were adhered to in the above exercise; however, Euclid immediately went on to show that this restriction on the use of the compasses is unnecessary:

- Ex.2.2:   *Given a line segment AB and another point C with $|AB| < |BC|$, construct a point D so that $|CD| = |AB|$. Hint: use the last exercise to construct E so that $\triangle BCE$ (read 'triangle BCE') is equilateral. Now use the compasses with centre B to transfer the length $|AB|$ onto one of the lines BC, BE; think carefully before you decide which. There is then one more step.*

The restriction $|AB| < |BC|$ is unnecessary, but is put in simply to make the problem more straightforward, by reducing the number of different diagrams that need to be drawn to *one*.

We have shown, then, that lengths can be moved around using ruler and compasses, of the floppy variety. So there is no harm in assuming from now on that the compasses are *not* floppy and can be used like a pair of dividers to move lengths around, for any construction thus accomplished could also have been done with Euclid's floppy compasses.

There are practical problems you may encounter when trying some of the constructions in this chapter. You may find that your ruler or compasses will not reach as far as you need them to, or that a construction goes off the edge of your piece of paper. The practical solution is to start again, and reduce the size of your drawing, but this already involves the assumption that a change of scale will somehow not affect or distort the geometrical properties you are investigating. An alternative approach might be to give certain constructions for overcoming the problems. For example, can we construct the join *AB* even when the points *A*, *B* are further apart than the length of our ruler? This is the *short ruler problem*, and its solution is non-obvious. We shall leave discussing it until Section 4.9, where we shall develop the necessary technique. In the meantime, we simply assume that our ruler, compasses, and paper are as large as necessary.

2.3 BISECTION, PARALLELS, AND SUBDIVISION

- Ex.2.3: *Draw a line and mark two points A, C on it. With equal radius, more than $\frac{1}{2}|AC|$, draw two circles (or arcs of circles) with centres A and C, to meet each other in B and D. Join AB, AD, CB, CD and BD, and let AC, BD meet at O. The figure ABCD is called a* **rhombus***; it is a particular sort of parallelogram (that is, AB ∥ DC—read 'AB is parallel to DC'—and BC ∥ AD) with all four sides of equal length: $|AB| = |BC| = |CD| = |DA|$. Check, by measurement, that your rhombus has the following additional properties: $|AO| = |OC|$, $|BO| = |OD|$, AC ⊥ BD (read 'AC is perpendicular to BD'—that is to say, ∠AOB = 90°), and ∠BAC = ∠CAD. To see why this happens, consider the effect on B, D of a reflection in AC, and vice versa.*

The line *BD* is called the **perpendicular bisector** of *AC*, because it is perpendicular to *AC* and meets it at its **mid-point**, *O*. Essentially the same construction will allow us to bisect angles:

- Ex.2.4: *Given lines AP, AQ, use the compasses to mark B on AP and D on AQ with $|AB| = |AD|$. Then draw arcs of equal radius with centres B, D to meet at C. Join up, and verify that ∠BAC = ∠CAD, as before. (If you make all the arcs of the same radius, you will have another rhombus, but the construction works just as long as both $|AB| = |AD|$ and also $|CB| = |CD|$.) The line AC is said to* **bisect** *∠PAQ.*

- Ex.2.5: *Using the same diagram, produce PA to P', so that A is between P and P'. Construct the bisector AC' of ∠P'AQ, check that ∠CAC' = 90°, and explain why this should be so.*

So we know now how to bisect both lengths and angles, and how to construct a line perpendicular to a given line. There are some other useful constructions involving right angles:

- Ex.2.6: *Given a line AB, to construct C so that ∠BAC = 90°. One method: construct equilateral triangles BAD, DAE, and then construct AC to bisect ∠DAE. (In practice, it is not necessary actually to draw the sides of those triangles: try it.)*

- Ex.2.7: *Another method: choose O anywhere not on AB, and draw a circle centre O, radius $|OA|$, to meet AB in A and another point, P say. Join PO and let this line (diameter) meet the circle again in C. Check that ∠BAC = 90°. (The* *construction is easiest to do accurately, in practical terms, if you choose O so that ∠OAB is about 45°. But it always works.)*

- Ex.2.8:   *Dropping a perpendicular: given a line ℓ and a point A ∉ ℓ (read 'not lying on ℓ'), to construct O ∈ ℓ with AO ⊥ ℓ. Draw any arc, centre A, to meet ℓ in C, D. Then, with the same radius, draw arcs with centres C, D to meet in B (where B ≠ A). Then AB ⊥ ℓ, so O = AB · CD (the **meet**, that is, the point where the lines AB, CD meet).*

To construct a pair of parallel lines, you could just construct two equilateral triangles, *ABC* and *BCD*, which (provided $A \neq D$) would give you $AB \parallel CD$, or indeed you could construct a rhombus, as above. But usually one wants to construct a parallel to a given line not just anywhere, but through a given point:

- Ex.2.9:   *Given a line ℓ and A ∉ ℓ, construct a line parallel to ℓ, through A. Method 1: drop a perpendicular AB from A onto ℓ, and then construct C with ∠BAC = 90°.*

- Ex.2.10:   *Method 2: construct any line m (not through A) with m ⊥ ℓ, and then drop a perpendicular from A onto m.*

- Ex.2.11:   *Method 3: draw an arc, centre A, to meet ℓ in B, C, and with the same radius draw a circle centre C; draw also a circle, centre A, of radius |BC|. If the two circles meet in D, D', then one of AD, AD' is parallel to ℓ.*

- Ex.2.12:   *Method 4: draw any line through A, and let it meet ℓ in B; mark P on BA produced, so that A is the mid-point of PB. Draw another line, through P, and let it meet ℓ in C; bisect PC and call the mid-point D. Then AD ∥ ℓ.*

- Ex.2.13:   *Method 5: On ℓ, mark any B, C, P with |BP| = |PC|. (Put your compasses to one side now, as the rest of this construction uses ruler only.) Join AB, and choose a third point, Q, on this line. Let AC and PQ meet in R, and let BR and CQ meet in D. Then AD ∥ ℓ.*

Sometimes, if you are unlucky, lines which you hope will meet turn out to be parallel, or to meet off the edge of your paper; so you will have to start again and use more forethought. Try the last construction various ways, such as (i) with *Q* on *BA* produced, i.e., *A* between *B* and *Q*; (ii) with *Q* between *A* and *B*; and (iii) with *Q* on *AB* produced.

Once you are convinced you can construct perpendiculars and parallels with ruler and compasses, you may wish to cheat a little in future exercises by holding your ruler firmly with one hand while sliding a set square along it to drop perpendiculars or draw parallels. You really need three hands to do this easily, but it can be managed with only two, with practice, provided your piece of paper is firmly anchored. The advantage, apart from saving time, is that you don't then clutter up future diagrams with too many construction lines; the purpose, after all, is to *understand* what is going on, rather than religiously to go through the same ritual of construction time after time. For those who want to make a hobby out of it, you could try to get hold of (or make) a pair of *parallel rulers*, such as are used by navigators. These consist of two equal rulers connected with two equal cross pieces, hinged to make an adjustable parallelogram.

There is quite a lot of language to be learned at this point, in connection with angles. Two angles that add up to $180°$ are said to be **supplementary**, and the one is the **supplement** of the other. When two lines meet, they form four angles at a point; choosing two from four in the $\binom{4}{2} = 6$ ways, we find two pairs equal, called **vertically opposite** angles, and four supplementary pairs, called **adjacent** angles. Draw it!

Suppose $AB \parallel CD$. If B, D are on the same side of the line AC, then we say AB and CD are *directly* parallel; otherwise they are *oppositely* parallel. For example, if $ABCD$ is a parallelogram, then AB and CD are oppositely parallel, but AB and DC are directly parallel.

- Ex.2.14: *Let AB and CD be directly parallel, and let PQ meet AB in R (between A and B), and meet CD in S (between C and D). Suppose also that P, R, S, Q occur in that order on PQ. (i) Check that $\angle ARP = \angle CSR$: these are called* **corresponding** *angles, and there are three other such pairs in the diagram, so find them. (ii) Check that $\angle ARS = \angle RSD$: these are called* **alternate** *angles, and there is one other such pair in the diagram, so find it. (iii) Check that $\angle ARS$ and $\angle RSC$ are supplementary: these are called* **interior** *or* **allied** *angles, and there is one other such pair in the diagram, so find it.*

There are other equalities, such as $\angle ARP$ and $\angle QSD$, and other supplementary pairs, such as $\angle ARP$ and $\angle RSD$ (or $\angle CSQ$), but these do not have special names.

- Ex.2.15: *Draw a* **parallelogram** *ABCD, that is, make $AB \parallel DC$ and $AD \parallel BC$. (Don't make it a rhombus this time: make $|AB| \neq |AD|$.) Join AC and BD (the* **diagonals** *of the parallelogram), and let them meet at O. This time AC and BD will not meet at right angles (unless you have drawn a rhombus, after all), but every one of the angles in your diagram is equal to at least one other, so mark them appropriately. The lengths $|AB|$, $|AO|$, $|AD|$, $|BO|$, $|BC|$, $|CO|$, $|CD|$, $|DO|$ are equal in pairs, so mark them appropriately also.*

Angles which add up to $90°$ are said to be **complementary**, and the one is the **complement** of the other.

- Ex.2.16: *Draw a triangle ABC with $\angle ABC = 90°$, and verify that $\angle CAB$ and $\angle ACB$ are complementary. For a proof, construct $CD \parallel BA$, and note that (i) $\angle ABC$ and $\angle BCD$ are allied angles, and (ii) $\angle DCA$ and $\angle CAB$ are alternate angles.*

- Ex.2.17: *Generalize this to prove the well-known theorem about the sum of the angles of a triangle. Draw any triangle ABC, construct $CD \parallel BA$, and produce DC to E. Find two pairs of alternate angles in this diagram, and deduce that $\angle ABC + \angle BCA + \angle CAB = 180°$. Prove also that the external angle of a triangle is equal to the sum of the opposite interior angles: produce AB to F, and show that $\angle CBF = \angle BCA + \angle CAB$.*

- Ex.2.18: *Draw a triangle ABC with $|AB| = |AC|$, an* **isosceles** *triangle. Check that $\angle ABC = \angle ACB$. Can you prove that this will always happen? Deduce that if $\angle BAC = \theta$, then $\angle ABC = 90° - \theta/2$.*

- Ex.2.19: *The angle in a semicircle is a right angle. Draw a circle, centre O, with AB a diameter, and C elsewhere on the circle. Verify that $\angle ACB = 90°$, and prove that this always happens. (Join everything in sight, including OC, and find some isosceles triangles to play with.) Converse: given $\triangle ABC$ with $\angle ACB = 90°$, and O the mid-point of AB, does it follow that $|OC| = |OA|$ ($= |OB|$)?*

Since we know how to bisect a line segment (by constructing the perpendicular bisector), it is easy to divide a line segment into four, or eight, or sixteen equal parts, etc., by repeated bisection. But suppose we want to divide it into three equal parts, or seven, say?

- Ex.2.20: *To divide AB into n equal parts. Method 1: draw a second line, ℓ, through A, and mark $A_1, A_2, \ldots, A_n$ on ℓ so that $|AA_1| = |A_1A_2| = \ldots = |A_{n-1}A_n|$. Join A_nB, and then through $A_1, A_2, \ldots, A_{n-1}$ construct lines parallel to A_nB, to meet ℓ in $B_1, B_2, \ldots, B_{n-1}$, respectively. Then $|AB_1| = |B_1B_2| = \ldots = |B_{n-1}B|$.*

- Ex.2.21: *Method 2: Draw any line ℓ parallel to AB, and on ℓ mark points $C_0, C_1, \ldots, C_n$ with $|C_0C_1| = |C_1C_2| = \cdots = |C_{n-1}C_n|$. Let AC_0 and BC_n meet in P, and let PC_i meet AB in B_i, for $1 \le i < n$. Then $|AB_1| = |B_1B_2| = \cdots = |B_{n-1}B|$, as before.*

- Ex.2.22: *To trisect AB. Put $n = 3$ above, or else choose any P not on AB and join PA, PB. On AB produced, mark C with $|BC| = |AB|$, and draw any line through C to meet PA, PB in Q, R respectively. Let $AR \cdot BQ = O$ (the meet), and $PO \cdot AB = D$. Then $|DB| = \frac{1}{3}|AB|$. (What happens if instead we take C with $|BC| = 2|AB|$? Or $|BC| = 3|AB|$? Or $|BC| = n|AB|$? What if we let $n \to \infty$?)*

Lengths can be trisected, then; but angles cannot, in general. Of course, it is easy to trisect an angle of $90°$, for instance, because we can construct an angle of $30°$, by bisecting an angle of $60°$, constructed first. But it is impossible with ruler and compasses, used as specified above, to construct an angle of $20°$, for example, and so it is impossible to trisect an angle of $60°$. This is, roughly, because the formula $\cos 3\theta = 4\cos^3 \theta - 3\cos\theta$ shows that, if $x = \cos 20°$, then $4x^3 - 3x = \cos 60° = \frac{1}{2}$. The solution of this cubic equation involves taking a cube root,[1] whereas our ruler and compasses constructions correspond algebraically to solving linear equations (e.g., seeing where two lines meet) or quadratic equations (e.g., Pythagoras, or seeing where line

---

[1] It is the cube root of a *complex* number which is involved, actually: $x = \frac{1}{2}(\omega + \omega^{-1})$, where $\omega^3 = \frac{1}{2}(1 + i\sqrt{3})$. Of course, this last quantity is just $\cos 60° + i\sin 60°$, and it has *three* cube roots, one of which is $\omega = \cos 20° + i\sin 20°$, which immediately gives $x = \cos 20°$. This seems to be begging the question, then; to see exactly why the presence of that cube root makes the construction impossible, you will have to learn about fields and field extensions, which is beyond the scope of this book. For details, see [30], Chapter 5.

and circle meet, or two circles). To trisect $90°$, on the other hand, you have to solve $4x^3 - 3x = \cos 90° = 0$, which is rather easier!

Of course, it is perfectly easy to trisect angles if you are allowed to bend the rules. (An engineer or surveyor would simply measure the angle and divide by three, but that is not what we have in mind.) First, since $\frac{1}{3} = \frac{1}{4} + \frac{1}{16} + \frac{1}{64} + \ldots$ (or, in binary, $(11)^{-1} = 0.010101\ldots$), continued bisection and adding pieces together will get you as close as you please to the answer; what a pity that to get the exact answer would take for ever.

- **Ex.2.23:** *Try it, for a little while. How many steps do you need to get within (say) $\frac{1}{2}°$ of $20°$?*

Here are two exact (and finite) methods that depend on using (or making) marks on the ruler:

- **Ex.2.24:** *To trisect a given angle AOB. We may as well assume $|OA| = |OB|$. Join AB, and draw the circle centre A, radius $|OA|$. (This will pass through O but not B in general, unless $\angle AOB = 60°$, for instance.) Make two marks on the ruler a distance $|OA|$ apart, and jiggle the position of the ruler so that you can draw a line through O, with one of the marks at C, say, on AB, and the other at D, say, on the circle. Verify that $\angle AOC = \frac{1}{3}(\angle AOB)$. For a proof, note that $|OA| = |OB| = |AD| = |CD|$, so you have lots of isosceles triangles. Chase angles.*

The reader might like to plot a polar curve to see what is going on in the last exercise:

- **Ex.2.25:** *Take O as origin, and OA as the x-axis, with $|OA| = 1$. Thus B is on the unit circle, centre O, with equation $r = 1$, and D is on the unit circle, centre A, whose equation is $r = 2\cos\theta$. Since $|CD| = 1$, the locus of C, as B varies, is $r = (2\cos\theta) - 1$. Plot this curve. It has two loops, one inside the other, and is called (naturally enough) the **trisectrix**. Don't let's have any nonsense about r having to be positive: the equations $x = r\cos\theta$ and $y = r\sin\theta$ don't give a hoot whether r is positive or negative, and neither should you—live a little! Note that, for any position of B, the line AB meets the trisectrix in three points, one of which is C. What is the significance of the other two?*

- **Ex.2.26:** *To trisect a given angle AOB (again). We may as well assume $\angle OAB = 90°$. Draw the line ℓ through B, parallel to OA, and make two marks on the ruler a distance $2|OB|$ apart. Now jiggle the position of the ruler so that you can draw a line through O, with one of the marks at C, say, on AB, and the other at D, say, on ℓ. Verify that $\angle AOC = \frac{1}{3}(\angle AOB)$. For a proof, join B to the mid-point E of CD, and note that $|OB| = |CE| = |ED| = |BE|$. (OK?) This time, you have isosceles triangles and parallel lines to play with, so go to it!*

There are more details of these and other ways of trisecting angles in [20], and various tools and mechanical linkages for doing the same are described in [11].

2.4   TRIANGLES

We have met some special types of triangle already: *isosceles* (two sides, and hence two angles, equal), *equilateral* (all three sides equal, and hence three angles of $60°$), and *right angled* (one **right** angle, that is, equal to $90°$). A **scalene** triangle is one in which *no* two sides are equal. To these we should add that an **acute-angled** triangle has all three angles **acute**, that is, less than $90°$, and an **obtuse-angled** triangle has one **obtuse** angle, that is, exceeding $90°$ (but less than $180°$, obviously). For the record, a **reflex** angle is an angle exceeding $180°$, but this cannot occur in a triangle, since the angle sum is always $180°$.

Some of the constructions that follow are rather trivial for some of the special triangles, e.g., isosceles or equilateral. To see anything interesting happen, you probably need to draw a scalene triangle, and perhaps make it acute-angled. Then try the obtuse-angled case to see what, if anything, is different about it; sometimes you will need to produce (extend) one or more sides to find the points you are looking for.

- Ex.2.27: *Draw a line ℓ and mark on it two points, A and B. If ABC is a triangle, then C cannot be on ℓ. Where can it be? If $|AC| \neq |BC|$, then C is not on the perpendicular bisector of AB, so draw this line, also. There are two circles C must avoid if $|AC| \neq |AB|$ and $|BC| \neq |BA|$, so draw these, and also the circle and two lines C must avoid if our triangle is not to be right angled. If we now insist $\triangle ABC$ is positively oriented (that is, the labels A, B, C go around the triangle in the counter-clockwise direction), and $|AB| > |BC| > |CA|$, then shade the region where C must lie (i) if $\triangle ABC$ is acute-angled; (ii) if $\triangle ABC$ is obtuse-angled.*

- Ex.2.28: *Choose, at random, three lengths a, b, c less than (say) 20 cm. Can you always then construct a triangle ABC with $|AB| = a$, $|BC| = b$ and $|CA| = c$? For example, what if $a = 10$, $b = 6$, and $c = 16$? Or $a = 10$, $b = 6$, and $c = 3$? And when it does work, can you tell in advance whether your chosen lengths are going to give an acute-angled triangle or not?—for instance, if $a = 8$, $b = 9$, and $c = 12$, say?*

Two triangles ABC, $A'B'C'$ are said to be **congruent** (written $\triangle ABC \equiv \triangle A'B'C'$) if $|AB| = |A'B'|$, $|BC| = |B'C'|$, and $|CA| = |C'A'|$. It then follows that $\angle ABC = \angle A'B'C'$, $\angle BCA = \angle B'C'A'$, and $\angle CAB = \angle C'A'B'$. This is a total of six equations, the last three following from the first three.

- Ex.2.29: *Which other subsets of three of these equations are sufficient to deduce that the remaining three equations hold also? (This can be regarded as an exercise in trying to construct a triangle from minimal information about it, in various ways.)*

- Ex.2.30: *Show that the last three equations do* not *determine the others, but that we then have*

$$|AB| : |A'B'| = |BC| : |B'C'| = |CA| : |C'A'|.$$

  (The triangles are then said to be **similar***.)*

- Ex.2.31: *Draw a triangle ABC, and put in the mid-points D, E, F of BC, CA, AB respectively. Join DE, EF, FD to form the* **mid-point triangle***, △DEF, of △ABC. Find the various parallels, and compare lengths where you can. Observe that you have four congruent triangles, each similar to △ABC.*

- Ex.2.32: *With the same diagram, put in the three* **medians** *AD, BE, CF, and observe that they meet in a point, G, the* **centroid** *of △ABC. Measure |AD| and |GD|, and compare; and do the same for the other medians. Where is the centroid of △DEF?*

- Ex.2.33: *Draw a triangle ABC, and put in the three perpendicular bisectors of the sides. You will find they meet in a point, O. Draw the circle with centre O, through A, and you will find it goes through B and C also. This is the* **circumscribed circle** *of △ABC, and O is the* **circumcentre** *of △ABC.*

- Ex.2.34: *Draw a triangle ABC, and put in the three angle bisectors. You will find they meet in a point, I. Drop a perpendicular IX from I onto BC, and draw the circle with centre I, through X. This circle* **touches** *BC at X (that is, BC is the tangent to it at X), and you will find it touches CA and AB also. It is the inscribed circle, or* **incircle***, of △ABC, and I is the* **incentre** *of △ABC.*

- Ex.2.35: *With the same diagram, put in the external angle bisectors of △ABC. That is, produce AC to P and bisect ∠PCB; this is just the line through C orthogonal (that is, perpendicular) to CI. Do the same at A and B, and let these three external bisectors meet pairwise in I_A, I_B, I_C, where A is on $I_B I_C$, and so on. If you produce AI, you will find it passes through I_A, and similarly for the others. Drop a perpendicular $I_A X'$ from I_A onto BC, and draw the circle with centre I_A, through X'. This circle touches BC at X', and you will find it touches AB and AC (both produced) also. Repeat the construction at I_B and I_C, and you now have a total of four circles all touching all three sides of △ABC. The last three drawn are the escribed circles, or* **excircles***, of △ABC, and I_A, I_B, I_C are the* **excentres** *of △ABC.*

- Ex.2.36: *In the same diagram, measure |BX| and |CX'|. Why are they the same? Mark the points where the other two excircles touch BC; call them X″ and X‴, say. Can you find more equal lengths on BC now?*

- Ex.2.37: *Draw a triangle ABC, preferably acute-angled, and drop perpendiculars AP, BQ, CR from A, B, C onto BC, CA, AB respectively. These three lines are the* **altitudes** *of △ABC. You will find that the altitudes meet in a point, H, the* **orthocentre** *of △ABC. This amounts to saying that, if four points A, B, C, H are such that AH ⊥ BC and BH ⊥ CA, then also CH ⊥ AB. (This is even true if the four points do not lie in a plane, but to* *think about this you will first have to decide what might be meant by saying that two lines are orthogonal, when they don't lie in a plane—**skew** lines—and so don't meet.) Notice that A is the orthocentre of △HBC, and B is the orthocentre of △AHC, and C is the orthocentre of △ABH. Such a set of four points, each the orthocentre of the triangle formed by the other three, is called an **orthocentric tetrad**.*

● Ex.2.38:   *In the same diagram, join PQ, QR, RP to form △PQR, the **pedal** triangle of △ABC (or HBC, or AHC, or ABH, or indeed of the orthocentric tetrad A, B, C, H). Measure the angles at P (and at Q and R), and observe that, if △ABC is acute-angled, then H is the incentre and A, B, C are the excentres of △PQR. (What if △ABC is obtuse-angled?)*

● Ex.2.39:   *The nine-point circle. If you have not done so recently, sharpen your pencil! Draw a triangle ABC and put in its circumcircle. (This diagram is going to get quite involved, so perhaps cheat by drawing a good big circle first, and then mark A, B, C on it. Make △ABC scalene and acute-angled.) Label the circumcentre, O. Join AO, and produce it until it meets the circle again, in D, so that AD is a diameter. Similarly, put in the diameters BE and CF. Next, draw the altitudes—perhaps use a set square—through A, B, C, and produce them until they meet the circumcircle again, in P, Q, R respectively. Label the orthocentre, H, and join HO. (If these points are rather close together, then your triangle is rather close to being equilateral. If H or O is not inside the triangle, then your triangle is obtuse-angle, or right angled.) Join H to D, E, F.*

   Put in, and label, the following mid-points: A', B', C' are the mid-points of HA, HB, HC respectively, and N is the mid-point of HO. Label the following meets: D' = HD · BC, E' = HE · CA, F' = HF · AB, P' = HP · BC, Q' = HQ · CA, and R' = HR · AB. Observe (check by measurement) that D', E', F' are the mid-points of BC, CA, AB respectively. (So OD', OE', OF' are the perpendicular bisectors of BC, CA, AB, respectively.) Observe also that D', E', F', P', Q', R' are the mid-points of HD, HE, HF, HP, HQ, HR, respectively.

   *Suppose we had set up coordinates in our plane, with the origin at H. (Not O—sorry!) Consider the effect of the map $(x, y) \mapsto (\frac{x}{2}, \frac{y}{2})$, which moves every point in the plane halfway towards the origin. We have nine points marked on the circumcircle: A, B, C, D, E, F, P, Q, R, and they get mapped to the nine mid-points A', B', C', D', E', F', P', Q', R'. The circumcircle gets mapped to a circle half its size (that is, half its radius), with its centre at N, the mid-point of HO, and we thus have that the nine points A', B', C', D', E', F', P', Q', R' lie on this circle. So now you can check how good your draughtsmanship is: take a deep breath, place the point of your compasses on N and draw the circle through (say) A'; you should find that this circle also goes through the other eight points listed. This is the **nine-point circle**, and N is the **nine-point centre**, of △ABC.*

ANSWERS TO EXERCISES

2.2: With E constructed, let the circle centre B through A cut the line segment BE at F, and let the circle centre E through F cut the line segment CE at D. Then $|CD| = |CE| - |DE| = |BE| - |FE| = |BF| = |BA|$.

2.24: Let $\angle AOD = \theta$; so $\angle ADO = \theta$ also, since $|AO| = |AD|$, and $\angle DAO = 180° - 2\theta$. Then $|DC| = |DA|$, so $\angle DAC = 90° - \frac{\theta}{2} = \angle DCA$, and $\angle BAO = \angle CAO = (180° - 2\theta) - (90° - \frac{\theta}{2}) = 90° - \frac{3\theta}{2}$. But now $|OA| = |OB|$, so $\angle AOB = 180° - 2\angle BAO = 3\theta$.

2.25: The point C has polar coordinates (r, θ) where $r = (2\cos\theta) - 1$, and $\theta = \frac{1}{3}\angle AOB$. The other points C', C'' where AB meets the trisectrix give the *other* trisections, namely $\frac{1}{3}(\angle AOB \pm 360°) = \frac{1}{3}\angle AOB \pm 120°$. In fact $\angle COC' = \angle C''OC = 60°$, not $120°$, but that is because r is negative at C' and C''. If you want to check that this is not a swindle, you will have to do some heavy algebra. Note that at (r, θ) we have $x = r\cos\theta = 2\cos^2\theta - \cos\theta$, and $y = r\sin\theta = 2\cos\theta\sin\theta - \sin\theta$. So, looking ahead at Proposition 4.24, all you need to show (!) is that

$$\begin{vmatrix} 1 & 2\cos^2\theta - \cos\theta & 2\cos^2\varphi - \cos\varphi \\ 0 & 2\cos\theta\sin\theta - \sin\theta & 2\cos\varphi\sin\varphi - \sin\varphi \\ 1 & 1 & 1 \end{vmatrix} = 0$$

when $\varphi = \theta \pm 120°$; and the best of luck.

2.26: Let $\angle AOC = \theta$. Then $\angle BDE = \theta$ (alternate angles), and since $|ED| = |EB|$, we have $\angle DBE = \theta$, so that $\angle OEB = 2\theta$. But $|BO| = |BE|$, so $\angle EOB = 2\theta$, whence $\angle AOB = 3\theta$.

2.28: To construct a triangle, you will need the sum of two side lengths to be greater than the other: $a + b > c$, $b + c > a$, and $c + a > b$. To test for acute/obtuse, use Pythagoras. For a right angled triangle, the hypotenuse is the longest side. When $a = 8$, $b = 9$, and $c = 12$, then $a^2 + b^2 = 64 + 81 = 145 > 144 = c^2$, so we have an acute-angled triangle in this case. (OK?)

2.36: Let the incircle touch BC, CA, AB at X, Y, Z respectively. Then $|BX| = |BZ|$, $|AZ| = |AY|$, and $|CY| = |CX|$ (equal tangents). Also $|BX| + |CX| = |BC|$, $|CY| + |AY| = |CA|$, and $|AZ| + |BZ| = |AB|$. It is not too hard to show from these equations that $|BX| = \frac{1}{2}(|AB| - |BC| + |CA|)$. Now let the excircle opposite A touch BC, CA, AB (produced as necessary) at X', Y', Z' respectively. Then $|BX'| = |BZ'|$, $|AZ'| = |AY'|$ and $|CY'| = |CX'|$ (equal tangents). Also $|BX'| + |CX'| = |BC|$, $|AY'| - |CY'| = |CA|$, and $|AZ'| - |BZ'| = |AB|$. Solve for $|CX'|$, and compare.

3

Plane geometry

3.1 POINTS

We shall assume that the set of real numbers, $\mathbb{R}$, is a familiar object on which we can perform the usual arithmetic operations. Our object of study is the *real plane* $\mathbb{R}^2$, or $\mathbb{R} \times \mathbb{R}$, which is the set whose members are all ordered pairs (x, y), where $x \in \mathbb{R}$ and $y \in \mathbb{R}$: in symbols,

$$\mathbb{R}^2 = \{ (x, y) : x, y \in \mathbb{R} \}.$$

The significance of the word *ordered* is that for instance $(3, 2)$ is not the same as $(2, 3)$: the order matters, and indeed we insist

$$(x, y) = (x', y') \text{ if and only if both } x = x' \text{ and } y = y'.$$

Incidentally, we shall often write 'iff' or '$\Longleftrightarrow$' for 'if and only if'.

Definition 3.1. *A* **point** *is a member* (x, y) *of the set* $\mathbb{R}^2$.

We shall often denote points by capital letters $A, B, \ldots$, and write A *is a point* to mean the same as $A \in \mathbb{R}^2$, and A *is the point* $(2, 3)$, for example, or even $A = (2, 3)$, when we want to specify which particular point we mean. If $A = (x, y)$, then the real numbers x, y are the *coordinates* of A, with x being the *first coordinate* or *x-coordinate*, and y being the *second coordinate* or *y-coordinate*. (It is probably better to say *first* and *second* as, for example, the x-coordinate of the point (y, x) is y, not x; but old habits die hard.) The special point $O = (0, 0)$ is called the *origin*.

3.2 LINES

A line is given by a linear equation:

Definition 3.2. *Given* $p, q, r \in \mathbb{R}$, *with at least one of* p, q *non-zero, then the set of points* (x, y) *such that* $px + qy + r = 0$ *is called a (straight)* **line**, *or (when we wish to be more specific) the line* $px + qy + r = 0$.

A line is just a certain set of points; if A is a point and ℓ is a line, we shall say A *lies on* ℓ or ℓ *passes through* A to mean the same as $A \in \ell$. The line $y = 0$ is (confusingly) called the *x-axis*, as it consists of all points $(x, 0)$; likewise $x = 0$ is called the *y-axis*, and consists of all points $(0, y)$.

Note that a line has many equations: for example, the equations $2x + 2y - 1 = 0$ and $-x - y + \frac{1}{2} = 0$ give the same line. To obtain a unique equation for the line $px + qy + r = 0$, we proceed in one of two ways. If $q \neq 0$, then divide through by q and rearrange to get the familiar equation $y = mx + c$, where $m = -p/q$ and $c = -r/q$. If $q = 0$ then necessarily $p \neq 0$, so divide by p and rearrange to get the equation $x = d$, where $d = -r/p$. In the first case, the constant m is called the *gradient* or *slope* of the line; the second case can be obtained from the first by letting $q \to 0$, which (with p fixed, $p \neq 0$) entails $m \to \infty$, so we shall say that the gradient is *infinite* in this case.

Definition 3.3. *Two lines (not necessarily distinct) having the same gradient, finite or infinite, are said to be* **parallel**.

We now collect together some elementary facts about points and lines:

Proposition 3.4. (i) *Two distinct points lie on one and only one line.*
(ii) *Two distinct lines which are not parallel intersect in one and only one point.*
(iii) *Two distinct lines are parallel if and only if they have no point in common.*
(iv) *Through a given point there is one and only one line parallel to a given line.*

Proof. This can safely be left to the reader. □

Definition 3.5. *Given points A, B, the unique line through them is called their* **join**, *denoted AB. Given two lines $\ell = AB$ and $m = CD$, if they are parallel we write $\ell \parallel m$, or $AB \parallel CD$, and if they are not parallel the point where they intersect is called their* **meet**, *and is denoted ℓm, or $AB \cdot CD$.*

3.3 VECTOR NOTATION

It is helpful to have various different ways of handling points and lines, and vector notation makes certain calculations simpler. Recall then that elements $(x_i, y_i) \in \mathbb{R}^2$ can be thought of as *vectors* which can be added and subtracted, and multiplied by any *scalar* $\lambda \in \mathbb{R}$, thus:

$$(x_1, y_1) + (x_2, y_2) = (x_1 + x_2, y_1 + y_2),$$
$$(x_1, y_1) - (x_2, y_2) = (x_1 - x_2, y_1 - y_2),$$
$$\lambda(x_1, y_1) = (\lambda x_1, \lambda y_1).$$

The confusion between the two meanings of (x, y) is quite deliberate: if $A = (x, y)$, a pair of coordinates, then $\overrightarrow{OA} = (x, y)$, a vector. We shall often use the corresponding bold lower-case letters $\mathbf{a}, \mathbf{b}, \ldots$ to stand for the vectors $\overrightarrow{OA}, \overrightarrow{OB}, \ldots$ respectively, and $\mathbf{0}$ will stand for the zero vector $(0, 0)$. We shall try not to fuss unduly about whether $\mathbf{b} - \mathbf{a}$ stands for the vector $\overrightarrow{AB}$ (a 'free' vector, neither end of which is at O) or for some vector $\overrightarrow{OC}$ (a 'position' vector, giving the position of C relative to the origin O).

● Ex.3.1:   *Show that* $AB \parallel CD$ *iff* $\overrightarrow{AB} = \lambda \overrightarrow{CD}$ *for some* $\lambda \in \mathbb{R}$. *(If we need to distinguish, we say* AB, CD *are* **directly parallel** *if* $\lambda > 0$, *and* **oppositely parallel** *if* $\lambda < 0$.)

The vector **a** is a *unit* vector if its length is 1, that is, $|\mathbf{a}| = 1$. (The length $|\mathbf{a}|$ of **a** will be defined properly in the next section.) Given a vector $\mathbf{a} \neq \mathbf{0}$, we can choose a scalar λ such that $\lambda \mathbf{a}$ is a unit vector simply by putting $\lambda = |\mathbf{a}|^{-1}$, so that $|\lambda \mathbf{a}| = |\lambda||\mathbf{a}| = |\mathbf{a}|^{-1}|\mathbf{a}| = 1$. Thus we obtain a unit vector parallel to **a**, a process known as *normalizing* **a**. The normalization of **a** is thus $|\mathbf{a}|^{-1}\mathbf{a}$, or $\mathbf{a}/|\mathbf{a}|$.

The unit vectors $\mathbf{i} = (1, 0)$ and $\mathbf{j} = (0, 1)$ are the *standard basis* of $\mathbb{R}^2$: for any (x, y), we have

$$(x, y) = (x, 0) + (0, y) = x(1, 0) + y(0, 1) = x\mathbf{i} + y\mathbf{j}.$$

Using vectors, lines can be written in a parametric form. Let $\mathbf{a}_1, \mathbf{a}_2$ be given vectors, with $\mathbf{a}_2 \neq \mathbf{0}$, and consider what happens to the point

$$\mathbf{r} = \mathbf{a}_1 + \xi \mathbf{a}_2$$

as the scalar parameter ξ varies. Putting $\mathbf{r} = (x, y)$ and $\mathbf{a}_i = (x_i, y_i)$, we have

$$x = x_1 + \xi x_2 \text{ and } y = y_1 + \xi y_2$$

whence $y_2(x - x_1) = x_2(y - y_1)$, which is a line; and conversely, as the next proposition shows, every line can be put in such a vector form. Let A_1, A_2 be distinct points, with $\mathbf{a}_i = \overrightarrow{OA_i}$, each i. The vector equation of the join $A_1 A_2$ is particularly easy to write down:

Proposition 3.6.   *The line joining the (distinct) points* $\mathbf{a}_1, \mathbf{a}_2$ *has equation*

$$\mathbf{r} = \mathbf{a}_1 + \xi(\mathbf{a}_2 - \mathbf{a}_1), \text{ or } \mathbf{r} = (1 - \xi)\mathbf{a}_1 + \xi \mathbf{a}_2.$$

Proof.   The equation $\mathbf{r} = \mathbf{a}_1 + \xi(\mathbf{a}_2 - \mathbf{a}_1)$ *is* a line; putting $\xi = 0$ gives $\mathbf{r} = \mathbf{a}_1$, so the line goes through $\mathbf{a}_1$, and putting $\xi = 1$ gives $\mathbf{r} = \mathbf{a}_2$, so the line goes through $\mathbf{a}_2$ also. Thus we have the required line, by Proposition 3.4(i).   □

● Ex.3.2:   *Show that the line joining* (x_1, y_1) *and* (x_2, y_2) *has equation* $(y_1 - y_2)x - (x_1 - x_2)y + (x_1 y_2 - x_2 y_1) = 0$.

Points on a line occur, not just anyhow, but in a particular order, and the last proposition allows us to make this explicit. Let B be a point of the line $A_1 A_2$, with $\mathbf{b} = \overrightarrow{OB}$. Then $\mathbf{b} = \mathbf{a}_1 + \xi(\mathbf{a}_2 - \mathbf{a}_1)$ for some $\xi \in \mathbb{R}$. Then:

Definition 3.7. B *is* **between** A_1 *and* A_2 *if* $0 < \xi < 1$. *The* **line segment** $A_1 A_2$ *is the set of all points between* A_1 *and* A_2, *together with the points* A_1 *and* A_2, *that is, it consists of all* B *with* $0 \leq \xi \leq 1$. *When* $\xi = \frac{1}{2}$, *then* B *is the* **mid-point** *of* $A_1 A_2$, *given by* $\mathbf{b} = \frac{1}{2}(\mathbf{a}_1 + \mathbf{a}_2)$.

Note that $A_1 A_2$ now has two meanings: the *line* $A_1 A_2$, and the *line segment* $A_1 A_2$. It should be clear from the context which is meant, if it matters.

- Ex.3.3: *Prove that if B is between A_1 and A_2, then B is between A_2 and A_1. (This may seem rather footling, but it had better be true, else we are going to get into a proper muddle!)*

- Ex.3.4: *Prove that B is between A_1 and A_2 iff $\overrightarrow{A_1 B} = \lambda \overrightarrow{B A_2}$ for some $\lambda > 0$.*

- Ex.3.5: *Prove that B is the mid-point of $A_1 A_2$ iff $|A_1 B| = |B A_2| = \frac{1}{2}|A_1 A_2|$.*

- Ex.3.6: *Given B_i, $i = 1, 2, 3$, with $\mathbf{b}_i = \overrightarrow{O B_i} = \mathbf{a}_1 + \xi_i(\mathbf{a}_2 - \mathbf{a}_1)$, $i = 1, 2, 3$, then B_2 is between B_1 and B_3 iff the real number ξ_2 is between ξ_1 and ξ_3, that is, iff $(\xi_1 - \xi_2)(\xi_2 - \xi_3) > 0$.*

- Ex.3.7: *Given distinct points B_1, B_2, B_3 on a line, exactly one of them is between the other two.*

3.4 DISTANCES AND ORTHOGONALITY

Definition 3.8. *Let $P = (x, y)$ and let O be the origin, $(0, 0)$. The **distance** from O to P is written $|OP|$, and is defined to be the non-negative quantity $\sqrt{x^2 + y^2}$. In vector language, $|OP|$ is the **length** of the vector $\overrightarrow{OP}$.*

Putting $Q = (x, 0)$, we see that $\triangle OPQ$ is a right-angled triangle, and we have used Pythagoras' theorem to define $|OP|$, as promised in Section 1.6.

More generally, let $P_1 = (x_1, y_1)$ and $P_2 = (x_2, y_2)$. Then $\overrightarrow{OP_1} - \overrightarrow{OP_2} = \overrightarrow{P_2 P_1} = (x_1 - x_2, y_1 - y_2)$. This motivates the next definition:

Definition 3.9. *The distance from the point $P_1 = (x_1, y_1)$ to the point $P_2 = (x_2, y_2)$ is $|P_1 P_2| = \sqrt{(x_1 - x_2)^2 + (y_1 - y_2)^2}$.*

- Ex.3.8: *Show that $|P_1 P_2| = 0$ iff $P_1 = P_2$.*

Definition 3.10. *The points $A, B, C, \ldots$ are **collinear** if they all lie on same line.*

Now let $P_i = (x_i, y_i)$, $i = 0, 1, 2$, be three points, not collinear. According to Pythagoras, if $\angle P_1 P_0 P_2$ is a right angle, then $|P_0 P_1|^2 + |P_0 P_2|^2 = |P_1 P_2|^2$. We have not yet defined angles, so we shall turn Pythagoras' theorem on its head, and use this last equation to *define* the notion of perpendicularity:

Definition 3.11. *$P_0 P_1$ and $P_0 P_2$ are **perpendicular** or **orthogonal** (in symbols, we write $P_0 P_1 \perp P_0 P_2$) if $|P_0 P_1|^2 + |P_0 P_2|^2 = |P_1 P_2|^2$.*

- Ex.3.9: *Let $ABCD$ be a rectangle. (So $|AB| = |CD|$, $|BC| = |DA|$, $AB \perp BC$, $BC \perp CD$, $CD \perp DA$, and $DA \perp AB$.) Let P be any other point. Prove that $|PA|^2 + |PC|^2 = |PB|^2 + |PD|^2$.*

- Ex.3.10: *Let Σ be a circle, centre P, and let A be a point inside Σ. Given B on Σ, find D on Σ such that $DA \perp AB$, and then find C (not on Σ!) such that $ABCD$ is a rectangle. Prove that, if Σ and A are fixed, and B moves around*

Σ, *then the locus of C is another circle, concentric with Σ, that is, having the same centre, P. Does this still work if we choose A outside Σ? Always, sometimes, or never?*

Lemma 3.12.   *Let $P_i = (x_i, y_i)$, $i = 0, 1, 2$. Then $P_0 P_1 \perp P_0 P_2$ iff $(x_0 - x_1)(x_0 - x_2) + (y_0 - y_1)(y_0 - y_2) = 0$.*

Proof.   For any $a, b \in \mathbb{R}$, we have $a^2 + b^2 - (a - b)^2 = 2ab$. Putting $a = x_0 - x_1$ and $b = x_0 - x_2$ gives

$$(x_0 - x_1)^2 + (x_0 - x_2)^2 - (x_1 - x_2)^2 = 2(x_0 - x_1)(x_0 - x_2).$$

Similarly,

$$(y_0 - y_1)^2 + (y_0 - y_2)^2 - (y_1 - y_2)^2 = 2(y_0 - y_1)(y_0 - y_2).$$

Adding these equations gives the result.   □

The reader may recognize the quantity $(x_0 - x_1)(x_0 - x_2) + (y_0 - y_1)(y_0 - y_2)$ as the dot product $\overrightarrow{P_1 P_0} \cdot \overrightarrow{P_2 P_0}$. Dot products will be introduced in Section 3.8.

Proposition 3.13.   *Let the line $P_0 P_i$ have gradient m_i, for $i = 1, 2$. Then $P_0 P_1 \perp P_0 P_2$ iff either (i) $m_1 m_2 = -1$, or (ii) $m_1 = 0$, $m_2 = \infty$, or (iii) $m_1 = \infty$, $m_2 = 0$.*

Proof.   Most of this is left to the reader. If $x_0 - x_1 \neq 0$, then show that $m_1 = (y_0 - y_1)/(x_0 - x_1)$, and if $x_0 - x_1 = 0$, show that $m_1 = \infty$. Similarly for m_2. Now suppose $P_0 P_1 \perp P_0 P_2$, so that $(x_0 - x_1)(x_0 - x_2) + (y_0 - y_1)(y_0 - y_2) = 0$. Case (i) follows on dividing through by $(x_0 - x_1)(x_0 - x_2)$, provided this is not zero. The other cases are left to the reader, who should also check that the argument reverses.   □

As a consequence of this proposition, orthogonality is a property of the *lines* $P_0 P_1$, $P_0 P_2$, and does not depend on the choice of position of P_1, P_2 on these lines.

- Ex.3.11:   *Suppose $P_0 P_i$ has equation $p_i x + q_i y + r_i = 0$, for $i = 1, 2$. Then $P_0 P_1 \perp P_0 P_2$ iff $p_1 p_2 + q_1 q_2 = 0$.*
- Ex.3.12:   *Given lines ℓ, m, n with $\ell \perp m$ and $m \perp n$, show that $\ell \parallel n$.*

Proposition 3.14.   *Let P_1, P_2 be distinct points. The set of points P such that $|P P_1| = |P P_2|$ is a line which passes through the mid-point of $P_1 P_2$ and is perpendicular to $P_1 P_2$.*

Definition 3.15.   *This line is the **perpendicular bisector** of $P_1 P_2$ (Fig. 3.1).*

Proof.   Let $P_i = (x_i, y_i)$, $i = 1, 2$, and $P = (x, y)$. We have

$$(x - x_1)^2 + (y - y_1)^2 = (x - x_2)^2 + (y - y_2)^2,$$

or

$$2(x_1 - x_2)x + 2(y_1 - y_2)y + (x_2^2 + y_2^2 - x_1^2 - y_1^2) = 0, \tag{3.1}$$

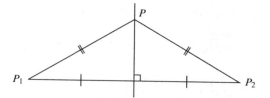

Fig. 3.1 Proposition 3.14

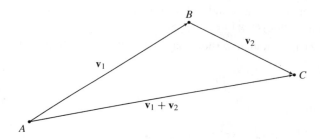

Fig. 3.2 Lemma 3.16

which, since $2(x_1 - x_2)$ and $2(y_1 - y_2)$ cannot both be zero, is the equation of a line. It is easy to check that $x = \frac{1}{2}(x_1 + x_2)$, $y = \frac{1}{2}(y_1 + y_2)$ satisfies (3.1). The line $P_1 P_2$ has equation

$$(y_1 - y_2)(x - x_1) - (x_1 - x_2)(y - y_1) = 0,$$

or

$$(y_1 - y_2)x - (x_1 - x_2)y + (x_1 y_2 - x_2 y_1) = 0,$$

and the fact that this is perpendicular to (3.1) follows from Exercise 3.11. □

Lemma 3.16 (The triangle inequality). *For any three points A, B, C,*

$$|AB| + |BC| \geq |AC|,$$

with $|AB| + |BC| = |AC|$ iff B lies on AC, between A and C.

Proof. Translate into vector language: put $\mathbf{v}_1 = \overrightarrow{AB}$ and $\mathbf{v}_2 = \overrightarrow{BC}$ (Fig. 3.2). We must show that $|\mathbf{v}_1| + |\mathbf{v}_2| \geq |\mathbf{v}_1 + \mathbf{v}_2|$, with equality iff $\mathbf{v}_2 = \lambda \mathbf{v}_1$ for some $\lambda > 0$. Note that we are assuming that A, B, C are distinct, so we are assuming that $\mathbf{v}_1, \mathbf{v}_2 \neq \mathbf{0}$.

Let $\mathbf{v}_1 = (x_1, y_1)$ and $\mathbf{v}_2 = (x_2, y_2)$, so that $\mathbf{v}_1 + \mathbf{v}_2 = (x_1 + x_2, y_1 + y_2)$. We want to show that

$$\sqrt{x_1^2 + y_1^2} + \sqrt{x_2^2 + y_2^2} \geq \sqrt{(x_1 + x_2)^2 + (y_1 + y_2)^2}.$$

Since both sides are positive, this is equivalent to showing that

$$(x_1^2 + y_1^2) + 2\sqrt{x_1^2 + y_1^2}\sqrt{x_2^2 + y_2^2} + (x_2^2 + y_2^2) \geq (x_1 + x_2)^2 + (y_1 + y_2)^2,$$

or

$$\sqrt{x_1^2 + y_1^2}\sqrt{x_2^2 + y_2^2} \geq x_1 x_2 + y_1 y_2. \tag{3.2}$$

(The right-hand side of this is $\mathbf{v}_1 \cdot \mathbf{v}_2$, the dot product. So we are trying to show that $|\mathbf{v}_1||\mathbf{v}_2| \geq \mathbf{v}_1 \cdot \mathbf{v}_2$. If we had already defined angles, and if we knew that $\mathbf{v}_1 \cdot \mathbf{v}_2 = |\mathbf{v}_1||\mathbf{v}_2|\cos\theta$, then the result would follow from the fact that $1 \geq \cos\theta$, with equality iff $\theta = 0$. But we have not yet defined angles.)

To prove (3.2), it is enough to show that $\sqrt{x_1^2 + y_1^2}\sqrt{x_2^2 + y_2^2} \geq |x_1 x_2 + y_1 y_2|$, or

$$(x_1^2 + y_1^2)(x_2^2 + y_2^2) \geq (x_1 x_2 + y_1 y_2)^2.$$

Expand: we want

$$x_1^2 x_2^2 + x_1^2 y_2^2 + y_1^2 x_2^2 + y_1^2 y_2^2 \geq x_1^2 x_2^2 + 2x_1 x_2 y_1 y_2 + y_1^2 y_2^2,$$

or

$$x_1^2 y_2^2 - 2x_1 x_2 y_1 y_2 + y_1^2 x_2^2 \geq 0,$$

or

$$(x_1 y_2 - y_1 x_2)^2 \geq 0. \tag{3.3}$$

The reader should have no difficulty believing (3.3), and so our inequality is proved. (A warning about logic: this proof has been written in reverse order, because it seems more natural: the more conventional alternative would be to start mysteriously with 'Consider the quantity $(x_1 y_2 - x_2 y_1)^2$.' The danger with writing proofs backwards, as we have just done, is that we may fall into the trap of thinking it is enough to start from a desired statement and deduce something correct from it: but this would *not* constitute a proof. For example, to 'prove' that $1 = 2$, one might start with $1 = 2$, multiply each side by zero (a legitimate operation), and observe that the resulting equation, $0 = 0$, is most certainly true. But few would claim that this proves that $1 = 2$! So for this lemma, the reader should check that, starting from the statement (3.3), each line of the argument follows from the line *below*.)

We still have to examine the case where equality occurs. This clearly happens iff both $x_1 y_2 - x_2 y_1 = 0$ and also $x_1 x_2 + y_1 y_2 = |x_1 x_2 + y_1 y_2|$, that is, $x_1 x_2 + y_1 y_2 \geq 0$.
Case 1: $x_1 = 0$. Then $x_2 y_1 = 0$, but $y_1 \neq 0$, so $x_2 = 0$ and $y_2 = \lambda y_1$ for some λ.
Case 2: $x_1 \neq 0$. So $x_2 = \lambda x_1$ for some λ, and $x_1 y_2 - \lambda x_1 y_1 = 0$, whence $y_2 = \lambda y_1$.
In each case, $(x_2, y_2) = \lambda(x_1, y_1)$, and also $0 \leq x_1 x_2 + y_1 y_2 = \lambda(x_1^2 + y_1^2)$, so that $\lambda > 0$. The argument reverses easily. $\qquad\square$

The triangle inequality says that the sum of the lengths of any two sides of a triangle is greater than the length of the other side. The last part does not quite say that the shortest route between two points is a straight line—for that, you need calculus—but it does say that, if you wish to travel from A to C using a succession of line segments, then the shortest route is along the single line segment AC.

3.5 ISOMETRIES

One of our main tools in geometry will be groups of transformations, of one kind or another, and we now introduce the notion of a *transformation*, and the first kind of transformation, called an *isometry*.

First recall some terminology from set theory. Let X, Y be sets, and let $f : X \to Y$ be a map. We say f is *injective*, or *one-to-one*, if whenever $x_1, x_2 \in X$ with $f(x_1) = f(x_2)$, then in fact $x_1 = x_2$. Equivalently, if $x_1, x_2 \in X$ with $x_1 \neq x_2$, then we require $f(x_1) \neq f(x_2)$: distinct elements of X are mapped by f to distinct elements of Y. Next, we say f is *surjective*, or *onto*, if given any $y \in Y$ there exists $x \in X$ with $f(x) = y$. Equivalently, if we write $f(X) = \{ f(x): x \in X \}$, then we require $f(X) = Y$. If f is both injective and surjective, we say it is *bijective*. For such a map, there is a unique *inverse* map $g : Y \to X$, also a bijection, such that $fg = 1_Y$ and $gf = 1_X$. Here fg means the composite map, given by $(fg)(y) = f(g(y))$, all $y \in Y$, and $1_Y : Y \to Y$ is the trivial (identity) map, given by $1_Y(y) = y$, all $y \in Y$. The construction of g is easy: given $y \in Y$, the conditions ensure the existence of a unique $x \in X$ with $f(x) = y$ (the existence because f is surjective, and the uniqueness because f is injective), and so we define $g(y) = x$. We write $g = f^{-1}$, so that $f(x) = y$ iff $x = f^{-1}(y)$.

Definition 3.17. *A* **transformation** *of* $\mathbb{R}^2$ *is a bijective map* $f : \mathbb{R}^2 \to \mathbb{R}^2$.

Given such f, the inverse f^{-1} is another transformation of $\mathbb{R}^2$.

Definition 3.18. *An* **isometry** *(of* $\mathbb{R}^2$*) is a transformation* f *(of* $\mathbb{R}^2$*) which is distance-preserving, that is, if* $f(P) = P'$ *and* $f(Q) = Q'$*, then* $|P'Q'| = |PQ|$.

In fact a distance-preserving map f is necessarily one-to-one, for if $P' = Q'$ (where $f(P) = P'$ and $f(Q) = Q'$), then $|P'Q'| = 0$, so that $|PQ| = 0$ also, and thus $P = Q$. It is also true that such a map is necessarily onto, though the proof is trickier and we shall not give it here. It makes life easier, and does no harm, just to assume at the outset that all our maps are transformations, as defined above. So, for example, we do not have to worry about whether an isometry has an inverse: the existence of the inverse is immediate from the definition.

- Ex.3.13: *Show that the inverse of an isometry is an isometry, and the composite of two isometries is an isometry.*

We shall denote the set of all isometries of $\mathbb{R}^2$ by $\mathcal{I}(\mathbb{R}^2)$. Readers who know about group theory will recognize that the last exercise, and the obvious fact that the identity map $1 : \mathbb{R}^2 \to \mathbb{R}^2$ is an isometry, show that $\mathcal{I}(\mathbb{R}^2)$ is a *group*, where the binary operation is map composition. For more details, see Section 5.2.

Example 1. Let $\mathbf{v} \in \mathbb{R}^2$. Then the map $f : \mathbb{R}^2 \to \mathbb{R}^2$ given by $f(\mathbf{u}) = \mathbf{u} + \mathbf{v}$, or $f(x, y) = (x + a, y + b)$, where $\mathbf{v} = (a, b)$, is an isometry, called **translation**

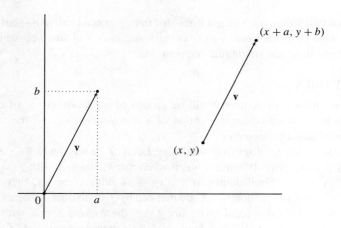

Fig. 3.3   Translation

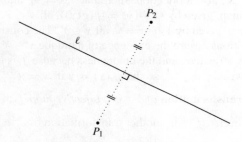

Fig. 3.4   Reflection

through **v** (Fig. 3.3). This map f is clearly a transformation. (OK? What is its inverse?) The fact that it is an isometry then follows from the calculation

$$f(\mathbf{u}_1) - f(\mathbf{u}_2) = (\mathbf{u}_1 + \mathbf{v}) - (\mathbf{u}_2 + \mathbf{v}) = \mathbf{u}_1 - \mathbf{u}_2.$$

Example 2.   Given a line ℓ, with equation $px + qy + r = 0$, and a point $P_1 = (x_1, y_1)$, we define the **reflection** of P_1 in ℓ to be the point $P_2 = (x_2, y_2)$ such that ℓ is the perpendicular bisector of $P_1 P_2$ (Fig. 3.4). We have

$$p(x_1 + x_2) + q(y_1 + y_2) + 2r = 0, \tag{3.4}$$

since the mid-point of $P_1 P_2$ is on ℓ. Also, the equation of $P_1 P_2$ is $(y_1 - y_2)x - (x_1 - x_2)y + (x_1 y_2 - x_2 y_1) = 0$, by Exercise 3.2; and since $P_1 P_2 \perp \ell$, we have

$$q(x_1 - x_2) - p(y_1 - y_2) = 0, \tag{3.5}$$

by Exercise 3.11. These two equations can be put together, in matrix form, thus:

$$\begin{pmatrix} p & q \\ -q & p \end{pmatrix} \begin{pmatrix} x_2 \\ y_2 \end{pmatrix} = \begin{pmatrix} -p & -q \\ -q & p \end{pmatrix} \begin{pmatrix} x_1 \\ y_1 \end{pmatrix} - \begin{pmatrix} 2r \\ 0 \end{pmatrix},$$

which can be solved for x_2, y_2, since

$$\det \begin{pmatrix} p & q \\ -q & p \end{pmatrix} = p^2 + q^2 \neq 0.$$

This allows us to define the **reflection** $f : \mathbb{R}^2 \to \mathbb{R}^2$, by $f(P_1) = P_2$. The line of reflection, or mirror, ℓ, is called the **axis** of the reflection. (It consists of all points P such that $f(P) = P$.) It is clear from the symmetry that $f(P_2) = P_1$, so that f is a transformation; in fact, $f^{-1} = f$, or $f^2 = 1$. (Here f^2 stands for the composite ff, of course.)

Proposition 3.19.   *The reflection f, as above, is an isometry.*

Proof.   Let $f(P_1) = P_2$, and $f(P_1') = P_2'$. From (3.4) and (3.5), and similar equations for P_1', P_2', we have

$$p(x_1 - x_1') + q(y_1 - y_1') = p(x_2' - x_2) + q(y_2' - y_2),$$
$$q(x_1 - x_1') - p(y_1 - y_1') = -q(x_2' - x_2) + p(y_2' - y_2).$$

Squaring and adding,

$$(p^2 + q^2)((x_1 - x_1')^2 + (y_1 - y_1')^2) = (p^2 + q^2)((x_2' - x_2)^2 + (y_2' - y_2)^2),$$

from which the result follows, since $p^2 + q^2 \neq 0$.   □

Example 3.   (Fig. 3.5.) The map f given by $f(x, y) = (-x, -y)$ is an isometry—OK?—called the **half-turn** about O. So if $f(P) = P'$, then $\overrightarrow{OP'} = -\overrightarrow{OP}$. Note that $f = f^{-1}$ and $f^2 = 1$, again. The map g given by $g(x, y) = (-y, x)$ is an isometry—OK?—called the **quarter-turn** about O, in the positive or counter-clockwise sense. Check that, if $g(P) = P'$, then $OP \perp OP'$. Here $g^2 = f$, the half-turn about O, and $g^4 = 1$. The isometry g^{-1} is given by $g^{-1}(x, y) = (y, -x)$, and is also a quarter-turn about O, but in the negative or clockwise sense. These are all examples of **rotations** (about O), but we cannot define rotation through an arbitrary angle until we have defined *angle*.

Example 4.   Let h be the translation given by $h(x, y) = (x + a, y + b)$, and let f be the half-turn about O. Then hfh^{-1} is the half-turn about (a, b).

• Ex.3.14:   *Write down the explicit formula for the half-turn about (a, b)—that is, evaluate $hfh^{-1}(x, y)$— and also for the two quarter-turns about (a, b).*

Proposition 3.20.   *If A, B, C are collinear and f is an isometry, then $f(A), f(B), f(C)$ are collinear also.*

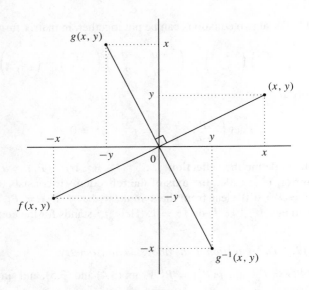

Fig. 3.5   Quarter-turns and half-turn about the origin

Proof.   Put $f(A) = A'$, $f(B) = B'$ and $f(C) = C'$. By Exercise 3.7 we may suppose, without loss of generality, that B lies between A and C, so that $|AB| + |BC| = |AC|$, by Lemma 3.16. Since f is an isometry, we have $|A'B'| + |B'C'| = |A'C'|$, whence A', B', C' are collinear, by Lemma 3.16, again.   □

Given a line ℓ and a vector $\mathbf{v} = \overrightarrow{AB}$, the statement $\ell \parallel \mathbf{v}$ will mean that the lines ℓ and AB are parallel.

- Ex.3.15:   *Let f be translation through $\mathbf{v}$, and let ℓ be any line, with $f(\ell) = \ell'$ (another line, by Proposition 3.20). Prove that $\ell \parallel \ell'$. Prove also that, if $\ell \parallel \mathbf{v}$, then $\ell = \ell'$.*

- Ex.3.16:   *Let f be the half-turn about O, and let ℓ be any line, with $f(\ell) = \ell'$. Prove that $\ell \parallel \ell'$. Prove also that, if ℓ passes through O, then $\ell = \ell'$. Generalize to half-turns about other points.*

- Ex.3.17:   *Let f be a quarter-turn about O, and let ℓ be any line, with $f(\ell) = \ell'$. Prove that $\ell \perp \ell'$. Generalize to quarter-turns about other points.*

Definition 3.21.   *Triangles ABC, $A'B'C'$ are **congruent** (written $\triangle ABC \equiv \triangle A'B'C'$) if there is an isometry f with $f(A) = A'$, $f(B) = B'$, and $f(C) = C'$. (Note that the order in which the triangles are labelled is a crucial part of this definition.) More generally, two subsets X, Y of $\mathbb{R}^2$ are congruent if there is an isometry f with $f(X) = Y$.*

Proposition 3.22. $\triangle ABC \equiv \triangle A'B'C'$ *iff* $|AB| = |A'B'|$, $|BC| = |B'C'|$, *and* $|CA| = |C'A'|$.

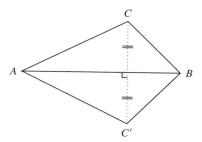

Fig. 3.6 Proof of Proposition 3.22, case 2

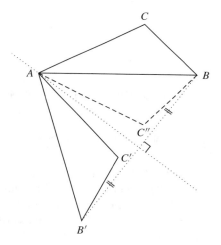

Fig. 3.7 Proof of Proposition 3.22, case 3

Proof. If the triangles are congruent, it is immediate from the definition that the various corresponding lengths are equal. For the converse, if the lengths are equal, we must construct a suitable isometry f:

Case 1: $A = A'$, $B = B'$, $C = C'$. There is nothing to do: just put $f = 1$.

Case 2: $A = A'$, $B = B'$, $C \neq C'$ (Fig. 3.6). Here $|AC| = |A'C'| = |AC'|$, and similarly $|BC| = |BC'|$, so that AB is the perpendicular bisector of CC'. Putting $f =$ reflection in AB achieves the desired effect.

Case 3: $A = A'$, $B \neq B'$, $C \neq C'$ (Fig. 3.7). Here $|AB| = |AB'|$, so A lies on the perpendicular bisector of BB'. Let f_1 be the reflection in this perpendicular bisector, so that f_1 fixes A and interchanges B, B', and let $C'' = f_1(C')$. Thus $f_1(A) = A = A'$, $f_1(B) = B'$, and also $f_1(C'') = C'$. (OK?) Since f_1 is an isometry, it follows that $\triangle A'B'C' \equiv \triangle ABC''$, and now $\triangle ABC$, $\triangle A'B'C'$, and $\triangle ABC''$ have corresponding side-lengths equal. By case 1 or 2 applied to $\triangle ABC$ and $\triangle ABC''$, there is an isometry

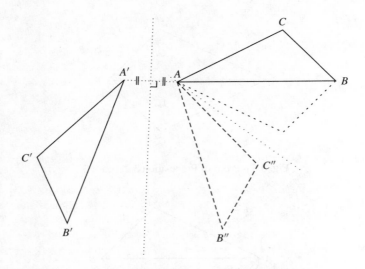

Fig. 3.8   Proof of Proposition 3.22, case 4

f_2 with $f_2(A) = A$, $f_2(B) = B$, and $f_2(C) = C''$, and so we just need to put $f = f_1 f_2$. (Check that it works.)

Case 4: $A \neq A'$, $B \neq B'$, $C \neq C'$ (Fig. 3.8). Let f_1 be reflection in the perpendicular bisector of AA', and let $B'' = f_1(B')$ and $C'' = f_1(C')$. Thus $f_1(A) = A'$, and also $f_1(B'') = B'$ and $f_1(C'') = C'$. (OK?) Since f_1 is an isometry, we have $\triangle A'B'C' \equiv \triangle AB''C''$, and now $\triangle ABC$, $\triangle A'B'C'$ and $\triangle AB''C''$ have corresponding side lengths equal. By case 1, 2 or 3 applied to $\triangle ABC$ and $\triangle AB''C''$, there is an isometry f_2 with $f_2(A) = A$, $f_2(B) = B''$, and $f_2(C) = C''$, so we put $f = f_1 f_2$. (Again, check that it works.)

In each case, we have a suitable isometry f, so the proof is complete.   □

Note (for future use) that in each of the above cases the isometry f that we constructed was a product (composite) of at most three reflections. Later on, we shall use this to prove that *every* isometry of $\mathbb{R}^2$ is a product of at most three reflections.

Corollary 3.23.   *Given $\triangle ABC$ and $\triangle A'B'C'$, suppose that $|AB| = |A'B'|$, and $|AC| = |A'C'|$, and also $AB \perp BC$ and $A'B' \perp B'C'$. Then $\triangle ABC \equiv \triangle A'B'C'$.*

Proof.   By Definition 3.11,

$$|BC|^2 = |AC|^2 - |AB|^2 = |A'C'|^2 - |A'B'|^2 = |B'C'|^2,$$

so that $|BC| = |B'C'|$, and the result follows by Proposition 3.22.   □

3.6 ANGLES

3.6.1 Preamble

Euclid defined *angle* to be 'the inclination of one line to another', but gave no indication as to how we might *measure* angles. Also, two lines which meet define *four* angles, α, β, γ, δ (Fig. 3.9), with $\alpha = \gamma$ (vertically opposite angles) and $\beta = \delta$ (ditto). If we define our measurement of angles so that one complete revolution is $360°$ (360 degrees), then $\alpha + \beta + \gamma + \delta = 360°$, so $2\alpha + 2\beta = 360°$, or $\alpha + \beta = 180°$ (adjacent angles are supplementary), and so on.

It leads to less confusion if we define the angle, not between two lines, but between two *half*-lines, or *rays*, with a common end-point. This gives two angles, α, β (Fig. 3.10), with $\alpha + \beta = 360°$, and if $0 < \alpha < \beta$, then $\beta > 180°$, a *reflex* angle. If we insist that all our angles θ lie in the range $0° \leq \theta \leq 180°$ (i.e., ban reflex angles), then two half-lines do now determine a unique angle, except perhaps when $\theta = 180°$. There are still difficulties, because we want to be able to do arithmetic with angles: we want to be able to write things like $\angle AOB + \angle BOC = \angle AOC$. This works fine if the angles are small enough, but when (for instance) $\angle AOB = \angle BOC = 120°$, then we arrive at the illegal value $240°$ for $\angle AOC$. Looking at Fig. 3.11, it is clear that $\angle AOC = 120°$, also, so we have the ridiculous statement that $120° + 120° = 120°$. At the very least, we now have to allow negative angles, and agree that in this case $\angle AOC = -120°$. But a better solution is to allow angles to take any value, and to work modulo 360. Our equation $\angle AOB + \angle BOC = \angle AOC$, putting in the values, now becomes $120 + 120 \equiv -120$ (mod 360), which we can write without blushing. (The statement $a \equiv b$ (mod c) is read

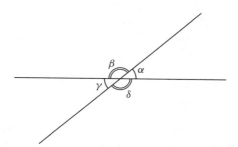

Fig. 3.9 Two lines determine *four* angles

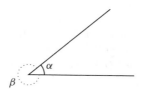

Fig. 3.10 Two *half*-lines determine *two* angles

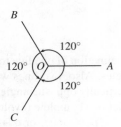

Fig. 3.11 $\angle AOB + \angle BOC = \angle AOC$?

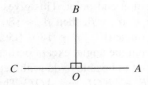

Fig. 3.12   Right angles

'a is congruent to b modulo c', and means that $a = b + nc$ for some integer n, or in other words $(a - b)/c \in \mathbb{Z}$.)

We still have not decided how to attach a number to each angle. Before we do this, notice that isometries allow us to compare angles: we could say that two angles are equal (ignoring problems about *sense*, or direction of measurement) if there is an isometry taking the pair of half-lines defining one of the angles onto the pair of half-lines defining the other. This would allow us, for instance, to deduce that corresponding angles of congruent triangles are equal: if $\triangle ABC \equiv \triangle A'B'C'$, then $\angle ABC = \angle A'B'C'$, $\angle BCA = \angle B'C'A'$, and $\angle CAB = \angle C'A'B'$. (The necessary isometry is provided by Definition 3.21.)

It would also allow us to define a *right* angle to be half of a 'flat' angle: if $\angle AOB = \angle BOC$, and A, O, C are collinear (with O between A and C), then $\angle AOB$ is a right angle, and so is $\angle BOC$ (Fig. 3.12). Here the isometry that does the trick is the reflection in OB. Extending this idea would allow us to define a measurement for any angle arrived at by addition (or subtraction) or by equal subdivision of previously known angles, and so we could define a measurement for any angle which is a *rational* multiple of one complete revolution. We should then be faced with defining an angle of (for instance) $\sqrt{2}°$ as a *limit* of angles of $1°, 1.4°, 1.41°, \ldots$, which would be rather messy to work with.

A right angle could also be defined via Pythagoras and orthogonality: we could say that $\angle AOB$ is a right angle if $|AO|^2 + |OB|^2 = |AB|^2$, that is, if $OA \perp OB$. We have seen that this depends only on the *lines* OA, OB and not on the lengths $|OA|, |OB|$. With $\overrightarrow{OA} = \mathbf{v}_1 = (x_1, y_1)$ and $\overrightarrow{OB} = \mathbf{v}_2 = (x_2, y_2)$, the condition becomes $|\mathbf{v}_1|^2 + |\mathbf{v}_2|^2 = |\mathbf{v}_1 + \mathbf{v}_2|^2$, or $(x_1^2 + y_1^2) + (x_2^2 + y_2^2) = (x_1 + x_2)^2 + (y_1 + y_2)^2$, or $0 = x_1 x_2 + y_1 y_2$, that is, $\mathbf{v}_1 \cdot \mathbf{v}_2 = 0$. This leads to a very respectable way to define angle measurement: assume

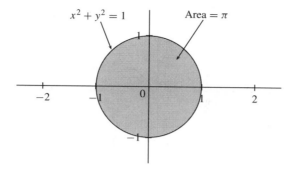

Fig. 3.13 Definition of π

that the cosine function is 'known' (or define it from its series $\cos x = 1 - \frac{x^2}{2!} + \frac{x^4}{4!} - \cdots$, where x is in radians) and, instead of having two interpretations of $\mathbf{v}_1 \cdot \mathbf{v}_2$ (as $x_1 x_2 + y_1 y_2$ or as $|\mathbf{v}_1||\mathbf{v}_2|\cos\theta$), *define* the angle θ between $\mathbf{v}_1$ and $\mathbf{v}_2$ by

$$\cos\theta = \frac{\mathbf{v}_1 \cdot \mathbf{v}_2}{|\mathbf{v}_1||\mathbf{v}_2|},$$

where $\mathbf{v}_1 \cdot \mathbf{v}_2 = x_1 x_2 + y_1 y_2$, and where we take the principal value of θ (in radians), that is, $0 \le \theta \le \pi$. This is exactly the approach used in [26]. Its advantage is that it gives a quick, neat, no-nonsense definition of angle. Its disadvantages are (1) we need to know all about trigonometry 'in the abstract', that is, properties of the sine and cosine functions, and (2) the addition formula for angles, $\angle AOB + \angle BOC = \angle AOC$, is non-obvious and becomes a theorem, requiring proof. (See [26], proposition (4.3.5).)

Our approach will be via *areas*. Let A be a point with $|OA| = 1$, so that, if $A = (x, y)$, then $x^2 + y^2 = 1$. This is the equation of the *unit circle*, centre O, and OA is a *radius* of the circle. The set of points (x, y) with $x^2 + y^2 \le 1$ is called the *unit disc*, centre O. We shall assume that we already know about areas, and we *define* the number π to be the area of the unit disc (Fig. 3.13). Given two radii OA, OB of the unit circle, they divide the disc into two regions, and we define the angle AOB (written $\angle AOB$) to be $2 \times$ the smaller of these two areas (Fig. 3.14). (This is called *circular measure*, and the units now are *radians*, not degrees. One complete turn is 2π radians, a flat is π radians, and a right angle is $\pi/2$ radians.) The advantages of this approach are that it is intuitively right, and makes the addition formula $\angle AOB + \angle BOC = \angle AOC$ easy and obvious (Fig. 3.15), provided we are prepared to allow negative angles (to take care of sense of rotation) and to work modulo 2π. The disadvantages are that we need a well-developed theory of areas, i.e. integration, and we *still* need the abstract theory of sines and cosines. But, unlike geometry, integration still features heavily in school syllabuses; and every mathematician needs to see the properties of the trigonometric functions developed from scratch at least once, so why not now?

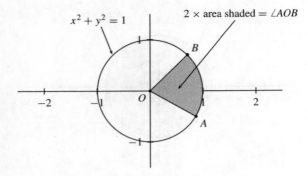

Fig. 3.14   Definition of $\angle AOB$

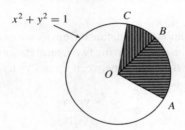

Fig. 3.15 $\angle AOB + \angle BOC = \angle AOC$

3.6.2   Abstract trigonometry

Let us define

$$s(x) = x - \frac{x^3}{3!} + \frac{x^5}{5!} - \cdots,$$

$$c(x) = 1 - \frac{x^2}{2!} + \frac{x^4}{4!} - \cdots.$$

In fact, $s(x)$ is $\sin x$ and $c(x)$ is $\cos x$, but we give the functions neutral names at the start, so as not to be in danger of using familiar properties of sine and cosine without proof. The two series are absolutely convergent for all x, and so the functions s, c are continuous, differentiable, and can be differentiated term by term. It follows that $s(0) = 0$, $c(0) = 1$, $s'(x) = c(x)$, and $c'(x) = -s(x)$, where the dashes denote differentiation with respect to x. Note also that $s(-x) = -s(x)$ and $c(-x) = c(x)$, for all x.

Put $f(x) = s(x)^2 + c(x)^2$, and observe that

$$f'(x) = 2s(x)s'(x) + 2c(x)c'(x) = 2s(x)c(x) - 2c(x)s(x) = 0,$$

so that $f(x) = $ constant, and therefore $f(x) = f(0) = 0^2 + 1^2 = 1$. So $s(x)^2 + c(x)^2 = 1$, for all x.[1] It follows that $|s(x)| \le 1$ and $|c(x)| \le 1$, for all x.

---

[1] This corresponds to the familiar identity $\sin^2 x + \cos^2 x = 1$. We write $s(x)^2$, *not* $s^2(x)$, because the latter ought to mean $s(s(x))$, whereas we mean $(s(x))^2$. The conventional notation for sines and cosines, where

We now need to find a zero of $c(x)$, and we do this by showing that $c(2) < 0$. Indeed,

$$c(2) = 1 - \frac{2^2}{2!} + \frac{2^4}{4!} - \frac{2^6}{6!} + \frac{2^8}{8!} - \frac{2^{10}}{10!} + \frac{2^{12}}{12!} - \frac{2^{14}}{14!} + \frac{2^{16}}{16!} - \cdots.$$

We bracket the first three terms together, and then the rest in pairs, thus:

$$c(2) = \left(1 - \frac{2^2}{2!} + \frac{2^4}{4!}\right) + \left(-\frac{2^6}{6!} + \frac{2^8}{8!}\right) + \left(-\frac{2^{10}}{10!} + \frac{2^{12}}{12!}\right) + \left(-\frac{2^{14}}{14!} + \frac{2^{16}}{16!}\right) + \cdots.$$

We claim that each bracketed expression is negative. For $1 - \frac{2^2}{2!} + \frac{2^4}{4!} = -\frac{1}{3} < 0$; and provided $n \geq 1$ then $4 < (2n + 1)(2n + 2)$, so that $2^{2n+2}/2^{2n} < (2n + 2)!/(2n)!$, and $2^{2n+2}/(2n + 2)! < 2^{2n}/(2n)!$, or $-2^{2n}/(2n)! + 2^{2n+2}/(2n + 2)! < 0$. Thus $c(2) < 0$, as claimed.

It follows that there is a real number k with $0 < k < 2$ such that $c(k) = 0$, and $c(x) > 0$ for $0 \leq x < k$. Since $s'(x) = c(x)$, we have $s'(x) > 0$ for $0 \leq x < k$, and so $s(x)$ is increasing on this interval, and thus $s(x) > 0$ for $0 < x \leq k$.

Next we obtain the angle-sum formulae. Let a be constant, and put

$$f(x) = s(x)c(a - x) + c(x)s(a - x).$$

It follows that

$$f'(x) = s'(x)c(a - x) - s(x)c'(a - x) + c'(x)s(a - x) - c(x)s'(a - x)$$
$$= c(x)c(a - x) + s(x)s(a - x) - s(x)s(a - x) - c(x)c(a - x)$$
$$= 0,$$

so $f(x)$ is constant, and thus $f(x) = f(0) = s(a)$. This is true for all a and all x, so put $a = \theta + \varphi$ and $x = \theta$ to get

$$s(\theta + \varphi) = s(\theta)c(\varphi) + c(\theta)s(\varphi).$$

A similar argument yields

$$c(\theta + \varphi) = c(\theta)c(\varphi) - s(\theta)s(\varphi).$$

(Details are left to the reader.) Putting $\theta = \varphi$ gives $s(2\theta) = 2s(\theta)c(\theta)$, and $c(2\theta) = c(\theta)^2 - s(\theta)^2 = 2c(\theta)^2 - 1 = 1 - 2s(\theta)^2$.

Next, recall $s(k)^2 + c(k)^2 = 1$, so that, since $c(k) = 0$ and $s(k) > 0$, we have $s(k) = 1$. Thus $s(2k) = 0$ and $c(2k) = -1$, whence $s(4k) = 0$ and $c(4k) = 1$. Thus $s(\theta + 4k) = s(\theta)$, and $c(\theta + 4k) = c(\theta)$, for all θ. Note also that $s(2k - x) = s(x)$, so that $s(x) > 0$ for $0 < x < 2k$, and also $c(x)$ is decreasing for $0 < x < 2k$.

$\sin^2 x$ means $(\sin x)^2$, but $\sin^{-1} x$ does *not* mean $(\sin x)^{-1}$, is appalling, but is so ingrained in most of us that we are probably stuck with it. It is better to use $\arcsin x$ for the inverse function, and we probably ought to write $(\sin x)^2$ instead of $\sin^2 x$. By the same token, we probably ought to write $(s(x))^2$, not $s(x)^2$, just in case anyone confuses the latter with $s(x^2)$; but too many brackets can be very confusing also, so we'll take the risk.

3.6.3 Angle as area

Now let $A = (x_0, y_0)$ be a point on the unit circle $x^2 + y^2 = 1$, and in the upper half-plane $y > 0$ (Fig. 3.16). So $x_0^2 + y_0^2 = 1$, and $y_0 > 0$, and also $-1 < x_0 < 1$. It follows that there is a real number α, with $0 < \alpha < 2k$, such that $c(\alpha) = x_0$, and $s(\alpha) > 0$. But $s(\alpha)^2 = 1 - c(\alpha)^2 = 1 - x_0^2 = y_0^2$, whence $s(\alpha) = y_0$. Put $B = (1, 0)$, and we have, by definition,

$$\angle BOA = 2 \times (\text{area of sector } BOA)$$

$$= 2 \int_0^{y_0} \left(\sqrt{1 - y^2} - \frac{x_0 y}{y_0} \right) dy$$

$$= 2 \int_0^{y_0} \sqrt{1 - y^2}\, dy - x_0 y_0.$$

Substitute $y = s(\theta)$, so that $dy/d\theta = c(\theta)$. We obtain:

$$\angle BOA = 2 \int_0^{\alpha} \sqrt{1 - s(\theta)^2}\, c(\theta)\, d\theta - x_0 y_0$$

$$= 2 \int_0^{\alpha} c(\theta)^2\, d\theta - x_0 y_0$$

$$= \int_0^{\alpha} (1 + c(2\theta))\, d\theta - x_0 y_0$$

$$= \left[\theta + \frac{s(2\theta)}{2} \right]_0^{\alpha} - x_0 y_0$$

$$= \alpha + \frac{s(2\alpha)}{2} - s(\alpha)c(\alpha)$$

$$= \alpha,$$

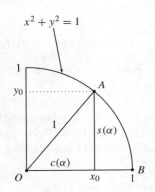

Fig. 3.16 Calculation of $\angle BOA$

which is just what we wanted to happen. If we put $C = (x_0, 0)$, then $\triangle OCA$ is a right-angled triangle with hypotenuse OA of length 1, and $s(\alpha) = |CA| = |CA|/|OA|$, $c(\alpha) = |OC| = |OC|/|OA|$. We have at last obtained the two formulae we wanted, giving $s(\alpha)$ and $c(\alpha)$ as ratios of lengths of sides of a right-angled triangle, and so we can finally give the functions s and c their familiar names: $s(\alpha) = \sin \alpha$ and $c(\alpha) = \cos \alpha$. More generally, let $\lambda > 0$ and put $A' = \lambda(x_0, y_0)$ and $C' = \lambda(x_0, 0)$, so that $\triangle OC'A'$ is a right-angled triangle with hypotenuse of length λ, and (by definition) $\angle C'OA' = \angle COA$. Then $|C'A'|/|OA'| = \lambda|CA|/\lambda = |CA| = \sin \alpha$, and similarly $|O'C'|/|OA'| = \cos \alpha$.

- Ex.3.18: *Show that the gradient of OA is $\tan \alpha$ (defined to be $\sin \alpha / \cos \alpha$, of course).*

As a special case, when $x_0 = -1$ and $y_0 = 0$, we have $\angle BOA = 2k$. But this is twice the area of the semicircle, which by definition is $2(\pi/2) = \pi$, and so $k = \pi/2$, as claimed earlier.

As another example, if $A = (0, 1)$ (so that $OA \perp OB$), then $\angle BOA = \pi/2$. If we take $A = (x_0, y_0)$ with $y_0 < 0$, then the value for $\angle BOA$, calculated as above, will be negative. Sometimes this will matter, and sometimes we shall take the absolute value of the angle and not bother about signs. So, for example, if $A = (0, -1)$, we get $\angle BOA = -\pi/2$, and so, properly, $\angle AOB = \pi/2$. But if the direction of measurement is unimportant we shall cheerfully write $\angle BOA = \pi/2$. There is also the business of working modulo 2π to worry about; normally we shall work with the principal values of angles: $-\pi < \alpha \leq \pi$, or sometimes (if it is more convenient) $0 \leq \alpha < 2\pi$. (In this case, with $A = (0, -1)$, still, $\angle BOA = 3\pi/2$, a reflex angle.) If we are ignoring signs, we can insist that our angles α satisfy $0 \leq \alpha \leq \pi$; this requires some care, as we still need to work modulo 2π, not modulo π.

Strictly, we have only defined the angle between the x-axis and another line through the origin, so far. The angle between two arbitrary lines through the origin can be expressed in terms of such angles; and the angle between two lines meeting at a point P is defined to be the same as the angle between the images of these lines under translation through $\overrightarrow{PO}$, that is, between two lines through the origin, parallel to the given lines. Indeed, since angle is defined in terms of area, which is defined in terms of distance, isometries must preserve angles:

Proposition 3.24. *Let $\triangle ABC \equiv \triangle A'B'C'$. Then $\angle ABC = \angle A'B'C'$, $\angle BCA = \angle B'C'A'$, and $\angle CAB = \angle C'A'B'$.* □

Corollary 3.25. (Pons asinorum[2]) *Let $|AB| = |AC|$, so that $\triangle ABC$ is isosceles. Then $\angle ABC = \angle ACB$ (Fig. 3.17).*

Proof. We shall use Proposition 3.22 to show that $\triangle ABC \equiv \triangle ACB$ (!), and the result will then follow from Proposition 3.24, with $A' = A$, $B' = C$, and $C' = B$.

We have $|AB| = |AC|$ (that is, $|AB| = |A'B'|$), and $|AC| = |AB|$ (that is, $|AC| = |A'C'|$), and also $|BC| = |CB|$ (that is, $|BC| = |B'C'|$). The result follows! □

---

[2] The asses' bridge. This probably refers to the bridge-like shape of the isosceles triangle.

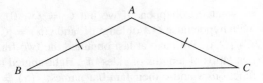

Fig. 3.17   Pons asinorum

The above strange but delightfully economical proof should bring home the importance of the note preceding Proposition 3.22. There are easier proofs (conceptually easier, that is): for example, since $|AB| = |AC|$, the point A must lie on the perpendicular bisector AX of BC, where X is the mid-point of BC. Reflection in AX now shows that $\angle ABX = \angle ACX$, as required.

- Ex.3.19: *With the notation of Exercise 2.14, use the translation through $\overrightarrow{RS}$, and the half-turns about R, S, and the mid-point of RS, to prove that vertically opposite angles are equal, corresponding angles are equal, alternate angles are equal, adjacent angles are supplementary, and interior angles are supplementary. (The terminology is all explained in and just before Exercise 2.14.)*

- Ex.3.20: *Given $\triangle PQR$ with $PQ \perp QR$, define $C = (|PQ|, 0)$ and $A = (|PQ|, |QR|)$. Use Proposition 3.22 to show that $\triangle PQR \equiv \triangle OCA$, whence $\angle COA = \angle QPR$, and hence $\sin\angle QPR = |QR|/|PR|$ and $\cos\angle QPR = |PQ|/|PR|$.*

- Ex.3.21: *Work out*

$$\int_0^1 \frac{dx}{1+x^2}$$

  two ways: (i) by substituting $x = \tan\theta$, and (ii) by first expanding $(1+x^2)^{-1}$ as an infinite series. Hence show that

$$\frac{\pi}{4} = 1 - \frac{1}{3} + \frac{1}{5} - \frac{1}{7} + \cdots.$$

- Ex.3.22: *Work out*

$$\int_0^1 \frac{x^4(1-x)^4}{1+x^2}\, dx$$

  and use your answer to show that $\pi \approx 22/7$. (You should be able to show easily that $0 < \frac{22}{7} - \pi < \frac{1}{256}$, though the approximation is a little better than this suggests. An even better rational approximation is $355/113$, which agrees with π to six places of decimals, but no proof is offered here.)

3.7   POLAR COORDINATES AND ROTATIONS

Let $A = (x, y)$. Here x, y are the *cartesian* coordinates of A. The *polar* coordinates of A are the quantities r, θ, where $r = |OA|$ and $\theta = \angle BOA$, with $B = (1, 0)$ (Fig. 3.18).

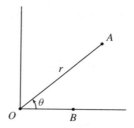

Fig. 3.18 Polar coordinates

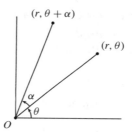

Fig. 3.19 Rotation through α

Thus (r, θ) can be obtained from (x, y) by $r^2 = x^2 + y^2$ and $\tan \theta = y/x$, and (x, y) can be obtained from (r, θ) by $x = r \cos \theta$ and $y = r \sin \theta$.

Definition 3.26. **Rotation** *about the centre O through the angle α is the map $f : \mathbb{R}^2 \to \mathbb{R}^2$ given in terms of polar coordinates by $f : (r, \theta) \mapsto (r, \theta + \alpha)$ (Fig. 3.19).*

Proposition 3.27. *Rotation is an isometry.*

Proof. Let $f, \alpha, A, x, y, r, \theta$ be as above, and note that f has an inverse, f^{-1}, which is just the rotation through $-\alpha$ about O. So f is a transformation of $\mathbb{R}^2$.

Next, if $f(A) = A' = (x', y')$, then

$$x' = r \cos(\theta + \alpha)$$
$$= r \cos \theta \cos \alpha - r \sin \theta \sin \alpha$$
$$= x \cos \alpha - y \sin \alpha, \tag{3.6}$$

and similarly

$$y' = x \sin \alpha + y \cos \alpha. \tag{3.7}$$

For $i = 1, 2$, let $A_i = (x_i, y_i)$, with $f(A_i) = A'_i = (x'_i, y'_i)$. We obtain equations like (3.6) and (3.7) relating x'_i, y'_i to x_i, y_i, and on subtracting and putting $x = x_1 - x_2$ and

$y = y_1 - y_2$, we obtain *exactly* (3.6) and (3.7). It follows from these that $(x')^2 + (y')^2 = r^2 = x^2 + y^2$, or

$$(x_1' - x_2')^2 + (y_1' - y_2')^2 = (x_1 - x_2)^2 + (y_1 - y_2)^2,$$

as required.   □

- Ex.3.23:   *Give an alternative proof of Proposition 3.27 by congruent triangles: show that, if f is a rotation about O with $f(A) = A'$, $f(B) = B'$, then provided O, A, B are not collinear, we have $\triangle AOB \equiv \triangle A'OB'$, so that $|AB| = |A'B'|$. (You may need to use Corollary 4.13.) What if O, A, B are collinear?*

- Ex.3.24:   *Use the technique of Example 4, page 33, to define a rotation through α about an arbitrary point P, and prove that the map obtained is an isometry.*

3.8   DOT PRODUCTS

We have met the dot product informally, but now let us introduce it officially:

Definition 3.28.   *Let $(x_1, y_1), (x_2, y_2) \in \mathbb{R}^2$. The* **dot product** *of these vectors is*

$$(x_1, y_1) \cdot (x_2, y_2) = x_1 x_2 + y_1 y_2.$$

Note that, if $\mathbf{u}, \mathbf{v} \in \mathbb{R}^2$, then $\mathbf{u} \cdot \mathbf{v} \in \mathbb{R}$, that is, $\mathbf{u} \cdot \mathbf{v}$ is a scalar, not a vector. For this reason the dot product is also known as the *scalar* product. (We have previously used dots for *meets*: $AB \cdot CD$ means the point where the line AB meets the line CD, but then AB and CD are not vectors, so no confusion should arise between the two usages.)

We leave it to the reader to check that the following hold for all $\mathbf{u}, \mathbf{v}, \mathbf{w} \in \mathbb{R}^2$ and $\lambda \in \mathbb{R}$:

$$\mathbf{u} \cdot \mathbf{v} = \mathbf{v} \cdot \mathbf{u},$$

$$\mathbf{u} \cdot \mathbf{u} = |\mathbf{u}|^2,$$

$$(\lambda \mathbf{u}) \cdot \mathbf{v} = \lambda (\mathbf{u} \cdot \mathbf{v}),$$

$$\mathbf{u} \cdot (\mathbf{v} + \mathbf{w}) = \mathbf{u} \cdot \mathbf{v} + \mathbf{u} \cdot \mathbf{w}.$$

The definition of orthogonality (Definition 3.11) was given in terms of line segments. We apply this to vectors in the obvious way: $\overrightarrow{AB} \perp \overrightarrow{BC}$ will mean the same as $AB \perp BC$. It seems sensible at the same time to extend the definition and agree that the zero vector is orthogonal to *every* vector. This allows us, for example, to *avoid* having to insist that $\mathbf{u}, \mathbf{v}$ are non-zero in the next lemma:

Lemma 3.29.   *Let $\mathbf{u}, \mathbf{v} \in \mathbb{R}^2$. Then $\mathbf{u} \cdot \mathbf{v} = 0 \Leftrightarrow \mathbf{u} \perp \mathbf{v}$.*

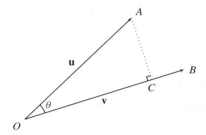

Fig. 3.20 Proposition 3.30

Proof. This is really Lemma 3.12 again, but we can now give the proof in vector notation:

$$\mathbf{u} \perp \mathbf{v} \Leftrightarrow |\mathbf{u}|^2 + |\mathbf{v}|^2 = |\mathbf{u} - \mathbf{v}|^2,$$

by Definition 3.11, as extended, above. But

$$|\mathbf{u} - \mathbf{v}|^2 = (\mathbf{u} - \mathbf{v}) \cdot (\mathbf{u} - \mathbf{v}) = \mathbf{u} \cdot \mathbf{u} - \mathbf{u} \cdot \mathbf{v} - \mathbf{v} \cdot \mathbf{u} + \mathbf{v} \cdot \mathbf{v} = |\mathbf{u}|^2 - 2\mathbf{u} \cdot \mathbf{v} + |\mathbf{v}|^2,$$

whence the result. □

Proposition 3.30. *Let* $\mathbf{u}, \mathbf{v} \in \mathbb{R}^2$. *Then* $\mathbf{u} \cdot \mathbf{v} = |\mathbf{u}||\mathbf{v}| \cos \theta$, *where* θ *is the angle between* $\mathbf{u}$ *and* $\mathbf{v}$.

Proof. The result is easy if $\theta = 0, \pi/2,$ or π, so assume θ does not take any of these values. Let $\mathbf{u} = \overrightarrow{OA}$ and $\mathbf{v} = \overrightarrow{OB}$, and let the line through A, orthogonal to OB, meet OB in C (Fig. 3.20). So $OC \perp CA$, and $|OA| = |\mathbf{u}|$, and $\angle COA = \theta$. By Exercise 3.20, $|OC| = |\mathbf{u}| \cos \theta$, provided θ is acute (that is, $0 < \theta < \pi/2$), and so we multiply the normalization of $\mathbf{v}$ by this length to get

$$\overrightarrow{OC} = |\mathbf{u}| \cos \theta \frac{\mathbf{v}}{|\mathbf{v}|}.$$

Since $\cos(\pi - \theta) = -\cos \theta$, this last equation still holds if θ is obtuse, that is, $\pi/2 < \theta < \pi$. Then

$$\overrightarrow{CA} = \overrightarrow{OA} - \overrightarrow{OC} = \mathbf{u} - \frac{|\mathbf{u}| \cos \theta}{|\mathbf{v}|}\mathbf{v},$$

so, by Lemma 3.29,

$$\left(\mathbf{u} - \frac{|\mathbf{u}| \cos \theta}{|\mathbf{v}|}\mathbf{v} \right) \cdot \mathbf{v} = 0,$$

or

$$\mathbf{u} \cdot \mathbf{v} = \frac{|\mathbf{u}| \cos \theta}{|\mathbf{v}|}\mathbf{v} \cdot \mathbf{v} = |\mathbf{u}||\mathbf{v}| \cos \theta. \qquad \square$$

The construction just used is called *dropping a perpendicular* from A onto OB. (See also Exercise 2.8.) The line segment AC is the *perpendicular*, C is its *foot*, and the line-segment OC, or sometimes its length $|OC|$, is the *orthogonal projection* of OA onto OB.

- Ex.3.25: *If $|\mathbf{u}| = |\mathbf{v}|$ then $(\mathbf{u} - \mathbf{v}) \cdot (\mathbf{u} + \mathbf{v}) = 0$. Interpret this geometrically.*

3.8.1   Distance from point to line

Proposition 3.31.   *The perpendicular distance from the point (x_0, y_0) to the line $px + qy + r = 0$ is*

$$\frac{|px_0 + qy_0 + r|}{\sqrt{p^2 + q^2}}.$$

Proof.   Let (x_1, y_1) and (x_2, y_2) be any two points of the line, so that $px_1 + qy_1 = -r = px_2 + qy_2$. Thus $p(x_1 - x_2) + q(y_1 - y_2) = 0$, so the vector (p, q) is orthogonal to the line. The required distance is the orthogonal projection of $(x_0 - x_1, y_0 - y_1)$ onto (p, q), which is

$$\frac{|p(x_0 - x_1) + q(y_0 - y_1)|}{|(p, q)|} = \frac{|px_0 + qy_0 + r|}{\sqrt{p^2 + q^2}},$$

as required.   □

If the absolute value signs are omitted in the above formula, then the sign of the formula distinguishes points on one side of the line from points on the other side. Points *on* the line give the value zero, of course.

Since the equation of a line may be multiplied by any non-zero scalar without affecting the line, it is sometimes convenient to divide through by $\sqrt{p^2 + q^2}$ and so assume that the line $px + qy + r = 0$ has been *normalized*, that is, that $p^2 + q^2 = 1$. The distance from (x_0, y_0) to this line then takes the particularly simple form $|px_0 + qy_0 + r|$.

3.8.2   Angle bisectors

Proposition 3.32.   *Given two non-parallel lines, the set of points equidistant from the two lines forms a pair of orthogonal lines.*

Proof.   Let the lines be $p_1x + q_1y + r_1 = 0$, and $p_2x + q_2y + r_2 = 0$, assumed normalized. Then (x, y) is equidistant from these lines iff

$$|p_1x + q_1y + r_1| = |p_2x + q_2y + r_2|,$$

by Proposition 3.31. This happens iff either

$$p_1x + q_1y + r_1 = p_2x + q_2y + r_2 \quad \text{or} \quad p_1x + q_1y + r_1 = -(p_2x + q_2y + r_2),$$

that is,

$$(p_1 - p_2)x + (q_1 - q_2)y + (r_1 - r_2) = 0 \quad \text{or} \quad (p_1 + p_2)x + (q_1 + q_2)y + (r_1 + r_2) = 0.$$

This is a pair of straight lines, and they are orthogonal by Exercise 3.11, because

$$(p_1 - p_2)(p_1 + p_2) + (q_1 - q_2)(q_1 + q_2) = p_1^2 + q_1^2 - p_2^2 - q_2^2 = 1 - 1 = 0.   □$$

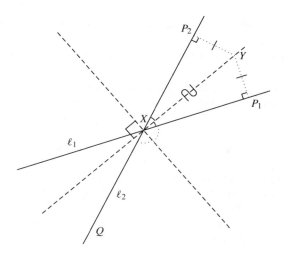

Fig. 3.21 Angle bisectors

The lines just constructed are the *angle bisectors* of the given lines. To see what this means, let the given lines be ℓ_1, ℓ_2, meeting at X, and let Y be on one of the lines constructed in Proposition 3.32, with $Y \neq X$. Let P_1, P_2 be the feet of the perpendiculars from Y onto ℓ_1, ℓ_2 respectively (Fig. 3.21). Then by construction we have $|YP_1| = |YP_2|$. Since the side XY is common to the triangles XYP_1, XYP_2, and $\angle XP_1Y = \pi/2 = \angle XP_2Y$, it follows that $\triangle XP_1Y \equiv \triangle XP_2Y$, by Corollary 3.23. Thus $\angle YXP_1 = \angle YXP_2 = \frac{1}{2}\angle P_1XP_2$, or in other words, XY *bisects* the angle P_1XP_2; XY is the *(internal) bisector* of $\angle P_1XP_2$. If we choose Q on P_2X produced, then the other line constructed in Proposition 3.32 bisects $\angle P_1XQ$; we also say that this line bisects $\angle P_1XP_2$ *externally*, or is the *external bisector* of $\angle P_1XP_2$.

ANSWERS TO EXERCISES

3.1: The line through (x_1, y_1) and (x_2, y_2) is $(x - x_1)(y_2 - y_1) = (y - y_1)(x_2 - x_1)$, and its gradient is $(y_2 - y_1)/(x_2 - x_1)$. OK? Similarly the line through (x_3, y_3) and (x_4, y_4) has gradient $(y_4 - y_3)/(x_4 - x_3)$, so the two lines are parallel iff $(y_2 - y_1)/(x_2 - x_1) = (y_4 - y_3)/(x_4 - x_3)$, or $(y_2 - y_1)/(y_4 - y_3) = (x_2 - x_1)/(x_4 - x_3) = \lambda$, say, or $(x_2 - x_1, y_2 - y_1) = \lambda(x_4 - x_3, y_4 - y_3)$, as required. (We have blithely assumed that no denominators vanish, so please attend to the exceptions.)

3.2: The equation *is* a line, and it is satisfied by $x = x_1, y = y_1$ and also by $x = x_2, y = y_2$.

3.3: $\mathbf{b} = \mathbf{a}_1 + \xi(\mathbf{a}_2 - \mathbf{a}_1) = \mathbf{a}_2 + (1 - \xi)(\mathbf{a}_1 - \mathbf{a}_2)$, and if $0 < \xi < 1$, then $0 < 1 - \xi < 1$.

3.4: We have $(1 - \xi)\mathbf{a}_1 + \xi\mathbf{a}_2 = \mathbf{b} = ((1 - \xi) + \xi)\mathbf{b}$, so $(1 - \xi)(\mathbf{b} - \mathbf{a}_1) = \xi(\mathbf{a}_2 - \mathbf{b})$. Put $\lambda = \xi/(1 - \xi)$, and note $0 < \xi < 1$ iff $\lambda > 0$.

3.5: In the notation of the last answer, $\xi = 1/2$ iff $\lambda = 1$.

3.6: $\overrightarrow{B_iB_j} = \mathbf{b}_j - \mathbf{b}_i = (\xi_j - \xi_i)(\mathbf{a}_2 - \mathbf{a}_1)$, so $\overrightarrow{B_1B_2} = \lambda\overrightarrow{B_2B_3}$ where $\lambda = (\xi_2 - \xi_1)/(\xi_3 - \xi_2)$. Thus $\lambda > 0 \Leftrightarrow (\xi_1 - \xi_2)(\xi_2 - \xi_3) > 0 \Leftrightarrow$ either $\xi_1 < \xi_2 < \xi_3$ or $\xi_1 > \xi_2 > \xi_3$.

3.7: Given distinct real numbers ξ_1, ξ_2, ξ_3, exactly one of them is between the other two. Now use the last exercise.

3.8: If $(x_1 - x_2)^2 + (y_1 - y_2)^2 = 0$, then $x_1 - x_2 = 0$ and $y_1 - y_2 = 0$, so $(x_1, y_1) = (x_2, y_2)$. The converse is obvious.

3.9: Let Q, R, S, T be the feet of the perpendiculars from P onto AB, BC, CD, DA (produced if necessary), respectively. Then $AQPT$ is a rectangle, so $|PA|^2 = |PQ|^2 + |QA|^2 = |PQ|^2 + |PT|^2$. Likewise, $|PC|^2 = |PR|^2 + |PS|^2$, so that $|PA|^2 + |PC|^2 = |PQ|^2 + |PR|^2 + |PS|^2 + |PT|^2$, and similarly for $|PB|^2 + |PD|^2$.

3.10: $|PA|$, $|PB|$, $|PD|$ are constant, so $|PC|$ is also constant, by the last exercise. When A is outside Σ, we get no locus at all unless A is close enough to Σ that its tangents to Σ form an obtuse angle. Then the locus of C is an *arc* of a circle. (Where are its end-points?) To get the whole circle, note that for each B there are *two* positions of D, giving two positions of C which *together* trace out the whole circle. (So now what about the second position of C when A is *inside* Σ?)

3.11: If $q_1 q_2 \neq 0$, then $-1 = m_1 m_2 = (-p_1/q_1)(-p_2/q_2) = p_1 p_2/q_1 q_2$, whence the result. If $q_1 = 0$ then $m_1 = \infty$, $m_2 = 0$ and so $p_2 = 0$, and the result follows again; similarly if $q_2 = 0$.

3.12: Writing ℓ, m, n for the *gradients*, we have $\ell m = -1 = mn$, so that $\ell = n$, provided no gradients are 0 or ∞; we leave these exceptional cases to the reader.

3.13: Given an isometry f, put $g = f^{-1}$; and given P, Q, put $P' = g(P)$ and $Q' = g(Q)$. We must show $|PQ| = |P'Q'|$. But this follows because $f(P') = P$ and $f(Q') = Q$, and f is an isometry. For the composite, you can write down some symbols, or you can say: if two maps each preserve distance, and if you apply one after the other, then you have preserved distance *twice*, which means you have preserved distance!

3.14: The image of (x, y) is $(2a - x, 2b - y)$, $(a + b - y, x - a + b)$, $(y + a - b, a + b - x)$, respectively.

3.15: Let ℓ be $px + qy + r = 0$; if $\mathbf{v} = (a, b)$ then ℓ' is $p(x - a) + q(y - b) + r = 0$, or $px + qy + r' = 0$, where $r' = r - pa - qb$. So $\ell \parallel \ell'$, since each has gradient $m = -p/q$. If $\ell \parallel \mathbf{v}$ then $m = b/a$ (OK?) so $pa + qb = 0$ and $r = r'$, so that $\ell = \ell'$. (As an alternative, let ℓ be $\mathbf{r} = \mathbf{a}_1 + \xi(\mathbf{a}_2 - \mathbf{a}_1)$. Then ℓ' is $\mathbf{r} = f(\mathbf{a}_1) + \xi(f(\mathbf{a}_2) - f(\mathbf{a}_1)) = \mathbf{a}_1 + \mathbf{v} + \xi(\mathbf{a}_2 - \mathbf{a}_1)$. If $\ell \parallel \mathbf{v}$, then $\mathbf{v} = \lambda(\mathbf{a}_2 - \mathbf{a}_1)$, say, so ℓ' is $\mathbf{r} = \zeta(\mathbf{a}_2 - \mathbf{a}_1)$, where $\zeta = \xi + \lambda$.)

3.16: $px + qy + r = 0$ becomes $-px - qy + r = 0$, or $px + qy - r = 0$, which is parallel to the original line, and the *same* iff $r = 0$. For the half-turn about (a, b), the line $px + qy + r = 0$ becomes $p(2a - x) + q(2b - y) + r = 0$, or $px + qy - (r + 2pa + 2qb) = 0$, which is parallel to the original line. The lines coincide iff $r = -(r + 2pa + 2qb)$, or $pa + qb + r = 0$, which is the condition that (a, b) lies on $px + qy + r = 0$.

3.17: The line $px + qy + r = 0$ becomes $-py + qx + r = 0$, or $qx - py + r = 0$, and the orthogonality follows by Exercise 3.11. The arguments for the other quarter-turn, and the general case, are left to the reader.

3.18: Easy: $O = (0, 0)$ and $A = (x_0, y_0)$, so the gradient of OA is $(y_0 - 0)/(x_0 - 0) = y_0/x_0 = s(\alpha)/c(\alpha) = \sin(\alpha)/\cos(\alpha) = \tan(\alpha)$.

3.19: It makes life a lot simpler (and makes no difference to the angles) if we position the various points (see Exercise 2.14) on the three lines so that $|AR| = |RB| = |CS| = |SD|$ and $|PR| = |RS| = |SQ|$. Then if $X \mapsto X'$ is translation by $\overrightarrow{RS}$, then $A' = C$, $R' = S$, and $P' = R$, so that $\angle ARP = \angle A'R'P' = \angle CSR$. This sorts out one pair of equal (corresponding) angles, and the other parts of the exercise are done more or less similarly.

3.20: We have $\overrightarrow{OC} = (|PQ|, 0)$, $\overrightarrow{CA} = (|PQ|, |QR|) - (|PQ|, 0) = (0, |QR|)$, and $\overrightarrow{OA} = (|PQ|, |QR|)$. So $|OC| = |PQ|$, $|CA| = |QR|$, and $|OA| = \sqrt{|PQ|^2 + |QR|^2} = |PR|$. Thus $\triangle PQR \equiv \triangle OCA$ and $\angle COA = \angle QPR$, so $\sin \angle QPR = \sin \angle COA = |CA|/|OA| = |QR|/|PR|$; and similarly for the cosine.

3.21: (i) is standard:

$$\int_0^1 \frac{dx}{1+x^2} = \int_0^{\pi/4} \frac{\sec^2 \theta \, d\theta}{1 + \tan^2 \theta} = [\theta]_0^{\pi/4} = \frac{\pi}{4}.$$

For (ii), we have

$$(1 + x^2)^{-1} = 1 - x^2 + x^4 - x^6 + \cdots,$$

which integrates to

$$\left[x - \frac{x^3}{3} + \frac{x^5}{5} - \frac{x^7}{7} + \cdots \right]_0^1 = 1 - \frac{1}{3} + \frac{1}{5} - \frac{1}{7} + \cdots.$$

For the legitimacy of equating the integral of the sum of an infinite series with the answer obtained by integrating the terms first and summing the series afterwards, ask an analyst.

3.22: We have

$$\frac{x^4(1-x)^4}{1+x^2} = x^6 - 4x^5 + 5x^4 - 4x^2 + 4 - 4\frac{1}{1+x^2},$$

so integrating gives $[\frac{1}{7}x^7 - \frac{2}{3}x^6 + x^5 - \frac{4}{3}x^3 + 4x - 4\arctan x]_0^1 = \frac{1}{7} - \frac{2}{3} + 1 - \frac{4}{3} + 4 - \pi = \frac{22}{7} - \pi$. Now for $0 < x < 1$ we have

$$0 < \frac{x^4(1-x)^4}{1+x^2} < x^4(1-x)^4 \leq (\tfrac{1}{2})^4 (1 - \tfrac{1}{2})^4 = \frac{1}{256}.$$

Integrating, $0 < \frac{22}{7} - \pi < \frac{1}{256}$.

3.23: Let $A = (r, \theta)$ and $B = (s, \phi)$, where $r > 0$ and $s > 0$. Then $A' = (r, \theta + \alpha)$ and $B' = (s, \phi + \alpha)$, where α is the angle of rotation. So now $|AO| = r = |A'O|$, and $|OB| = s = |OB'|$, and finally $\angle AOB = |\theta - \phi| = |(\theta + \alpha) - (\phi + \alpha)| = \angle A'OB'$. We deduce that $\triangle AOB \equiv \triangle A'OB'$, by the case SAS (see p. 60), and thus $|AB| = |A'B'|$, as required. OK? *No*, actually: we have ignored possible difficulties caused by the fact that the angle in polar coordinates is determined only up to an integer multiple of 2π. So for example if $A = (1, 3\pi/4)$ and $B = (1, -3\pi/4)$, then most people would want to say $\angle AOB = \pi/2$ (draw it!) whereas the above calculation gives $\angle AOB = |3\pi/4 - (-3\pi/4)| = 3\pi/2$, the reflex angle. The answer is that it is always possible to alter θ (or ϕ) by an integer multiple of 2π so as to ensure that $|\theta - \phi| \leq \pi$, and then everything works as it should. For the collinear case we have either $|\theta - \phi| = 0$, when $|AB| = |r - s| = |A'B'|$, or else $|\theta - \phi| = \pi$, when $|AB| = r + s = |A'B'|$.

3.24: Let h be the translation by $\overrightarrow{OP}$, and let f be the rotation about O through the given angle α. Put $g = hfh^{-1}$, which is an isometry by Exercise 3.13. We claim g is the required rotation. Note that $g(P) = hfh^{-1}(P) = hf(O) = h(O) = P$, and so if $C \neq P$ and $g(C) = D$, then $|PC| = |PD|$. Put $h^{-1}(C) = A$ and $f(A) = B$, so that $\angle AOB = \alpha$. Then $h(A) = C$, $h(O) = P$, and $h(B) = hf(A) = hfh^{-1}(C) = g(C) = D$, so that $\triangle AOB \equiv \triangle CPD$, whence $\angle CPD = \alpha$, and we are done.

3.25: $(\mathbf{u} - \mathbf{v}) \cdot (\mathbf{u} + \mathbf{v}) = \mathbf{u} \cdot \mathbf{u} + \mathbf{u} \cdot \mathbf{v} - \mathbf{v} \cdot \mathbf{u} - \mathbf{v} \cdot \mathbf{v} = |\mathbf{u}|^2 - |\mathbf{v}|^2 = 0$. This says that the diagonals of a rhombus (an equal-sided parallelogram) are orthogonal *or* it says that the angle in a semicircle is a right angle.

4

Triangles and triangle formulae

We shall now introduce some of the special points associated with a triangle, and obtain formulae relating the various lengths and angles involved. So, given $\triangle ABC$, which just means given three non-collinear points A, B, C, we shall use corresponding lower-case bold $\mathbf{a}$, $\mathbf{b}$, $\mathbf{c}$ to stand for the position vectors $\overrightarrow{OA}$, $\overrightarrow{OB}$, $\overrightarrow{OC}$ respectively, where O is the origin; and similarly for any further points introduced.

We shall use the following standard notation: $a = |BC|$, $b = |CA|$, $c = |AB|$, and also $A = \angle BAC$, $B = \angle CBA$, $C = \angle ACB$.

We have not given a definition of what it means for a point to be *inside* $\triangle ABC$. Formal arguments about insides and outsides of triangles can very easily get excessively technical, and for the time being at least we propose to treat such matters informally. So, given $\triangle ABC$, the line AB divides the plane into two regions, called *half-planes*, one on each side. The line BC, not being parallel to AB, divides each of these half-planes into two *angular* regions, four in all, and all *unbounded*, that is, containing points arbitrarily far from the origin. The third side, CA, meets only three of these angular regions, each of which it divides into two regions, and so the three sides of the triangle divide the plane into *seven* regions. All but one of these regions is unbounded; they (together) constitute the *outside* of the triangle, and the remaining region, which is bounded, is the *inside* of $\triangle ABC$. (A region is *bounded* if it is not unbounded, that is, there is a number $k > 0$ such that $|OX| < k$ for every point X of the region.)

There are various equivalent ways of defining *inside*, which we give without proof:

(a) X is inside $\triangle ABC$ if AX meets BC in Y, with Y between B and C, and X between A and Y.

(b) X is inside $\triangle ABC$ if AX meets BC in Y, and BX meets CA in Z, with Y between B and C, and Z between C and A.

(c) X is inside $\triangle ABC$ if there is a line through X meeting BC in Y and CA in Z, with Y between B and C, and Z between C and A, and X between Y and Z.

(d) X is inside $\triangle ABC$ if it is on the same side of BC as A, on the same side of CA as B, and on the same side of AB as C.

Over the next few pages, we shall build up a list of different sets of sufficient conditions for two triangles to be congruent. Proposition 3.22, in the previous chapter, says it is sufficient if the *three pairs of sides* are equal: we shall abbreviate this to SSS (standing for side, side, side). Corollary 3.23 says it is sufficient if *two* pairs of sides are equal, and if also a *non*-included angle (that is, one of the angles *not* between the chosen sides) is a right angle in each triangle. This is the case of *two sides and a non-included right angle*, or SSR (side, side, right angle) for short.

- Ex.4.1:   *Show that if* $|AB| = |A'B'|$, *and* $|BC| = |B'C'|$, *and also* $\angle ABC = \pi/2 = \angle A'B'C'$, *then* $\triangle ABC \equiv \triangle A'B'C'$.

This exercise would be the case of two sides and an *included* right angle, but for this to work there is no need for the included angles to be right angles, as we shall see below, in Corollary 4.13.

- Ex.4.2:   *Given triangles* ABC *and* $A'B'C'$, *it is not possible to have* $|AB| = |A'B'|$, *and* $|AC| = |A'C'|$, *and also* $\angle ABC = \pi/2 = \angle B'C'A'$.

So in the case SSR, we do not have to worry about *which* of the non-included angles are right angles: there is at most one possibility.

Definition 4.1.   *The lines* $\ell, m, n, \ldots$ *are* **concurrent** *if they have a point in common.*

4.1   THE CENTROID

Let the mid-points of BC, CA, AB be D, E, F respectively. The lines AD, BE, CF are the *medians* of $\triangle ABC$. In terms of vectors, we have $\mathbf{d} = \frac{1}{2}(\mathbf{b} + \mathbf{c})$, and so on. The median AD has two points of trisection; let the one nearer D be G (Fig. 4.1), so that $\overrightarrow{DG} = \frac{1}{3}\overrightarrow{DA}$, that is, $\mathbf{g} - \mathbf{d} = \frac{1}{3}(\mathbf{a} - \mathbf{d})$, or

$$\mathbf{g} = \tfrac{1}{3}\mathbf{a} + \tfrac{2}{3}\mathbf{d} = \tfrac{1}{3}\mathbf{a} + \tfrac{2}{3}\left(\tfrac{1}{2}(\mathbf{b} + \mathbf{c})\right) = \tfrac{1}{3}(\mathbf{a} + \mathbf{b} + \mathbf{c}).$$

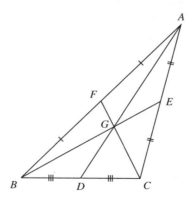

Fig. 4.1   Proposition 4.2: the centroid

The fact that this expression is symmetrical in **a**, **b**, **c** shows that G also lies on the other medians, BE and CF. We have proved:

Proposition 4.2.   *The medians of a triangle are concurrent.*   □

Definition 4.3.   *G is the* **centroid** *of* $\triangle ABC$.

Note that, by condition (a) above, G is always inside $\triangle ABC$. The centroid is the centre of mass of a system of three unit masses placed at A, B, C. For the masses at B, C can be replaced by a mass of magnitude 2 at the mid-point D, and the system of a unit mass at A and a mass of magnitude 2 at D clearly has its centre of mass at G, the point of trisection.

It is less clear, though nonetheless true, that G is also the centre of mass of the uniform triangular lamina ABC. Informally, we can slice the triangle into thin slices parallel to BC; each slice is approximately a uniform rod, and has its centre of mass at its mid-point, which is (the reader may wish to verify) on the median AD. So the centre of mass of the lamina must lie on this median, and hence (similarly) on the other medians also, and so it is at G. (The reader might like to try to prove this result by coordinate geometry, using integration to take moments.)

- Ex.4.3:   *Prove that, if $ABCD$ is a parallelogram, then the centre of mass of a system of four unit masses at A, B, C, D coincides with the centre of mass of the uniform lamina $ABCD$.*

4.2   THE ORTHOCENTRE

The perpendicular from a vertex of a triangle onto the opposite side is called an *altitude* of the triangle.

Proposition 4.4.   *The altitudes of a triangle are concurrent.*

Proof.   Let the altitudes of $\triangle ABC$ be AP, BQ, CR, and let AP, BQ meet at H (Fig. 4.2). Thus $AH \perp BC$ and $BH \perp CA$, that is, $(\mathbf{h} - \mathbf{a}) \cdot (\mathbf{c} - \mathbf{b}) = 0$ and

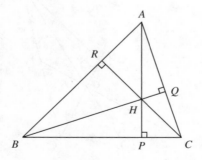

Fig. 4.2   Proposition 4.4: the orthocentre

$(\mathbf{h} - \mathbf{b}) \cdot (\mathbf{a} - \mathbf{c}) = 0$. Expanding and rearranging,

$$\mathbf{h} \cdot (\mathbf{c} - \mathbf{b}) = \mathbf{a} \cdot \mathbf{c} - \mathbf{a} \cdot \mathbf{b}$$

and

$$\mathbf{h} \cdot (\mathbf{a} - \mathbf{c}) = \mathbf{b} \cdot \mathbf{a} - \mathbf{b} \cdot \mathbf{c}.$$

Adding,

$$\mathbf{h} \cdot (\mathbf{a} - \mathbf{b}) = \mathbf{a} \cdot \mathbf{c} - \mathbf{b} \cdot \mathbf{c},$$

or $(\mathbf{h} - \mathbf{c}) \cdot (\mathbf{a} - \mathbf{b}) = 0$, whence $HC \perp BA$, and the result follows. □

Definition 4.5. *H is the* **orthocentre** *of* $\triangle ABC$.

Note that A is the orthocentre of $\triangle HBC$; B is the orthocentre of $\triangle AHC$; and C is the orthocentre of $\triangle ABH$. The set of points A, B, C, H, each being the orthocentre of the triangle formed by the other three, is called an *orthocentric tetrad*.

We may ask, when is H inside $\triangle ABC$? The point P is between B and C iff both B and C are acute, and similarly Q is between C and A iff both C and A are acute. By condition (b), page 52, H is inside $\triangle ABC$ iff $\triangle ABC$ is acute-angled.

4.3 THE SINE FORMULA

Let us denote the area of $\triangle ABC$ by Δ, and, with the same notation as in the last section, let us put (temporarily) $h = |AP|$.

- Ex.4.4: *Starting from the formula* base × height *for the area of a rectangle, prove the formula* $\frac{1}{2}$ × base × height *for the area of a triangle, that is, in the above notation,* $\Delta = \frac{1}{2}ah$. *Method: first deal with the case* $C = \pi/2$, *and then in the general case use the altitude* AP *to write* Δ *as the sum or difference of the areas of two right-angled triangles.*

Proposition 4.6 (The sine formula).

$$\frac{a}{\sin A} = \frac{b}{\sin B} = \frac{c}{\sin C} = \frac{abc}{2\Delta}.$$

(See also Exercise 4.5, below.)

Proof. From Exercise 4.4, we have $\Delta = \frac{1}{2}ah$, so that

$$\frac{abc}{2\Delta} = \frac{bc}{h}.$$

By Exercise 3.20, $h = c \sin B$, and thus

$$\frac{abc}{2\Delta} = \frac{b}{\sin B}.$$

The other parts of the formula follow similarly. □

If we know the values of B, a, C (two angles and the side between them), then $A = \pi - B - C$, and the sine formula then supplies the values of b and c. So this gives another criterion for congruence of triangles, the case ASA (angle, side, angle):

Corollary 4.7.   *Given* $\triangle ABC$ *and* $\triangle A'B'C'$, *suppose* $B = B'$, $a = a'$, *and* $C = C'$. *Then* $\triangle ABC \equiv \triangle A'B'C'$.   □

Of course, since any two angles of a triangle determine the third, there is also a case SAA (side, angle, angle) for congruence, but this is not usually thought worth listing as a separate case.

4.4   THE CIRCUMCENTRE

Proposition 4.8.   *The perpendicular bisectors of the sides of a triangle are concurrent.*

Proof.   Let the perpendicular bisectors of AB, BC meet[1] in O.[2] Then $|OA| = |OB|$, and also $|OB| = |OC|$, so that $|OA| = |OC|$, and O must lie on the perpendicular bisector of AC also.   □

It follows that the circle, centre O, radius $|OA|$, passes through A, B and C (Fig. 4.3):

Definition 4.9.   *The (unique) circle through the vertices* A, B, C *is called the* **circumcircle** *of* $\triangle ABC$, *and its centre,* O, *is called the* **circumcentre** *of* $\triangle ABC$. *Its radius,* $|OA|$, *is the* **circumradius** *of* $\triangle ABC$.

We write R for the circumradius of $\triangle ABC$.

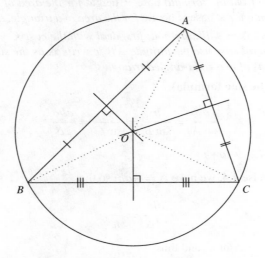

Fig. 4.3   Proposition 4.8: the circumcentre and circumcircle

[1] Could these lines *fail* to meet?

[2] It is perhaps unfortunate that the traditional name for the circumcentre is O, as we use this name also for the origin; but in practice no confusion should arise.

- Ex.4.5: *Let D be the mid-point of BC, as before. Prove that* $\angle DOB = A$ *or* $\pi - A$, *and then use Exercise 3.20 to show that*

$$\sin A = \frac{a/2}{R}.$$

Hence prove the following extension of the sine formula:

$$\frac{a}{\sin A} = \frac{b}{\sin B} = \frac{c}{\sin C} = \frac{abc}{2\Delta} = 2R.$$

In effect, to do this exercise you have to prove (or quote) the theorem about the angle at the centre being twice the angle at the circumference: $\angle BOC = 2A$, with the former being the *reflex* angle if A is obtuse. There are lots of isosceles triangles in the diagram, and it is just a matter of chasing angles around. You may find yourself worrying about whether O is inside $\triangle ABC$ or not; in fact, like the orthocentre, the circumcentre is inside the triangle iff the triangle is acute-angled. Sketch proof: if X is on AC, then X is between A and C iff $0 < \angle CBX < \angle CBA$ (in sense, as well as magnitude). So BO meets AC between A and C iff $0 < \pi/2 - A < B$, that is, iff both A and C are acute. Similarly, CO meets AB between A and B iff both A and B are acute. The result follows by condition (b), page 52.

4.5   THE INCENTRE AND EXCENTRES

Here we shall construct several circles that *touch* the sides of a triangle, that is, the sides of the triangle are *tangents* to the circles. To make this precise, let Σ be a circle, centre X, radius k, that is, Σ is the set of all points Y with $|XY| = k$. Let Y be a typical point of Σ, so that XY is a radius of Σ, and let ℓ be the line through Y orthogonal to XY. For any point $Z \in \ell$, $Z \neq Y$, we have $|XZ|^2 = |XY|^2 + |YZ|^2$, so that $|XZ| > |XY| = k$, and thus Z is not on Σ. Thus ℓ meets Σ in Y only, and this line ℓ is called the *tangent* to Σ at Y (Fig. 4.4).

- Ex.4.6: *Let m be a second line through Y, so that* $m \neq \ell$. *Prove that the perpendicular distance from X to m is* less *than k, and show that m meets* Σ *in (Y and) a second point. (So there is* precisely *one tangent to* Σ *at Y.)*

- Ex.4.7: *Tangents from a point to a circle are of equal length: explicitly, let Y′ be another point of* Σ, *with* ℓ' *the tangent at Y′, and suppose* ℓ *and* ℓ' *meet at P. Show that* $|PY| = |PY'|$. *(Method: show that* $\triangle PXY \equiv \triangle PXY'$; *or (possibly) reflect in PX.)*

Given $\triangle ABC$ then, let m be one of the angle bisectors of the lines AB, BC, let n be one of the angle bisectors of the lines BC, CA, and (temporarily) let m and n meet at W. (Could these lines *fail* to meet?) Put $\ell = WA$, so that ℓ, m, n are concurrent at W. Let the feet of the perpendiculars from W onto BC, CA, AB be X, Y, Z respectively. Then, because W is on n, we have $|WX| = |WY|$, and because W is on m, we have $|WZ| = |WX|$. Thus the circle, centre W, radius $|WX|$, not only passes through all three of X, Y, Z, but actually *touches* the three sides BC, CA, AB of $\triangle ABC$ (produced, if

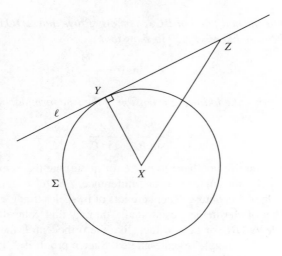

Fig. 4.4   The tangent ℓ to the circle Σ at Y

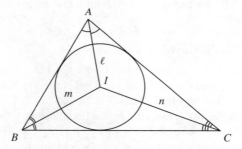

Fig. 4.5   The incircle and incentre

necessary). Also, since $|WZ| = |WY|$, the line ℓ is one of the angle bisectors of the lines CA, AB.

We shall refer to the angle bisectors of the lines AB, BC as *bisectors of B*. There is, of course, a choice of *two* bisectors of B, that is, two choices for m; and similarly two choices for n, giving $2 \times 2 = 4$ positions for W, and so *four* circles each touching all three sides of the triangle. Exactly one of the bisectors of B meets AC between A and C, by an argument like that at the end of the previous section. This is called the *internal* bisector of B, and the other is the *external* bisector of B, and contains *no* points inside the triangle.

First take the case where m, n are both internal bisectors. Let m and n meet at I which, by criterion (b), page 52, is inside $\triangle ABC$, and so ℓ is the internal bisector of A. The point I is called the *incentre* of $\triangle ABC$, and the circle centre I which touches the sides of the triangle is the *inscribed circle* or *incircle* of $\triangle ABC$ (Fig. 4.5). This circle lies

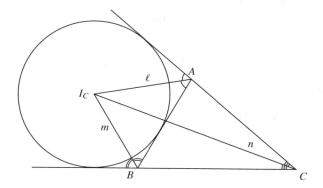

Fig. 4.6   The excircle and excentre opposite C

inside the triangle, and its point of contact with each side is between the vertices. Its radius is the *inradius* of $\triangle ABC$, denoted by r.

Now change m to be the external bisector of B, but leave n alone; so ℓ must change to the external bisector of A and if we let m and n meet at I_C, then I_C is called the *excentre* opposite C (Fig. 4.6). The corresponding circle is the *escribed circle* or *excircle* of $\triangle ABC$ opposite C. It lies outside the triangle, touches AB between A and B, but touches the other sides *produced*, i.e., *not* between the vertices. Similarly, there are two other excentres, I_B (where m is the internal bisector, and n, ℓ are external bisectors), and I_A (where m, n are external bisectors, and ℓ is the internal bisector). The radii of the three excircles are the *exradii* of $\triangle ABC$, denoted by r_A, r_B, r_C, in the obvious order.

Definition 4.10.   The **semi-perimeter** *of $\triangle ABC$ is the quantity* $s = \frac{1}{2}(a + b + c)$.

Proposition 4.11. $\Delta = sr = (s - a)r_A = (s - b)r_B = (s - c)r_C$.

Proof.   We have (see Fig. 4.5)

$$\Delta = \text{area}(BCI) + \text{area}(CAI) + \text{area}(ABI) = \tfrac{1}{2}ar + \tfrac{1}{2}br + \tfrac{1}{2}cr = sr.$$

Again (see Fig. 4.6),

$$\Delta = \text{area}(BCI_C) + \text{area}(CAI_C) - \text{area}(BAI_C) = \tfrac{1}{2}ar_C + \tfrac{1}{2}br_C - \tfrac{1}{2}cr_C = (s - c)r_C,$$

and similarly for the other parts.   □

- Ex.4.8:   *Let the feet of the perpendiculars from I onto BC, CA, AB be X, Y, Z respectively; let the feet of the perpendiculars from I_A onto BC, CA, AB be X_A, Y_A, Z_A respectively; and so on. Use Exercise 4.7 to prove that (i) $|BX| = s - b = |CX_A|$; (ii) $|CX| = s - c = |BX_A|$; $|BX_C| = s - a = |CX_B|$; and (iv) $|CX_B| = s = |CX_C|$.*

4.6 THE COSINE FORMULA

Proposition 4.12. **(The cosine formula)** *In any triangle* ABC,

$$a^2 = b^2 + c^2 - 2bc\cos A, \quad or \quad \cos A = \frac{b^2 + c^2 - a^2}{2bc},$$

with similar formulae involving $\cos B$ *and* $\cos C$.

Proof. Let $\mathbf{u} = \overrightarrow{AB}$ and $\mathbf{v} = \overrightarrow{AC}$, so that $\overrightarrow{CB} = \mathbf{u} - \mathbf{v}$. Then

$$a^2 = |\mathbf{u} - \mathbf{v}|^2 = (\mathbf{u} - \mathbf{v}) \cdot (\mathbf{u} - \mathbf{v}) = |\mathbf{u}|^2 + |\mathbf{v}|^2 - 2\mathbf{u} \cdot \mathbf{v} = b^2 + c^2 - 2bc\cos A,$$

by Proposition 3.30. □

If we know the values of b, A, c (two sides and the *included* angle, that is, the angle between them), then the cosine formula supplies the value of a. This gives the final criterion for congruence of triangles, the case SAS (side, angle, side):

Corollary 4.13. *Given* $\triangle ABC$ *and* $\triangle A'B'C'$, *suppose* $b = b'$, $A = A'$, *and* $c = c'$. *Then* $\triangle ABC \equiv \triangle A'B'C'$. □

To summarize, we now have four criteria for congruence:

Proposition 4.14. *Given* $\triangle ABC$ *and* $\triangle A'B'C'$, *if we know*

  (i) $a = a'$, $b = b'$, *and* $c = c'$, *or*
 (ii) $a = a'$, $b = b'$, *and* $B = B' = \pi/2$, *or*
(iii) $A = A'$, $b = b'$, *and* $C = C'$, *or*
(iv) $a = a'$, $B = B'$, *and* $c = c'$,

then in each case we deduce $\triangle ABC \equiv \triangle A'B'C'$.

Proof. (i) is SSS (Proposition 3.22) (Fig. 4.7), (ii) is SSR (Corollary 3.23) (Fig. 4.8), (iii) is ASA (or SAA) (Corollary 4.7) (Fig. 4.9), and (iv) is SAS (Corollary 4.13) (Fig. 4.10). □

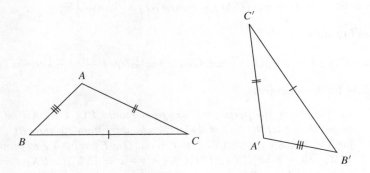

Fig. 4.7 Proposition 4.14(i): SSS

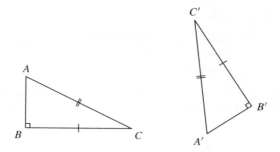

Fig. 4.8 Proposition 4.14(ii): SSR

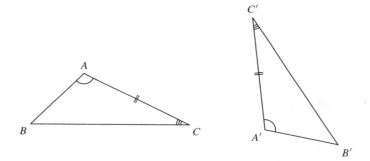

Fig. 4.9 Proposition 4.14(iii): ASA

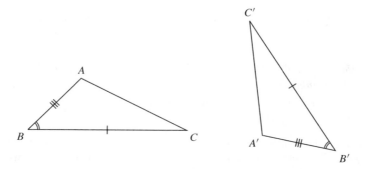

Fig. 4.10 Proposition 4.14(iv): SAS

Of the remaining cases one might be tempted to write down, SRS (two sides and an included right angle) works, being just a special case of SAS; AAA (three angles) does not work, being the criterion for *similarity* of triangles (of which more in Section 4.7), rather than congruence; and SSA (two sides and a non-included angle) does not work unless the angle is a right angle:

- Ex.4.9:   *Given $\triangle ABC$, with C obtuse, let AP be the altitude through A, let $A' = A$, let $B' = B$, and let C' be the reflection of C in AP. Show that $|BA| = |B'A'|$, $|AC| = |A'C'|$, and $\angle CBA = \angle C'B'A'$, but $\triangle ABC \not\equiv \triangle A'B'C'$.*

The next formula gives the area of $\triangle ABC$ purely in terms of its side lengths. It is usually attributed to Heron of Alexandria (AD 60), but was known to Archimedes (287–212 BC).

Proposition 4.15.   (Heron's formula)

$$\Delta = \sqrt{s(s - a)(s - b)(s - c)}.$$

Proof.   From the sine formula,

$$\begin{aligned}
\Delta^2 &= \tfrac{1}{4}b^2c^2 \sin^2 A \\
&= \tfrac{1}{4}b^2c^2(1 - \cos^2 A) \\
&= \tfrac{1}{4}b^2c^2 \left(1 - \left(\frac{b^2 + c^2 - a^2}{2bc} \right)^2 \right) \quad \text{(by the cosine formula)} \\
&= \tfrac{1}{16}(4b^2c^2 - (b^2 + c^2 - a^2)^2) \\
&= \tfrac{1}{16}(2bc + (b^2 + c^2 - a^2))(2bc - (b^2 + c^2 - a^2)) \\
&= \tfrac{1}{16}((b + c)^2 - a^2)(a^2 - (b - c)^2) \\
&= \tfrac{1}{16}(a + b + c)(-a + b + c)(a - b + c)(a + b - c) \\
&= s(s - a)(s - b)(s - c). \qquad \square
\end{aligned}$$

- Ex.4.10:   *Show that, for fixed s and a, the area Δ is greatest when $b = c$, i.e. when the triangle is isosceles.*
- Ex.4.11:   *Show that, for fixed s, the area Δ is greatest when $a = b = c$, i.e. when the triangle is equilateral.*

The reader may have noticed the way properties and concepts tend to occur twice, with the words *point*, *line* interchanged at the second occurrence. Thus two points determine a line, and in general[3] two lines determine a point. If three points lie on a line they are collinear, and if three lines pass through a point they are concurrent. The three vertices determine the three sides of a triangle, and in general three lines determine a triangle of which they are the sides. This partial symmetry between point and line

---

[3] The phrase 'in general' is a let-out, much loved by geometers and often provoking fury in others, and usually it means something like 'provided nothing special happens'. For example, we might say that *in general a quadratic equation has two roots* (but of course they just might be equal); here we mean we are ignoring the case of parallel lines.

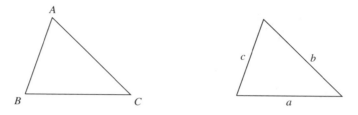

Fig. 4.11 Triangle *ABC* and trilateral *abc*

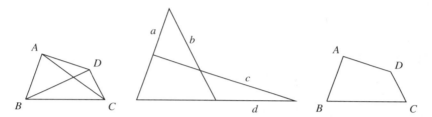

Fig. 4.12 Quadrangle *ABCD*, quadrilateral *abcd*, and tetragram *ABCD*

will become full symmetry when we set up projective geometry, in a later chapter. The idea there is to invent 'points at infinity' where parallel lines 'meet', so that we can then say that in all cases, two lines meet in a point. This will be done in a perfectly formal way, and points at infinity will have coordinates, just like any other points, and all without ever doing arithmetic involving the mysterious symbol '∞'. (Well, hardly ever.) The symmetry between points and lines is called *duality*, and once it is established, definitions, statements and even theorems have *duals*, obtained by interchanging point with line, join with meet, and so on.

A triangle, thus, is equally a three-cornered figure or a three-sided figure. If we should ever need to distinguish, then a *triangle* (three angles) is specified by its three vertices, and a *trilateral* (three lines) is the 'same' as a triangle but specified by its three sides instead (Fig. 4.11). The word 'quadrilateral' is traditionally taken to mean a four-sided figure *ABCD*, that is, four points *A*, *B*, *C*, *D* (no three of which are collinear) joined by the four line segments *AB*, *BC*, *CD*, *DA*. But four points *A*, *B*, *C*, *D* can in general be joined pairwise in a total of six ways, giving six lines *AB*, *AC*, *AD*, *BC*, *BD*, *CD*, and (dually!) four lines meet pairwise in six points. To avoid the confusion, the figure with four vertices (and six sides) is called a *quadrangle*, and the dual figure with four sides (and six vertices) is called a *quadrilateral* (Fig. 4.12). Sometimes the latter is referred to as a 'complete' quadrilateral, to distinguish it from the (incomplete?) four-vertex four-sided figure, but it would seem safer to have a separate name for the latter, and J. A. Tyrrell has suggested:

Definition 4.16. *A **tetragram** (Fig. 4.12) is an ordered set of four points ABCD (no three of which are collinear). Its **sides** are the line segments AB, BC, CD, and DA, and*

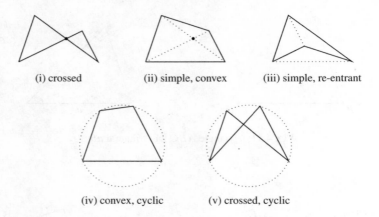

(i) crossed   (ii) simple, convex   (iii) simple, re-entrant

(iv) convex, cyclic   (v) crossed, cyclic

Fig. 4.13   Various types of tetragram

its **diagonals** *are the line segments AC and BD. Each of the vertices A, C is* **opposite** *the other, and similarly for the vertices B, D. Each of the sides AB, CD is* **opposite** *the other, and similarly for the sides BC, DA.*

Definition 4.17.   *A tetragram is* **crossed** *(Fig. 4.13(i)) if two opposite sides have a point in common (noting that the sides were defined to be line segments, not lines), and if it is not crossed it is* **simple**. *A simple tetragram is* **convex** *(Fig. 4.13(ii)) if the two diagonals have a point in common, and if it is not convex it is* **re-entrant** *(Fig. 4.13(iii)).*

It is only in the simple case that the tetragram has an 'inside' and an 'outside', and so it is only in this case that we can assign a simple-minded (!) notion of *area* to a tetragram.

Definition 4.18.   *A tetragram ABCD is* **cyclic** *if its vertices lie on a circle (Fig. 4.13(iv), (v)).*

Here is a nice generalization of Heron's formula to a convex cyclic tetragram $ABCD$, of area Δ. Put $a = |AB|$, $b = |BC|$, $c = |CD|$, and $d = |DA|$, and let $s = \frac{1}{2}(a + b + c + d)$. Then:

- Ex.4.12:

$$\Delta = \sqrt{(s - a)(s - b)(s - c)(s - d)}.$$

(Method: let $\theta = \angle ABC$ and $\varphi = \angle CDA$, so that

$$\Delta = \tfrac{1}{2}ab \sin \theta + \tfrac{1}{2}cd \sin \varphi.$$

Proceed as in the proof of Proposition 4.15. You will need to use the cosine formula twice to equate two values of $|AC|^2$, and you will need to prove (or quote) a theorem about the value of $\theta + \varphi$.)

- Ex.4.13: *Given a simple tetragram $ABCD$, with a, b, c, d as above and fixed in value, show that the area Δ is greatest when $ABCD$ is cyclic. (Assume throughout that B, D are on opposite sides of AC, if it helps.)*

Note that Heron's formula can be obtained from Exercise 4.12 by allowing D to coincide with A, so that $d = 0$.

- Ex.4.14: *Let $ABCD$ be a simple tetragram, and let $O = AC \cdot BD$ (the meet), and $\theta = \angle AOB$. Prove that $\text{area}(ABCD) = \frac{1}{2}|AC||BD|\sin\theta$. What does this formula represent if $ABCD$ is a crossed tetragram? Can it ever yield the answer zero? Or a negative answer?*

4.7 SIMILARITIES

We now introduce a new type of transformation of $\mathbb{R}^2$.

Definition 4.19. *A **similarity** (of $\mathbb{R}^2$) is a transformation f (of $\mathbb{R}^2$) which multiplies all distances by the same constant. That is, there is a positive real number k such that, if $f(P) = P'$ and $f(Q) = Q'$, then $|P'Q'| = k|PQ|$. The number k is called the **scale factor** of f.*

It should be clear that every isometry is a similarity (with scale factor $= 1$), that the composite of two similarities is a similarity, and so is the inverse of a similarity. If we denote the set of all similarities of $\mathbb{R}^2$ by $\mathcal{E}(\mathbb{R}^2)$, then group theorists will recognize that we have just said that $\mathcal{E}(\mathbb{R}^2)$ is a group, containing $\mathcal{I}(\mathbb{R}^2)$, the set of all isometries, as a subgroup. $\mathcal{E}(\mathbb{R}^2)$ is called the *Euclidean group*.

Definition 4.20. *Triangles ABC, $A'B'C'$ are **similar** (written $\triangle ABC \sim \triangle A'B'C'$) if there is a similarity f with $f(A) = A'$, $f(B) = B'$, and $f(C) = C'$. More generally, two subsets X, Y of $\mathbb{R}^2$ are similar if there is a similarity f with $f(X) = Y$.*

Proposition 4.21. *Suppose $\triangle ABC \sim \triangle A'B'C'$. Put $\alpha = \angle CAB$, $\beta = \angle ABC$, $\gamma = \angle BCA$, $\alpha' = \angle C'A'B'$, $\beta' = \angle A'B'C'$, and $\gamma' = \angle B'C'A'$, Then $a : a' = b : b' = c : c'$, and also $\alpha = \alpha'$, $\beta = \beta'$, and $\gamma = \gamma'$.*

Proof. There is a similarity f, as above, and if k is its scale factor, then $|B'C'| = k|BC|$, or $a' = ka$; and likewise $b' = kb$ and $c' = kc$. So $a : a' = b : b' = c : c' = 1 : k$. Then, by the cosine formula,

$$\cos\alpha' = \frac{(kb)^2 + (kc)^2 - (ka)^2}{2(kb)(kc)} = \frac{b^2 + c^2 - a^2}{2bc} = \cos\alpha.$$

Since $\cos : [0, \pi] \to [-1, 1]$ is bijective (OK?), we deduce $\alpha' = \alpha$, and likewise $\beta' = \beta$ and $\gamma' = \gamma$. $\qquad\square$

Just as with congruence of triangles, there are various criteria sufficient to assert that two given triangles are similar, and we collect these together in the next proposition:

Proposition 4.22.   *Given $\triangle ABC$ and $\triangle A'B'C'$, let us write $\alpha = \angle CAB$, $\beta = \angle ABC$, $\gamma = \angle BCA$, $\alpha' = \angle C'A'B'$, $\beta' = \angle A'B'C'$, and $\gamma' = \angle B'C'A'$, If we know*

(i) *$a : a' = b : b' = c : c'$, or*
(ii) *$\alpha = \alpha'$ and $\beta = \beta'$, or*
(iii) *$\alpha = \alpha'$ and $b : b' = c : c'$, or*
(iv) *$\alpha = \alpha' = \pi/2$ and $a : a' = b : b'$,*

then in each case we deduce $\triangle ABC \sim \triangle A'B'C'$.

Remark: We can use the same notation for these cases as for congruent triangles. Thus (i) is SSS (three sides), (iii) is SAS (two sides and the included angle), and (iv) is SSR (two sides and a non-included right angle). Case (ii) is new; note that if $\alpha = \alpha'$ and $\beta = \beta'$ then also $\gamma = \gamma'$ (because the angle sum of a triangle is π), so we denote this case by AAA (three angles). The congruence case ASA (two angles and the included side) does not appear here, because as soon as we know that two angles are equal, we are in case (ii) AAA, and anyway you cannot give information about corresponding sides being in proportion if you are only going to mention *one* pair of corresponding sides!

Proof.   (i) Let $k = a'/a \ (= b'/b = c'/c)$, and let g be any similarity with scale factor $= k$. (For example, we could define g by $g(x, y) = (kx, ky)$, for all x, y; check that this is a similarity.) Let $g(A) = A''$, $g(B) = B''$, and $g(C) = C''$, and consider $\triangle A''B''C''$. We have $a''/a = b''/b = c''/c = k$, and so $a'' = a'$, $b'' = b'$, and $c'' = c'$. Thus $\triangle A''B''C'' \equiv \triangle A'B'C'$ (SSS, Proposition 3.22), so there is an *isometry* h with $h(A'') = A'$, $h(B'') = B'$, and $h(C'') = C'$. Both g and h are similarities, and hence so is $f = hg$; but $f(A) = A'$, $f(B) = B'$, and $f(C) = C'$, whence the result.

(ii) As remarked above, we have not only $\alpha = \alpha'$ and $\beta = \beta'$, but also $\gamma = \gamma'$. From the sine formula,

$$\frac{a}{b} = \frac{\sin \alpha}{\sin \beta} = \frac{\sin \alpha'}{\sin \beta'} = \frac{a'}{b'},$$

so that $a/a' = b/b'$, or $a : a' = b : b'$, and similarly $a : a' = c : c'$. Now use (i).

(iii) Put $k = b'/b \ (= c'/c)$. Then, by the cosine formula,

$$(a')^2 = (kb)^2 + (kc)^2 - 2(kb)(kc) \cos \alpha' = k^2(b^2 + c^2 - 2bc \cos \alpha) = k^2 a^2,$$

so we deduce $a'/a = k$ also. The result follows by (i) again.

(iv) Put $k = a'/a \ (= b'/b)$. Then, by Pythagoras,

$$(c')^2 = (ka)^2 - (kb)^2 = k^2(a^2 - b^2) = k^2 c^2,$$

so we deduce $c'/c = k$ also. Use (i) once more.   □

• Ex.4.15:   *In $\triangle ABC$, let D, E, F be the mid-points of BC, CA, AB respectively. Show that $\triangle DEF \sim \triangle ABC$. Show also that $\triangle DEF \equiv \triangle AFE \equiv \triangle FBD \equiv \triangle EDC$.*

- Ex.4.16: *Given $\triangle ABC$ with E on AC and F on AB, prove that if $BC \parallel FE$, then $|AF| : |FB| = |AE| : |EC|$. Is the converse true?*
- Ex.4.17: *Show that if A, B, C are collinear, and f is a similarity, then $f(A), f(B), f(C)$ are also collinear. (cf. Proposition 3.20.)*

4.8   BARYCENTRIC COORDINATES

In this section we introduce a new system of coordinates based on an arbitrary *triangle of reference* rather than on two (perpendicular) axes. First, we need conditions for collinearity and concurrency, expressed in matrix form.

Lemma 4.23.   *The line through (a_1, b_1) and (a_2, b_2) has equation*

$$\begin{vmatrix} a_1 & a_2 & x \\ b_1 & b_2 & y \\ 1 & 1 & 1 \end{vmatrix} = 0.$$

Proof.   The equation is linear in x, y, so it is a line, and it is clearly satisfied by $x = a_i, y = b_i$ for $i = 1, 2$, because any matrix with two equal columns has determinant equal to 0.   □

As a consequence:

Proposition 4.24.   *The points (a_1, b_1), (a_2, b_2), (a_3, b_3) are collinear iff*

$$\begin{vmatrix} a_1 & a_2 & a_3 \\ b_1 & b_2 & b_3 \\ 1 & 1 & 1 \end{vmatrix} = 0.$$ □

Proposition 4.25.   *The lines $p_1x + q_1y + r_1 = 0$, $p_2x + q_2y + r_2 = 0$, $p_3x + q_3y + r_3 = 0$ are either concurrent or all parallel iff*

$$\begin{vmatrix} p_1 & q_1 & r_1 \\ p_2 & q_2 & r_2 \\ p_3 & q_3 & r_3 \end{vmatrix} = 0.$$

Proof.   Suppose (x, y) lies on all three lines. Then

$$\begin{pmatrix} p_1 & q_1 & r_1 \\ p_2 & q_2 & r_2 \\ p_3 & q_3 & r_3 \end{pmatrix} \begin{pmatrix} x \\ y \\ 1 \end{pmatrix} = \begin{pmatrix} 0 \\ 0 \\ 0 \end{pmatrix},$$

so

$$\begin{vmatrix} p_1 & q_1 & r_1 \\ p_2 & q_2 & r_2 \\ p_3 & q_3 & r_3 \end{vmatrix} = 0.$$

Now suppose the lines are all parallel: so their gradients are equal, and $p_1 : q_1 = p_2 : q_2 = p_3 : q_3$. Thus the vectors

$$\begin{pmatrix} p_1 \\ p_2 \\ p_3 \end{pmatrix} \quad \text{and} \quad \begin{pmatrix} q_1 \\ q_2 \\ q_3 \end{pmatrix}$$

are linearly dependent, and once again,

$$\begin{vmatrix} p_1 & q_1 & r_1 \\ p_2 & q_2 & r_2 \\ p_3 & q_3 & r_3 \end{vmatrix} = 0.$$

For the converse, if the determinant is zero, then

$$\begin{pmatrix} p_1 & q_1 & r_1 \\ p_2 & q_2 & r_2 \\ p_3 & q_3 & r_3 \end{pmatrix} \begin{pmatrix} x \\ y \\ z \end{pmatrix} = \begin{pmatrix} 0 \\ 0 \\ 0 \end{pmatrix} \quad \text{for some} \quad \begin{pmatrix} x \\ y \\ z \end{pmatrix} \neq \begin{pmatrix} 0 \\ 0 \\ 0 \end{pmatrix}.$$

Case 1: $z = 0$. Then

$$x \begin{pmatrix} p_1 \\ p_2 \\ p_3 \end{pmatrix} + y \begin{pmatrix} q_1 \\ q_2 \\ q_3 \end{pmatrix} = 0$$

with x, y not both zero, so that $p_1 : q_1 = p_2 : q_2 = p_3 : q_3 = -y : x$, and the lines are all parallel.

Case 2: $z \neq 0$. Then

$$\begin{pmatrix} p_1 & q_1 & r_1 \\ p_2 & q_2 & r_2 \\ p_3 & q_3 & r_3 \end{pmatrix} \begin{pmatrix} x/z \\ y/z \\ 1 \end{pmatrix} = \begin{pmatrix} 0 \\ 0 \\ 0 \end{pmatrix},$$

so the point $(x/z, y/z)$ lies on all three lines.   □

Now for some applied mathematics. Given $\triangle A_1 A_2 A_3$, with particles of mass ξ, η, ζ at A_1, A_2, A_3 respectively, let us find the coordinates (x, y) of the centre of mass of the three particles. Let $A_i = (a_i, b_i)$, $i = 1, 2, 3$. Taking moments about each axis in turn, we obtain:

$$(\xi + \eta + \zeta)x = \xi a_1 + \eta a_2 + \zeta a_3,$$
$$(\xi + \eta + \zeta)y = \xi b_1 + \eta b_2 + \zeta b_3.$$

Putting $\mathbf{r} = (x, y)$ and $\mathbf{u}_i = (a_i, b_i)$, we get

$$\mathbf{r} = \frac{\xi}{\xi + \eta + \zeta}\mathbf{u}_1 + \frac{\eta}{\xi + \eta + \zeta}\mathbf{u}_2 + \frac{\zeta}{\xi + \eta + \zeta}\mathbf{u}_3. \tag{4.1}$$

It clearly makes no difference to **r** if we replace (ξ, η, ζ) by $(k\xi, k\eta, k\zeta)$, for any $k \neq 0$, so to reverse the process we could solve

$$\xi a_1 + \eta a_2 + \zeta a_3 = x$$
$$\xi b_1 + \eta b_2 + \zeta b_3 = y \qquad (4.2)$$
$$\xi \quad + \eta \quad + \zeta \quad = 1$$

or

$$\mathbf{M} \begin{pmatrix} \xi \\ \eta \\ \zeta \end{pmatrix} = \begin{pmatrix} x \\ y \\ 1 \end{pmatrix}, \quad \text{where } \mathbf{M} = \begin{pmatrix} a_1 & a_2 & a_3 \\ b_1 & b_2 & b_3 \\ 1 & 1 & 1 \end{pmatrix},$$

which is possible since $|\mathbf{M}| \neq 0$, by Proposition 4.24.

In the above, we have been assuming that ξ, η, ζ are positive, which means that (x, y) is inside $\triangle A_1 A_2 A_3$. But it makes no difference to the calculations if, instead of placing masses at the three points, we have *forces* of magnitude ξ, η, ζ acting there, perpendicular to the plane. These can be positive, negative or zero, and the above calculation gives the coordinates (x, y) of the *point of action* of the system of forces, provided that there *is* a resultant force, that is, provided $\xi + \eta + \zeta \neq 0$. Notice also that the equations (4.2) can be solved for ξ, η, ζ whatever values are chosen for x, y, which means there is a triple (ξ, η, ζ) corresponding to $\mathbf{r} = (x, y)$, for every $\mathbf{r} \in \mathbb{R}^2$.

Definition 4.26. *If* **r** *and* (ξ, η, ζ) *are related by* (4.1), *then* (ξ, η, ζ) *are* **barycentric coordinates** *for* **r** *with respect to* $\triangle A_1 A_2 A_3$.

Note that (ξ, η, ζ) and $(k\xi, k\eta, k\zeta)$ are the *same* point, for all $k \neq 0$; and also $\xi + \eta + \zeta \neq 0$.

Definition 4.27. $\triangle A_1 A_2 A_3$ *is the* **triangle of reference** *for this coordinate system.*

The vertex A_1 of the triangle of reference is $(1, 0, 0)$. This says that if a mass of 1 unit is placed at A_1 and nothing at all at A_2 or A_3, then the centre mass of this system will be at A_1! Similarly $A_2 = (0, 1, 0)$, and $A_3 = (0, 0, 1)$ (see Fig. 4.14, below).

Barycentric coordinates can also be defined in terms of the areas of certain triangles (and are then called *areal* coordinates): for details see Section 5.3.2, and in particular Exercise 5.43.

We now give the barycentric versions of Lemma 4.23, Proposition 4.24, and Proposition 4.25, but not in that order:

Proposition 4.28. $(\xi_1, \eta_1, \zeta_1), (\xi_2, \eta_2, \zeta_2), (\xi_3, \eta_3, \zeta_3)$ *are collinear iff*

$$\begin{vmatrix} \xi_1 & \xi_2 & \xi_3 \\ \eta_1 & \eta_2 & \eta_3 \\ \zeta_1 & \zeta_2 & \zeta_3 \end{vmatrix} = 0.$$

Proof.   If the corresponding cartesian coordinates are (x_1, y_1), (x_2, y_2), (x_3, y_3), then, with **M** as before,

$$\mathbf{M} \begin{pmatrix} \xi_1 & \xi_2 & \xi_3 \\ \eta_1 & \eta_2 & \eta_3 \\ \zeta_1 & \zeta_2 & \zeta_3 \end{pmatrix} = \begin{pmatrix} x_1 & x_2 & x_3 \\ y_1 & y_2 & y_3 \\ 1 & 1 & 1 \end{pmatrix},$$

provided $\xi_i + \eta_i + \zeta_i = 1$, $1 \le i \le 3$, or, more generally,

$$\mathbf{M} \begin{pmatrix} \xi_1 & \xi_2 & \xi_3 \\ \eta_1 & \eta_2 & \eta_3 \\ \zeta_1 & \zeta_2 & \zeta_3 \end{pmatrix} \begin{pmatrix} k_1 & 0 & 0 \\ 0 & k_2 & 0 \\ 0 & 0 & k_3 \end{pmatrix} = \begin{pmatrix} x_1 & x_2 & x_3 \\ y_1 & y_2 & y_3 \\ 1 & 1 & 1 \end{pmatrix},$$

where $k_i = (\xi_i + \eta_i + \zeta_i)^{-1}$, $1 \le i \le 3$. Now take determinants and use Proposition 4.24.
□

Corollary 4.29.   *The line ℓ joining (ξ_1, η_1, ζ_1) and (ξ_2, η_2, ζ_2) consists of all (ξ, η, ζ) such that*

$$\begin{vmatrix} \xi_1 & \xi_2 & \xi \\ \eta_1 & \eta_2 & \eta \\ \zeta_1 & \zeta_2 & \zeta \end{vmatrix} = 0.$$
□

This last equation is of the form $p'\xi + q'\eta + r'\zeta = 0$: a *homogeneous* linear equation in ξ, η, ζ. (A polynomial is *homogeneous* if all its terms have the same degree.) Note that, if $p'\xi + q'\eta + r'\zeta = 0$, then also $p'(k\xi) + q'(k\eta) + r'(k\zeta) = 0$, so that whether the equation is satisfied is a property of the *point* (ξ, η, ζ), rather than of the choice of coordinates representing it. The equation $p'\xi + q'\eta + r'\zeta = 0$ is the *barycentric* or *homogeneous* equation of ℓ. Barycentric coordinates are an example of a system of *homogeneous coordinates*; we shall meet these in full generality when we study projective geometry. For the meantime, note that the equation $\xi + \eta + \zeta = 0$ looks like a line—it is homogeneous linear, after all—but is in fact satisfied by *no* points, by definition. When we do projective geometry, we shall add new 'points' to our system, called *points at infinity*, and they will all lie on this 'line', $\xi + \eta + \zeta = 0$, the so-called *line at infinity*. But for the present, *no* point (ξ, η, ζ) can satisfy $\xi + \eta + \zeta = 0$.

We can now find the homogeneous equations of the sides of the triangle of reference. Since $A_2 = (0, 1, 0)$ and $A_3 = (0, 0, 1)$, the line $A_2 A_3$ has equation

$$\begin{vmatrix} 0 & 0 & \xi \\ 1 & 0 & \eta \\ 0 & 1 & \zeta \end{vmatrix} = 0, \text{ or } \xi = 0.$$

This says that if *no* mass is placed at A_1, i.e. $\xi = 0$, then the centre of mass of the resulting system (with masses at either or both of A_2, A_3) is somewhere on $A_2 A_3$; and every point of $A_2 A_3$ is the centre of mass for some such system. Similarly, $A_3 A_1$ has equation $\eta = 0$, and $A_1 A_2$ has equation $\zeta = 0$ (Fig. 4.14).

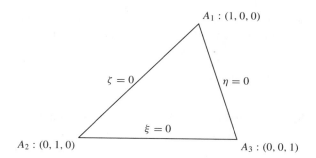

Fig. 4.14 Triangle of reference

Note that the line with cartesian equation $px + qy + r = 0$ can be written

$$(p \quad q \quad r) \begin{pmatrix} x \\ y \\ 1 \end{pmatrix} = 0, \text{ or } (p \quad q \quad r) \mathbf{M} \begin{pmatrix} \xi \\ \eta \\ \zeta \end{pmatrix} = 0.$$

That is,

$$(p' \quad q' \quad r') \begin{pmatrix} \xi \\ \eta \\ \zeta \end{pmatrix} = 0, \text{ where } (p' \quad q' \quad r') = (p \quad q \quad r) \mathbf{M}.$$

Proposition 4.30. *The lines with barycentric equations* $p'_1\xi + q'_1\eta + r'_1\zeta = 0$, $p'_2\xi + q'_2\eta + r'_2\zeta = 0$, *and* $p'_3\xi + q'_3\eta + r'_3\zeta = 0$ *are* either *concurrent or* all parallel *iff*

$$\begin{vmatrix} p'_1 & q'_1 & r'_1 \\ p'_2 & q'_2 & r'_2 \\ p'_3 & q'_3 & r'_3 \end{vmatrix} = 0.$$

Proof. From the last calculation,

$$\begin{pmatrix} p'_1 & q'_1 & r'_1 \\ p'_2 & q'_2 & r'_2 \\ p'_3 & q'_3 & r'_3 \end{pmatrix} = \begin{pmatrix} p_1 & q_1 & r_1 \\ p_2 & q_2 & r_2 \\ p_3 & q_3 & r_3 \end{pmatrix} \mathbf{M}.$$

Now take determinants and use Proposition 4.25. □

• Ex.4.18: *Prove that the lines with barycentric equations* $p'_1\xi + q'_1\eta + r'_1\zeta = 0$ *and* $p'_2\xi + q'_2\eta + r'_2\zeta = 0$ *are parallel iff*

$$\begin{vmatrix} p'_1 & q'_1 & r'_1 \\ p'_2 & q'_2 & r'_2 \\ 1 & 1 & 1 \end{vmatrix} = 0.$$

Comparing this exercise with the preceding proposition, our two lines are parallel iff they are concurrent with the 'line' $\xi + \eta + \zeta = 0$, that is, parallel lines meet at infinity.

Lemma 4.31. *Let $P_i = (\xi_i, \eta_i, \zeta_i)$ with $\xi_i + \eta_i + \zeta_i = 1$, $i = 1, 2$. Then the point dividing $P_1 P_2$ in the ratio $\lambda_2 : \lambda_1$ (note the order) has barycentric coordinates $\lambda_1(\xi_1, \eta_1, \zeta_1) + \lambda_2(\xi_2, \eta_2, \zeta_2)$.*

Proof. Let P_3 be the required point, and let the cartesian coordinates of P_i be (x_i, y_i), $1 \leq i \leq 3$. Then

$$(x_3, y_3) = \frac{\lambda_1}{\lambda_1 + \lambda_2}(x_1, y_1) + \frac{\lambda_2}{\lambda_1 + \lambda_2}(x_2, y_2),$$

which may be written

$$\begin{pmatrix} x_3 \\ y_3 \\ 1 \end{pmatrix} = \frac{\lambda_1}{\lambda_1 + \lambda_2} \begin{pmatrix} x_1 \\ y_1 \\ 1 \end{pmatrix} + \frac{\lambda_2}{\lambda_1 + \lambda_2} \begin{pmatrix} x_2 \\ y_2 \\ 1 \end{pmatrix}.$$

Multiplying each side on the left by $\mathbf{M}$ gives the barycentric coordinates of P_3 as

$$\begin{pmatrix} \xi_3 \\ \eta_3 \\ \zeta_3 \end{pmatrix} = \frac{\lambda_1}{\lambda_1 + \lambda_2} \begin{pmatrix} \xi_1 \\ \eta_1 \\ \zeta_1 \end{pmatrix} + \frac{\lambda_2}{\lambda_1 + \lambda_2} \begin{pmatrix} \xi_2 \\ \eta_2 \\ \zeta_2 \end{pmatrix}.$$

Multiplying the coordinates of P_3 through by $\lambda_1 + \lambda_2$ (as we may) gives the result. □

Remark: To return to the applied mathematical interpretation, the last lemma may be thought of as finding the point of action of a force of λ_1 at P_1 and λ_2 at P_2.

As an example (and for use in the next section), the point dividing $A_2 A_3$ in the ratio $\lambda : 1$ has coordinates $(0, 1, 0) + \lambda(0, 0, 1)$, that is, $(0, 1, \lambda)$. This works because $0 + 1 + 0 = 1$ and also $0 + 0 + 1 = 1$, so the conditions of Lemma 4.31 are satisfied.

4.9   MENELAUS AND CEVA

In this section we prove two famous theorems about collinearity and concurrency. First, some notation. Suppose B, C, D are collinear, with $BD : DC = \lambda : 1$, that is, $\overrightarrow{BD} = \lambda \overrightarrow{DC}$. We shall write $BD/DC = \lambda$, the ratio of *signed* lengths. This is positive if D lies between B and C, and negative otherwise. (And, of course, $|BD|/|DC| = |\lambda|$ in either case.)

For the next two theorems and the following exercises, let D, E, F lie on the sides BC, CA, AB (respectively) of $\triangle ABC$, where none of D, E, F coincides with any of A, B, C. Put $\lambda = BD/DC$, $\mu = CE/EA$, and $\nu = AF/FB$. With $\triangle ABC$ as triangle of reference, we have $D = (0, 1, \lambda)$, $E = (\mu, 0, 1)$, and $F = (1, \nu, 0)$, as in the example at the end of the last section.

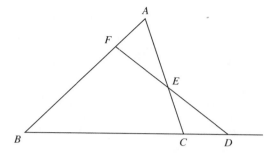

Fig. 4.15 Menelaus' theorem

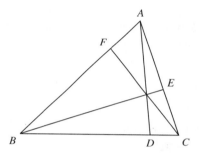

Fig. 4.16 Ceva's theorem

Theorem 4.32. **(Menelaus)** (See Fig. 4.15.) D, E, F *are collinear iff*

$$\frac{BD}{DC}\frac{CE}{EA}\frac{AF}{FB} = -1.$$

Proof. By Proposition 4.28, D, E, F are collinear iff

$$\begin{vmatrix} 0 & \mu & 1 \\ 1 & 0 & \nu \\ \lambda & 1 & 0 \end{vmatrix} = 0, \text{ or } \lambda\mu\nu + 1 = 0, \text{ or } \lambda\mu\nu = -1. \qquad \square$$

Theorem 4.33. **(Ceva)** (See Fig. 4.16.) AD, BE, CF *are concurrent or all parallel iff*

$$\frac{BD}{DC}\frac{CE}{EA}\frac{AF}{FB} = 1.$$

Proof. By Corollary 4.29, the equation of AD is

$$\begin{vmatrix} 1 & 0 & \xi \\ 0 & 1 & \eta \\ 0 & \lambda & \zeta \end{vmatrix} = 0,$$

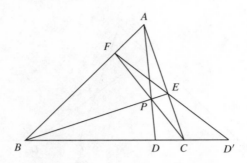

Fig. 4.17   Harmonic conjugates

or $\zeta = \lambda\eta$. Similarly, BE has equation $\xi = \mu\zeta$, and CF has equation $\eta = \nu\xi$. By Proposition 4.30, the required condition is

$$\begin{vmatrix} 0 & \lambda & -1 \\ -1 & 0 & \mu \\ \nu & -1 & 0 \end{vmatrix} = 0, \text{ or } \lambda\mu\nu - 1 = 0, \text{ or } \lambda\mu\nu = 1. \qquad \square$$

For the following exercises, assume that A, B, C, D, E, F are as above, and that AD, BE, CF are concurrent at P.

• Ex.4.19: *Suppose EF meets BC at D'. Prove that*

$$\frac{BD}{DC} = -\frac{BD'}{D'C}.$$

Definition 4.34.   *Here D and D' divide BC internally and externally—or externally and internally—in the same ratio. We say D, D' **separate** B, C **harmonically**, and that D' is the **harmonic conjugate** of D with respect to B, C (Fig. 4.17).*

• Ex.4.20: *Suppose also that AP meets EF in D''. Prove that D'', D' separate E, F harmonically.*

• Ex.4.21: *Suppose EF, FD, DE meet BC, CA, AB in D', E', F' respectively. Prove that D', E', F' are collinear.*

• Ex.4.22: *Taking $\triangle ABC$ as triangle of reference, show that if $P = (a, b, c)$, then $D'E'F'$ (as in the last exercise) has equation*

$$\frac{\xi}{a} + \frac{\eta}{b} + \frac{\zeta}{c} = 0.$$

Referring back to Exercise 4.19, suppose D is the mid-point of BC. Then $BD/DC = 1$, so $BD'/D'C = -1$. But this means that $\overrightarrow{D'B} = \overrightarrow{D'C}$, which entails $B = C$—impossible. We are forced to conclude that in this case EF does *not* meet BC, so $EF \parallel BC$ (Fig. 4.18).

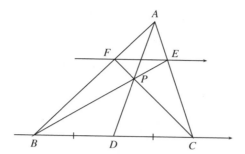

Fig. 4.18   If $BD/DC = 1$, then $EF \parallel BC$

- Ex.4.23: *Show that D is the mid-point of BC iff $AE/EC = AF/FB$ (cf. Exercise 4.16).*

- Ex.4.24: *Given a line BC with mid-point D, and a point F not on BC, give a ruler-only construction for the line through F parallel to BC.*

- Ex.4.25: *Given a line BC and a second line ℓ, parallel to BC, give a ruler-only construction for the mid-point of BC.*

- Ex.4.26: *Given a line BC and a second line ℓ, parallel to BC, give a ruler-only construction for the point D such that C is the mid-point of BD.*

- Ex.4.27: *Given a line BC, a second line ℓ, parallel to BC, and an integer $n \geq 2$, give a ruler-only construction for the point D on BC such that $BD/BC = 1/n$.*

It follows from these exercises that, if the points $(0, 0)$ and $(1, 0)$ are given, together with the line $y = 1$, then for every rational number q, the point $(q, 0)$ can be constructed by ruler alone. Again, if the points $(0, 0)$, $(1, 0)$, $(0, 1)$, and $(1, 1)$ are given, then for any rational q_1, q_2 the point (q_1, q_2) can be constructed by ruler alone.

- Ex.4.28: *Suppose a rectangle $ABCD$ is given, together with two more points, P and Q. Show that it is possible with ruler alone to construct the mid-point of PQ.*

- Ex.4.29: *Let $ABCD$ be a rectangle, and let P, Q lie on the line segments AB, AD respectively. Let R be such that $APRQ$ is another rectangle. Prove that the lines PD, QB, RC are concurrent. (The intended proof is via Menelaus; for several other proofs, including one by mechanics, see [5], pp. 175–7.)*

- Ex.4.30: *Let ℓ, m be lines meeting at A; let B, C, D lie on ℓ, let B', C', D' lie on m, and suppose BB', CC', DD' are concurrent at P. Apply Menelaus' theorem twice to $\triangle ABB'$ to show that*

$$\frac{AC}{CB} \bigg/ \frac{AD}{DB} = \frac{AC'}{C'B'} \bigg/ \frac{AD'}{D'B'}.$$

Definition 4.35.   *Given four collinear points* A, B, C, D, *the* **cross-ratio** $\{A, B; C, D\}$
is defined by

$$\{A, B; C, D\} = \frac{AC}{CB} \bigg/ \frac{AD}{DB}.$$

So Exercise 4.30 says $\{A, B; C, D\} = \{A, B'; C', D'\}$. The cross-ratio could perhaps
be written more easily as

$$\{A, B; C, D\} = \frac{AC}{CB} \frac{BD}{DA},$$

but this obscures its structure as a *ratio of ratios* (whence its name): it is the (signed)
ratio in which C divides AB, divided by the (signed) ratio in which D divides AB. In
symbols, we have $AC : CB = AC/CB : 1$, and $AD : DB = AD/DB : 1$, and finally
$AC/CB : AD/DB = \{A, B; C, D\} : 1$.

- Ex.4.31:   *Let ℓ, m be lines; let A, B, C, D lie on ℓ, let A', B', C', D' lie on m, and suppose AA', BB', CC', DD' are concurrent at O. Prove that $\{A, B; C, D\} =$
  $\{A', B'; C', D'\}$. (Use Exercise 4.30.)*

This exercise says that, just as four collinear points have a cross-ratio, so do four
concurrent lines:

Definition 4.36.   *If p, q, r, s are four lines meeting at O, and if a line ℓ (not through
O) meets p, q, r, s at A, B, C, D respectively, we define $\{p, q ; r, s\} = \{A, B ; C, D\}$.*

Exercise 4.31 says this is well defined, that is, it does not depend on the choice of ℓ.

- Ex.4.32:   *Given collinear points A, B, C, D, show that $\{A, B; C, D\} =$
  $\{B, A; D, C\} = \{C, D; A, B\} = \{D, C; B, A\}$ and also that
  $\{A, B; D, C\} = \{A, B; C, D\}^{-1}$. Show that if A, B, C, D are distinct, then
  $\{A, B; C, D\} \neq 1$, and deduce that $\{A, B; C, D\} = \{A, B; D, C\} \iff$
  $\{A, B; C, D\} = -1$.*

We have met this last condition before: $\{A, B; C, D\} = -1$ says that C, D separate
A, B harmonically, and that D is the harmonic conjugate of C with respect to A, B. It
is clear that, equally, C is the harmonic conjugate of D with respect to A, B; and also
A, B separate C, D harmonically, and so on.

Definition 4.37.   *A, B, C, D form a* **harmonic range** *if $\{A, B; C, D\} = -1$.*

We shall meet cross-ratios and harmonic ranges again in Section 9.2.

- Ex.4.33:   *Given collinear points A, B, C, D, let O be the mid-point of AB. Show
  that $\{A, B; C, D\} = -1$ iff $OA^2 = (OC)(OD)$ (the signed product of the
  lengths).*
- Ex.4.34:   *Let a circle touch the sides BC, CA, AB of $\triangle ABC$ at D, E, F respectively.
  (The circle is the incircle or one of the excircles of the triangle.) Suppose
  EF meets BC at D'. Prove that $\{B, C; D, D'\} = -1$.*

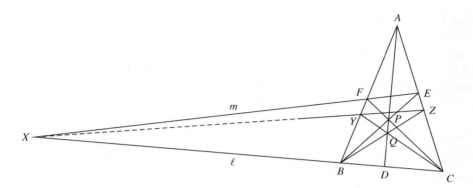

Fig. 4.19 The short ruler problem

• Ex.4.35: *Given three collinear points B, C, D, with D not the mid-point of BC, show how to construct D' such that {B, C ; D, D'} = −1, using ruler only.*

We return now to the *short ruler problem*, which was mentioned in Section 2.2. However short our ruler, it is easy (though tedious) to draw (straight) lines of arbitrary length, and likewise to produce a given line by an arbitrary amount. What is *not* easy, as a practical problem, is to draw the line joining two given points X and Y which are further apart than the length of our ruler. (Sending out for a longer ruler is regarded as cheating!)

So suppose X and Y are given, as above. The construction involves choosing and constructing extra points A, B, ..., which are *all* to be placed *near* to Y and to each other, i.e., within range of the ruler. This involves skill and judgement or (if you prefer) guesswork; if it goes wrong and you find you need to join points that are too far apart, begin again and plan more carefully.

Start by drawing two lines ℓ and m through X to come near Y (Fig. 4.19), and choose B on ℓ near Y. Let $F = m \cdot BY$, the meet. Choose C on ℓ near B and E on m near F, and let $A = BF \cdot CE$, $P = BE \cdot CF$, and $D = \ell \cdot AP$. By Exercise 4.19 (with X in place of D') we have $BD/DC = -BX/XC$. Now let $Q = YC \cdot AD$ and $Z = BQ \cdot AC$. We have $(BD/DC)(CZ/ZA)(AY/YB) = 1$, by Ceva. Thus $(BX/XC)(CZ/ZA)(AY/YB) = -1$, so that X, Y, Z are collinear, by Menelaus. So use the ruler to join YZ and produce, and this line will go through X, as required.

ANSWERS TO EXERCISES

4.1: $|AC|^2 = |AB|^2 + |BC|^2 = |A'B'|^2 + |B'C'|^2 = |A'C'|^2$, so $|AC| = |A'C'|$.

4.2: The hypotenuse is longer than either of the other sides.

4.3: $\triangle ABC$ has its centroid at $\mathbf{g}_1 = \frac{1}{3}(\mathbf{a} + \mathbf{b} + \mathbf{c})$, and $\triangle CDA$ has its centroid at $\mathbf{g}_2 = \frac{1}{3}(\mathbf{c} + \mathbf{d} + \mathbf{a})$. Since $\triangle ABC \equiv \triangle CDA$ (OK?), the centre of mass of the lamina $ABCD$ is at $\mathbf{g} = \frac{1}{2}(\mathbf{g}_1 + \mathbf{g}_2) = \frac{1}{3}\mathbf{a} + \frac{1}{6}\mathbf{b} + \frac{1}{3}\mathbf{c} + \frac{1}{6}\mathbf{d}$. But $ABCD$ is a parallelogram, so $\mathbf{a} + \mathbf{c} = \mathbf{b} + \mathbf{d}$, whence

$\mathbf{g} = \frac{1}{3}(\mathbf{a}+\mathbf{c}) + \frac{1}{6}(\mathbf{a}+\mathbf{c}) = \frac{1}{2}(\mathbf{a}+\mathbf{c}) = \frac{1}{4}(\mathbf{a}+\mathbf{b}+\mathbf{c}+\mathbf{d})$, which is the centre of mass of the system of four unit masses.

4.4: If $C = \frac{\pi}{2}$, position D so that $ACBD$ is a rectangle; then $\triangle ABC \equiv \triangle BAD$, so $\Delta = \frac{1}{2}\,\text{area}(ACBD) = \frac{1}{2}|BC||AC| = \frac{1}{2}ah$. Next, if B and C are acute, then $\Delta = \text{area}(ABP) + \text{area}(ACP) = \frac{1}{2}|BP|h + \frac{1}{2}|PC|h = \frac{1}{2}ah$. Finally, if C is obtuse, then $\Delta = \text{area}(ABP) - \text{area}(ACP) = \frac{1}{2}|BP|h - \frac{1}{2}|PC|h = \frac{1}{2}ah$ (and similarly if B is obtuse).

4.5: If A is acute, then $\angle BOC = 2A$ (angle at the centre), and since $\angle DOB = \angle DOC$, we have $\angle DOB = A$. Then, in $\triangle DOB$, we have $|DB| = a/2$ and $|BO| = R$, so $\sin A = (a/2)/R$, or $a/\sin A = 2R$. If A is obtuse, then $\pi < 2A$, which is thus the *reflex* angle BOC, so that $\angle BOC = 2\pi - 2A$, and $DOB = \pi - A$. Since $\sin(\pi - A) = \sin A$, we obtain the same result as before.

4.6: Let Z be the foot of the perpendicular from X to m; then $\triangle XYZ$ is a right triangle, so $|XZ| < |XY|$, the hypotenuse. Let Y' be the reflection of Y in XZ, so that $Y \neq Y'$, and Y' is on m (since Y is on m and m reflects to itself), and $|XY'| = |XY|$, so that Y' is on Σ.

4.7: Considering $\triangle PXY$ and $\triangle PXY'$, we have $|XY| = |XY'|$ (equal radii), PX is common, and $\angle XYP = \pi/2 = \angle XY'P$, so $\triangle PXY \equiv \triangle PXY'$ by the case SSR. Thus $|PY| = |PY'|$. Alternatively, reflection in XP sends Σ to itself (OK?) and swaps Y with Y' (OK?), so that $|PY| = |PY'|$.

4.8: Using equal tangents to the incircle, put $x = |AY| = |AZ|$, $y = |BX| = |BZ|$, and $z = |CX| = |CY|$. Then $y + z = a$, $x + z = b$, $x + y = c$, which solve to give $y = \frac{1}{2}(a-b+c) = \frac{1}{2}(a+b+c) - b$, so that $|BX| = s - b$. Now use equal tangents to the excircle opposite A: put $x_A = |AY_A| = |AZ_A|$, $y_A = |BX_A| = |BZ_A|$, and $z_A = |CX_A| = |CY_A|$. Then $y_A + z_A = a$, $x_A - z_A = b$, $x_A - y_A = c$, which solve to give $z_A = s - b$ also, which completes (i). The other parts are done similarly.

4.9: The equations $|BA| = |B'A'|$ and $\angle CBA = \angle C'B'A'$ are trivial, and $|AC| = |A'C'|$ follows from the reflection. However, $\angle AC'B = \angle AC'C = \angle ACC'$ (by the reflection), and this is the supplement of $\angle ACB$, and not equal to it.

4.10: $4\Delta^2 = s(s-a)(a-(b-c))(a+(b-c)) = s(s-a)(a^2-(b-c)^2)$, which (since $s(s-a)$ is fixed and positive) is greatest when $a^2 - (b-c)^2$ is greatest, that is, when $b - c = 0$.

4.11: By the last exercise, for fixed a the greatest area is Δ_a, where $4\Delta_a^2 = s(s-a)a^2$. By calculus, the greatest value of this for $0 < a < s$ occurs when $a = \frac{2}{3}s$, and then $b = c = s - (a/2) = \frac{2}{3}s$, again.

4.12: Since $ABCD$ is cyclic, we have $\theta + \varphi = \pi$, and so $\sin\varphi = \sin\theta$ and $\cos\varphi = -\cos\theta$. Next, $a^2 + b^2 - 2ab\cos\theta = |AC|^2 = c^2 + d^2 - 2cd\cos\varphi$, so that $a^2 + b^2 - c^2 - d^2 = 2(ab+cd)\cos\theta$. Then $\Delta^2 = \frac{1}{4}(ab+cd)^2\sin^2\theta = \frac{1}{4}(ab+cd)^2 - \frac{1}{4}(ab+cd)^2\cos^2\theta = \frac{1}{16}(4(ab+cd)^2 - (a^2 + b^2 - c^2 - d^2)^2) = \frac{1}{16}(2(ab+cd) - (a^2+b^2-c^2-d^2))(2(ab+cd) + (a^2+b^2-c^2-d^2)) = \frac{1}{16}((c+d)^2 - (a-b)^2)((a+b)^2 - (c-d)^2) = \frac{1}{16}(-a+b+c+d)(a-b+c+d)(a+b-c+d)(a+b+c-d) = (s-a)(s-b)(s-c)(s-d)$.

4.13: Again we have $a^2 + b^2 - 2ab\cos\theta = |AC|^2 = c^2 + d^2 - 2cd\cos\varphi$, so that $a^2 + b^2 - c^2 - d^2 = 2(ab\cos\theta - cd\cos\varphi)$. Then $\Delta^2 = (\frac{1}{2}ab\sin\theta + \frac{1}{2}cd\sin\varphi)^2 = \frac{1}{4}a^2b^2\sin^2\theta + \frac{1}{4}c^2d^2\sin^2\varphi + \frac{1}{2}abcd\sin\theta\sin\varphi = \frac{1}{4}(a^2b^2 + c^2d^2) - \frac{1}{4}(a^2b^2\cos^2\theta + c^2d^2\cos^2\varphi) + \frac{1}{2}abcd\sin\theta\sin\varphi = \frac{1}{4}(a^2b^2 + c^2d^2) - \frac{1}{4}(ab\cos\theta - cd\cos\varphi)^2 - \frac{1}{2}abcd(\cos\theta\cos\varphi - \sin\theta\sin\varphi) = \frac{1}{4}(a^2b^2 + c^2d^2) - \frac{1}{16}(a^2+b^2-c^2-d^2)^2 - \frac{1}{2}abcd\cos(\theta+\varphi)$. For fixed a, b, c, d, this is greatest when $\cos(\theta+\varphi) = -1$, that is, when $\theta + \varphi = \pi$.

4.14: If we write $\Delta = \text{area}(ABCD)$ and $a = |OA|$, $b = |OB|$, $c = |OC|$, $d = |OD|$, then in the convex case, $\Delta = \frac{1}{2}ab\sin\theta + \frac{1}{2}bc\sin\theta + \frac{1}{2}cd\sin\theta + \frac{1}{2}da\sin\theta = \frac{1}{2}(a+c)(b+d)\sin\theta = \frac{1}{2}|AC||BD|\sin\theta$. In the re-entrant case, we have $|AC| = a + c$ and $|BD| = |b - d|$ or else $|AC| = |a - c|$ and $|BD| = b + d$, and the above proof still works with appropriate sign changes. (It always gives a positive answer, since $0 < \theta < \pi$.) In the crossed case, the diagonals could be parallel (so that $\theta = 0$); but if they are not, then we have $|AC| = |a - c|$ *and* $|BD| = |b - d|$. If the sides AB and CD meet at X, the formula yields $|\,\text{area}(XAD) - \text{area}(XBC)|$; or, in the other case, if the sides BC and DA meet at Y, it yields $|\,\text{area}(YAB) - \text{area}(YDC)|$. This quantity cannot be negative, but it can be zero if the diagonals are parallel. (Please check all the details.)

4.15: First, $|AB| = 2|AF|$, $|AC| = 2|AE|$, and $\angle BAC = \angle FAE$, so that $\triangle ABC \sim \triangle AFE$ (SAS). Since the scale factor is 2, we have that $|BC| = 2|FE|$. (Alternatively: $\mathbf{e} - \mathbf{f} = \frac{1}{2}(\mathbf{a} + \mathbf{c}) - \frac{1}{2}(\mathbf{a} + \mathbf{b}) = \frac{1}{2}(\mathbf{c} - \mathbf{b})$ shows $|FE| = \frac{1}{2}|BC|$.) Likewise, $|CA| = 2|DF|$ and $|AB| = 2|ED|$, so that $\triangle ABC \sim \triangle DEF$ (SSS). For the last part, $|DE| = |AF| = |FB| = |ED|$ $(= \frac{1}{2}|AB|)$, and likewise for the other sides of the four triangles.

4.16: We have $\angle FAE = \angle BAC$, and if $BC \parallel FE$, then also $\angle AFE = \angle ABC$. So $\triangle AFE \sim \triangle ABC$ (AAA), from which $|AF| : |FB| = |AE| : |EC|$. Conversely, if $|AF| : |FB| = |AE| : |EC|$, then since $\angle FAE = \angle BAC$, we have $\triangle AFE \sim \triangle ABC$ (SAS), so that $\angle AFE = \angle ABC$, and $BC \parallel FE$.

4.17: Put $f(A) = A'$, etc. By Exercise 3.7 we may suppose, without loss of generality, that B lies between A and C, so that $|AB| + |BC| = |AC|$, by Lemma 3.16. Since f is a similarity, we have $|A'B'| + |B'C'| = k|AB| + k|BC| = k(|AB| + |BC|) = k|AC| = |A'C'|$, where k is the scale factor. So A', B', C' are collinear, by Lemma 3.16, again.

4.18: If the corresponding cartesian equations are $p_1 x + q_1 y + r_1 = 0$ and $p_2 x + q_2 y + r_2 = 0$, then the lines are parallel iff $p_1 q_2 = p_2 q_1$. But

$$\begin{pmatrix} p_1' & q_1' & r_1' \\ p_2' & q_2' & r_2' \\ 1 & 1 & 1 \end{pmatrix} = \begin{pmatrix} p_1 & q_1 & r_1 \\ p_2 & q_2 & r_2 \\ 0 & 0 & 1 \end{pmatrix} \mathbf{M},$$

and the result follows on taking determinants. Alternatively, the two barycentric equations, being two homogeneous linear equations in three unknowns, always have a solution $(\xi, \eta, \zeta) \neq (0, 0, 0)$; but if the lines are parallel, this triple cannot represent a point and so must also satisfy $\xi + \eta + \zeta = 0$, whence the vanishing of the given determinant. (Now reverse this argument to complete the proof.)

4.19: By Ceva,

$$\frac{BD}{DC}\frac{CE}{EA}\frac{AF}{FB} = 1,$$

and by Menelaus,

$$\frac{BD'}{D'C}\frac{CE}{EA}\frac{AF}{FB} = -1.$$

Compare!

4.20: There is no new calculation required here: it is just a matter of applying the result of Exercise 4.19 to $\triangle AFE$ instead of $\triangle ABC$. Thus, we have points D'', C, B on the sides FE, EA, AF of $\triangle AFE$, and the lines AD'', FC and EB meet at P. Further, the join BC meets the side FE at D'. So $(FD'')/(D''E) = -(FD')/(D'E)$, as required. (If, for Exercise 4.19, you drew $\triangle ABC$ with P inside, then P will be *outside* $\triangle AFE$; but you could equally well have drawn $\triangle ABC$ with P outside, and the proof would have worked just as well: indeed, the proof doesn't 'know' where P is at all, except that it lies on each of the three lines AD, BE, CF, wherever they happen to be.)

4.21: We have

$$\frac{BD'}{D'C} = -\frac{BD}{DC}, \quad \frac{CE'}{E'A} = -\frac{CE}{EA}, \quad \text{and} \quad \frac{AF'}{F'B} = -\frac{AF}{FB},$$

so

$$\frac{BD'}{D'C}\frac{CE'}{E'A}\frac{AF'}{F'B} = \left(-\frac{BD}{DC}\right)\left(-\frac{CE}{EA}\right)\left(-\frac{AF}{FB}\right) = -\frac{BD}{DC}\frac{CE}{EA}\frac{AF}{FB} = -1$$

by Ceva(!), and the result follows by Menelaus.

4.22: A is $(1, 0, 0)$ and P is (a, b, c), so AP is

$$\begin{vmatrix} 1 & a & \xi \\ 0 & b & \eta \\ 0 & c & \zeta \end{vmatrix} = 0,$$

or $b\zeta - c\eta = 0$, or $\eta/b = \zeta/c$. This meets $\xi = 0$ (i.e. BC) at D, which is therefore $(0, b, c)$: you just need *any* non-trivial solution of the simultaneous equations $\xi = 0$, $\eta/b = \zeta/c$, and certainly $\xi = 0$, $\eta = b$, $\zeta = c$ will do. Similarly $E = (a, 0, c)$ and $F = (a, b, 0)$. From Exercise 4.19, $D' = (0, b, -c)$ (or, more laboriously, EF is

$$\begin{vmatrix} a & a & \xi \\ 0 & b & \eta \\ c & 0 & \zeta \end{vmatrix} = 0,$$

or $(\xi/a) = (\eta/b) + (\zeta/c)$, which meets $\xi = 0$ at $(0, b, -c)$). Similarly $E' = (a, 0, -c)$ (and $F' = (a, -b, 0)$, but this is not needed). So $D'E'$ is

$$\begin{vmatrix} 0 & a & \xi \\ b & 0 & \eta \\ -c & -c & \zeta \end{vmatrix} = 0,$$

or (after a little juggling) $(\xi/a) + (\eta/b) + (\zeta/c) = 0$—done. If you wish, you can now give another proof of Exercise 4.21 simply by checking that F' lies on this line:

$$\begin{vmatrix} 0 & a & a \\ b & 0 & -b \\ -c & -c & 0 \end{vmatrix} = 0.$$

4.23: D is the mid-point of BC iff $\frac{BD}{DC} = 1$. But

$$\frac{BD}{DC}\frac{CE}{EA}\frac{AF}{FB} = 1$$

(Ceva), so D is the mid-point of BC iff

$$\frac{CE}{EA}\frac{AF}{FB} = 1, \quad \text{or} \quad \frac{AF}{FB} = \frac{EA}{CE} = \frac{AE}{EC}.$$

4.24: Join BF and pick a point A on BF (produced, if you like, but it doesn't matter). Join AD, and also CF, and put $P = AD \cdot CF$ (the meet). Join BP, and also AC, and put $E = BP \cdot AC$. Then $FE \parallel BC$.

4.25: Pick any point A not on either of the lines, let $F = AB \cdot \ell$, $E = AC \cdot \ell$, $P = BE \cdot CF$, and finally $D = AP \cdot BC$. Then D is the mid-point of BC.

4.26: This is surprisingly hard. Pick any point A not on either of the lines, and let $B' = AB \cdot \ell$, $D' = AC \cdot \ell$, $P = BD' \cdot CB'$, and $C' = AP \cdot \ell$. So now C' is the mid-point of $B'D'$. Let $A' = CC' \cdot AB$ and finally $D = A'D' \cdot BC$.

4.27: Choose E_0, E_1 on ℓ, and successively construct $E_2, E_3, \ldots, E_n$ on ℓ so that E_i is the mid-point of $E_{i-1}E_{i+1}$, for all i. Let $A = BE_0 \cdot CE_n$ (if the lines are parallel, or meet off the page, start again with more foresight!), and finally $D = AE_1 \cdot BC$.

4.28: This is based on Exercises 4.24 and 4.25, using the fact that we are given $AB \parallel DC$ and $BC \parallel AD$. If PQ is parallel to one of the sides of the rectangle, just use Exercise 4.25. Otherwise, construct lines through P and Q parallel to AB (Exercise 4.24), and let these meet AD in P', Q' respectively. Construct the mid-point of $P'Q'$ (Exercise 4.25), and call it R'. Construct the line through R' parallel to AB (Exercise 4.24 again), and this line will meet PQ at its mid-point. (Note that $ABCD$ does not have to be a rectangle: a parallelogram would do just as well.)

4.29: Let $PD \cdot QB = X$, $PR \cdot CD = P'$, and $QR \cdot BC = Q'$. Apply Menelaus to the line QXB and $\triangle APD$: we have

$$\frac{PX}{XD}\frac{DQ}{QA}\frac{AB}{BP} = -1.$$

But $|DQ| = |P'R|$, $|QA| = |RP|$, $|AB| = |DC|$, and $|BP| = |CP'|$, so

$$\frac{PX}{XD}\frac{DC}{CP'}\frac{P'R}{RP} = -1,$$

whence X, R, C are collinear, by Menelaus applied to $\triangle DPP'$.

4.30: We have

$$\left(\frac{AC}{CB}\frac{BP}{PB'}\frac{B'C'}{C'A}\right) \Big/ \left(\frac{AD}{DB}\frac{BP}{PB'}\frac{B'D'}{D'A}\right) = (-1)/(-1) = 1,$$

and the terms BP/PB' cancel—and we are done.

4.31: Join AD'; let $AD' \cdot BB' = B''$ and $AD' \cdot CC' = C''$. Then $\{A, B; C, D\} = \{A, B''; C'', D'\} = \{A', B'; C', D'\}$, by two applications of Exercise 4.30.

4.32: The first two parts just say that

$$\frac{AC}{CB}\frac{BD}{DA} = \frac{BD}{DA}\frac{AC}{CB} = \frac{CA}{AD}\frac{DB}{BC} = \frac{DB}{BC}\frac{CA}{AD} = \left(\frac{AD}{DB}\frac{BC}{CA}\right)^{-1},$$

all of which is obvious. If $\{A, B; C, D\} = 1$, then $AC/CB = AD/DB$, so if $A \neq B$, then C, D divide AB in the same ratio, so that $C = D$. Finally, $\{A, B; C, D\} = \{A, B; D, C\} \Leftrightarrow \{A, B; C, D\}^2 = 1 \Leftrightarrow \{A, B; C, D\} = \pm1 \Leftrightarrow \{A, B; C, D\} = -1$, since $\{A, B; C, D\} \neq +1$.

4.33: Let $OA = a$, $OB = b$, $OC = c$, $OD = d$, so that $b = -a$. Then $OA^2 = (OC)(OD) \Leftrightarrow a^2 = cd \Leftrightarrow (a - c)(d + a) = (a + c)(d - a) \Leftrightarrow (c - a)(b - d) = -(b - c)(d - a) \Leftrightarrow (AC)(DB) = -(CB)(AD) \Leftrightarrow \{A, B; C, D\} = -1$.

4.34: We have

$$\frac{BD}{DC}\frac{CE}{EA}\frac{AF}{FB} = 1,$$

since the lengths cancel in pairs, being equal tangents from A, from B, and from C. (Check that the signs are correct.) So AD, BE and CF are concurrent, by Ceva, and the result follows by Exercise 4.19.

4.35: Choose A not collinear with the given points, and join AB, AC, AD. Choose P on AD, join BP and CP, and let $E = BP \cdot AC$ (the meet), $F = CP \cdot AB$, and join EF. Then by Exercise 4.21, $D' = BC \cdot EF$.

5

Isometries of $\mathbb{R}^2$

We have met isometries in Section 3.5; recall that an isometry is a distance-preserving transformation of $\mathbb{R}^2$. One can also define isometries in $\mathbb{R}^3$: it is simply a matter of replacing $\mathbb{R}^2$ by $\mathbb{R}^3$ in Definitions 3.17 and 3.18. We shall continue to concentrate attention on $\mathbb{R}^2$, but the reader who feels sufficiently confident is encouraged to generalize statements to $\mathbb{R}^3$ where possible. We shall occasionally indicate where extra care or new terminology is needed to do this. There will be more about isometries of $\mathbb{R}^3$, and indeed about isometries of $\mathbb{R}^n$, in the next chapter.

5.1 FIXED POINTS AND LINES

Let f be an isometry (or indeed any transformation).

Definition 5.1. *P is a* **fixed point** *of f if $f(P) = P$.*

Example 1. If f is a reflection, then every point of its axis is a fixed point of f. (In $\mathbb{R}^3$, the mirror is a plane, not a line.)

Example 2. If f is a rotation about P, then P is a fixed point. (In $\mathbb{R}^3$, rotation is about a *line*, not a point. This line is called the *axis* of rotation, and all its points are fixed by the rotation.)

Example 3. A translation has *no* fixed points.

Example 4. The trivial or identity map 1 fixes *every* point.

Definition 5.2. *The line ℓ is an* **invariant line** *of f if $f(\ell) = \ell$, that is, $P \in \ell \iff f(P) \in \ell$.*

Example 1. If f is a translation, P is any point and $f(P) = P'$, then the line PP' is an invariant line of f.

Exampel 2. If f is a reflection, then its axis is an invariant line, and so is every line orthogonal to its axis.

Example 3. A rotation has no invariant lines unless it is a half-turn, in which case every line through the centre of rotation is invariant.

Perhaps this is the moment to point out that, if a statement is not obvious and if no proof is supplied, then the statement is an implied challenge to the reader to give a proof. So, are you *sure* about all these examples?

Definition 5.3. *The line ℓ is a* **line of fixed points** *of f if every point of ℓ is a fixed point of f.*

Example. If f is a reflection, its axis is a line of fixed points of f.

Note that a line of fixed points of f is necessarily an invariant line of f, but not conversely. (Find an example.) Some authors call an invariant line a *fixed line*; we shall avoid this because of the likelihood of confusion with *line of fixed points*. (In $\mathbb{R}^3$ one can also define invariant planes, and planes of fixed points. Find some examples.)

Lemma 5.4. *Let f be an isometry, let R be a point and let $f(R) = R'$, where $R \neq R'$. Let P be a fixed point of f. Then P lies on the perpendicular bisector of RR'.*

Proof. Immediate from Proposition 3.14. □

(Note: in $\mathbb{R}^3$, the perpendicular bisector of RR' is a *plane*.)

For the next few results, we shall put the three-dimensional adaptions in brackets, where required; but the proofs of the three-dimensional versions are for the most part left to the reader.

Corollary 5.5. *Let f be an isometry of $\mathbb{R}^2$ ($\mathbb{R}^3$), with $f \neq 1$. Then the fixed points of f, if any, lie in a line (plane).*

Proof. Just choose $R \neq R'$ as in the lemma, and the line (plane) in question is the perpendicular bisector of RR'. □

Note that we are *not* saying that this gives a line (plane) of fixed points, but merely that the line (plane) obtained contains any fixed points there might happen to be. We obtain immediately:

Corollary 5.6. *Let f be an isometry of $\mathbb{R}^2$ ($\mathbb{R}^3$), and suppose f has three (four) fixed points which are not collinear (coplanar). Then $f = 1$.* □

(Points $A, B, C, D, \ldots$ are *coplanar* if they lie in a plane.)

Corollary 5.7. *Every isometry of $\mathbb{R}^2$ ($\mathbb{R}^3$) is a product of at most three (four) reflections.*

Proof. Given f, choose any $\triangle ABC$ and put $A' = f(A)$, $B' = f(B)$, $C' = f(C)$. Then $\triangle ABC \equiv \triangle A'B'C'$, and, as in the proof of Proposition 3.22 (and as in the note following the proof of that Proposition), there is an isometry g, the product of at most three reflections, with $g(A) = A'$, $g(B) = B'$, and $g(C) = C'$. But now $g^{-1}f$ is an isometry (OK?) fixing A, B, C, so that $g^{-1}f = 1$, by Corollary 5.6. Thus $f = g$, a product of at most three reflections. (To do the three-dimensional version, you will need to prove a three-dimensional version of Proposition 3.22.)   □

Corollary 5.8.   *(i) Let f be an isometry fixing P, Q, where $P \neq Q$. Then PQ is a line of fixed points of f. (ii) Let f be an isometry of $\mathbb{R}^3$ fixing three non-collinear points P, Q, R. Then PQR is a plane of fixed points of f.*

Proof.   (i) Suppose R is on PQ: we must show $R = R'$ (where $R' = f(R)$). Now P, Q, R are collinear, so P', Q', R' are collinear, by Proposition 3.20. But P, Q are fixed, that is, $P = P'$ and $Q = Q'$, so R' is on PQ. If $R \neq R'$, then P is on the perpendicular bisector of RR', by Lemma 5.4. But since P, R, R' are collinear, P is the meet of RR' and its perpendicular bisector, that is, P is the mid-point of RR'. Similarly Q is the mid-point of RR', whence $P = Q$, a contradiction. Thus $R = R'$. (This argument works whether f is an isometry of $\mathbb{R}^2$ or of $\mathbb{R}^3$.)

(ii) Similarly to (i)—please try it!—or use the fact that, since f maps planes to planes (OK?), the plane PQR is an invariant plane of f. Thus the restriction of f to the plane PQR is a plane isometry with fixed points P, Q, R; now use the two-dimensional version of Corollary 5.6.   □

Corollary 5.9.   *(i) Let f be an isometry of $\mathbb{R}^2$ with a line of fixed points, ℓ. If $f \neq 1$, then f is reflection in ℓ. (ii) Let f be an isometry of $\mathbb{R}^3$ with a plane of fixed points, p. If $f \neq 1$, then f is reflection in p.*

Proof.   (i) If $f \neq 1$, there is a point R with $R \neq R'$, where $R' = f(R)$. But now ℓ must be the perpendicular bisector of RR', so if g is reflection in ℓ, we have that $g^{-1}f$ fixes R *and* every point of ℓ. By Corollary 5.6, $g^{-1}f = 1$, or $f = g$. (ii) Similarly.   □

Corollary 5.10.   *Let f be an isometry of $\mathbb{R}^2$ with $f^2 = 1$. If $f \neq 1$, then f is either a reflection or a half-turn (that is, a rotation through the angle π).*

Proof.   If f has two or more fixed points, it is a reflection, by Corollaries 5.8 and 5.9. So now suppose f has at most one fixed point. If A is *not* a fixed point, then put $A' = f(A)$ and we have $A \neq A'$; and if we put $A'' = f(A')$, then $A'' = A$, since $f^2 = 1$. Let P be the mid-point of AA', and put $P' = f(P)$: so $|AP| = |PA'| = \frac{1}{2}|AA'|$. But this means $|A'P'| = |P'A''| = \frac{1}{2}|A'A''|$, and P' is the mid-point of $A'A''$, that is, of $A'A$. So $P = P'$, and we have a fixed point. Since there is not more than one fixed point, we deduce that P is the mid-point of AA' for *every* choice of A, and thus f is the half-turn about P.   □

● Ex.5.1:   *State and prove the three-dimensional version of Corollary 5.10.*

5.2   GROUPS OF ISOMETRIES

Recall that a *transformation* of $\mathbb{R}^2$ is a map $f : \mathbb{R}^2 \to \mathbb{R}^2$ which is bijective, that is, it is both injective and surjective, and so has an inverse f^{-1}, satisfying $ff^{-1} = 1 = f^{-1}f$. Likewise, a transformation of $\mathbb{R}^3$ is a map $f : \mathbb{R}^3 \to R^3$ which is bijective. We have referred to the notion of a group of transformations informally; now is the time to give the proper definition:

Definition 5.11. *Let G be a set of transformations of* $\mathbb{R}^2$ *(or of* $\mathbb{R}^3$ *). Then G is a* **group of transformations**, *or* **transformation group**, *if (i)* $1 \in G$; *(ii)* $f, g \in G \Rightarrow fg \in G$; *and (iii)* $f \in G \Rightarrow f^{-1} \in G$.

Here 1 stands for the identity map $1 : P \mapsto P$, for all P, and fg means the composite transformation $fg : P \mapsto f(g(P))$, that is, apply g first, then f.

For the sake of completeness (and to save time later) we give the definition of the more general concept, that of (abstract) *group*. Let G be a set with a *binary operation*, that is, a map $G \times G \to G$, which we shall write as $(x, y) \mapsto x \circ y$. This operation is *associative* if $(x \circ y) \circ z$ is always the same as $x \circ (y \circ z)$ $(x, y, z \in G)$, so that we can cheerfully write $x \circ y \circ z$ to mean either. A *neutral* or *identity* element is an element $e \in G$ such that $e \circ x = x = x \circ e$, for all $x \in G$, and an *inverse* for x is an element $x^* \in G$ such that $x \circ x^* = e = x^* \circ x$. Then:

Definition 5.12. *The set G, with a given binary operation, is a* **group** *if (i) the operation is associative; (ii) G has a neutral element, and (iii) every element of G has an inverse in G.*

To save writing and to make calculations less clumsy, it is customary to write xy in place of $x \circ y$, 1 in place of e, and x^{-1} in place of x^*. This is called *multiplicative* notation, and G is then called a *multiplicative group*.

- Ex.5.2: *A group of transformations is a multiplicative group. (This is simply a matter of checking that the associative law holds for map composition. Just check that* $(fg)h : P \mapsto f(g(h(P)))$, *and similarly for* $f(gh)$.)
- Ex.5.3: *Give an example to show that, if* f, g *are transformations, then* fg *need not be the same as* gf.

Definition 5.13. *The group G is* **abelian** *if every pair of elements of G* **commute**, *that is,* $x \circ y = y \circ x$, *for all* $x, y \in G$.

In the case of an abelian group, we sometimes use *additive* notation, writing $x + y$ in place of $x \circ y$, 0 in place of e, and $-x$ in place of x^*. (Also, $x - y$ means $x \circ y^*$.) G is then an *additive group*; additive groups are *always* abelian (in this book, and in most others), but multiplicative groups may or may not be.

If this were a text on group theory, we should now spend several pages developing elementary properties of groups. We shall not do this, but instead just collect a few of these properties as exercises, so as to get back to geometry as quickly as possible. (The reader who wants a more thorough introduction to group theory should refer to one of the standard texts, such as [19], [1], or [6].)

- Ex.5.4: *A group cannot have more than one neutral element, and an element of a group cannot have more than one inverse.*

- Ex.5.5:   *Inverting a product:* $(xy)^{-1} = y^{-1}x^{-1}$. *(In the morning, you put on your socks and then your shoes; at night you reverse this process by taking off your shoes* before *you attempt to take off your socks!)*

- Ex.5.6:   *Cancellation:* $xy = xz \Rightarrow y = z$, *and* $xz = yz \Rightarrow x = y$. *Also* $xy = x \Rightarrow y = 1$ *and* $xy = y \Rightarrow x = 1$. *(But if* $xy = yz$, *it does* not *follow, in general, that* $x = z$: *give an example.)*

- Ex.5.7:   *Powers: write* x^2 *for* xx, x^3 *for* xxx, *and so on. What should* x^{-2} *mean:* $(x^2)^{-1}$ *or* $(x^{-1})^2$? *And what about* x^0? *Make sure that your definitions satisfy* $x^n x^m = x^{n+m}$ *for all* $n, m \in \mathbb{Z}$. *(In an* additive *group, one would write* $2x$ *for* $x + x$, *and so on, so write out this version of the exercise also.)*

- Ex.5.8:   *Let G be a set of transformations. Show that G is a group of transformations iff (i)* $G \neq \emptyset$, *and (ii)* $f, g \in G \Rightarrow fg^{-1} \in G$.

Definition 5.14.   *Let* G, H *be groups of transformations. If* $H \subseteq G$, *we say H is a* **subgroup** *of G. More generally, if G is a (multiplicative) group, the subset* $H \subseteq G$ *is a* **subgroup** *of G if (i)* $1 \in H$, *(ii)* $x, y \in H \Rightarrow xy \in H$, *and (iii)* $x \in H \Rightarrow x^{-1} \in H$.

This means H is itself a group, of course. If $H \neq G$, then H is a *proper* subgroup. Comparing with Definition 5.11, we see that to say G is a group of transformations is the same as to say G is a *sub*group of the group of *all* transformations of $\mathbb{R}^2$ (or of $\mathbb{R}^3$).

Examples. $\mathcal{I}(\mathbb{R}^2)$ is a group of transformations, and so is $\mathcal{E}(\mathbb{R}^2)$; indeed, $\mathcal{I}(\mathbb{R}^2)$ is a subgroup of $\mathcal{E}(\mathbb{R}^2)$, since every isometry is a similarity. The one-element set $\{1\}$, which by abuse of notation we shall denote by 1, is a subgroup of every group of transformations, and is called the *trivial* (sub)group; other subgroups are *non-trivial*.

Definition 5.15.   *The transformation* $f : \mathbb{R}^2 \to \mathbb{R}^2$ *is a* **collineation** *if, for every line* ℓ, *the image* $f(\ell)$ *is a line. That is, if A, B, C are collinear, then so are* $f(A), f(B), f(C)$. *The set of all collineations is denoted* $\mathcal{C}(\mathbb{R}^2)$.

- Ex.5.9:   *Prove that* $\mathcal{C}(\mathbb{R}^2)$ *is a group of transformations, and that* $\mathcal{E}(\mathbb{R}^2)$ *is a subgroup of* $\mathcal{C}(\mathbb{R}^2)$. *(Cf. Exercise 4.17.)*

- Ex.5.10:   *Show that the map* $f : \mathbb{R}^2 \to \mathbb{R}^2$ *given by* $f(x, y) = (2x, y)$, *for all* x, y, *is a collineation but not a similarity, and deduce that* $\mathcal{E}(\mathbb{R}^2)$ *is a* proper *subgroup of* $\mathcal{C}(\mathbb{R}^2)$.

We shall meet collineations again in Chapter 9.

- Ex.5.11:   *Prove that the set* $\mathcal{T}(\mathbb{R}^2)$ *of all translations of* $\mathbb{R}^2$ *is a subgroup of* $\mathcal{I}(\mathbb{R}^2)$. *(See Example 1 on p. 31 for the definition of* translation.*) Show also that* $\mathcal{T}(\mathbb{R}^2)$ *is abelian.*

Definition 5.16.   *Let* $X \subseteq \mathbb{R}^2$, *where* $X \neq \emptyset$. *A* **symmetry** *of X is an isometry f with* $f(X) = X$, *that is, such that* $P \in X \iff f(P) \in X$.

Note that 1 is a (trivial) symmetry of X, whatever peculiar shape X may be; this is at odds with what one might call the common-sense notion of symmetry, but is needed to ensure (as we shall show next) that the symmetries of X always form a group. If 1 is the only symmetry of X, so that the symmetry group is trivial, most people would say X has no symmetry. The mathematician says instead that the only symmetry of X is the trivial symmetry.

Proposition 5.17. *The set $\mathcal{S}(X)$ of symmetries of X is a group, the* symmetry group *of X.*

Proof. (i) $1 \in \mathcal{S}(X)$, as remarked above. (ii) If $f, g \in \mathcal{S}(X)$, then $(fg)(X) = f(g(X)) = f(X) = X$, so $fg \in \mathcal{S}(X)$. (iii) If $f \in \mathcal{S}(X)$, then $f(X) = X$, so $f^{-1}(f(X)) = f^{-1}(X)$, or $X = f^{-1}(X)$. So $f^{-1} \in \mathcal{S}(X)$. $\qquad\square$

Example 1. (Fig. 5.1.) Let X be the square with vertices $(\pm 1, \pm 1)$, that is, $X = \{ (x, y) \in \mathbb{R}^2 : |x + y| + |x - y| = 2 \}$. (OK?) This has four reflectional symmetries, namely reflections in the four *lines of symmetry* $x = 0$, $y = 0$, $x - y = 0$, $x + y = 0$. There are also some rotational symmetries, namely the rotations about the origin O through $\pi/2$, π, and $3\pi/2$ (or $-\pi/2$; it's where you arrive that matters, not how you got there). And we're not going to forget 1, are we?—so there are eight symmetries in all. (Make sure there are no more.) If we denote the rotation through $\pi/2$ by f, then the other rotations are f^2 and f^3, and $f^4 = 1$ (so $f^3 = f^{-1}$). Choose any one of the reflections and call it g, so that $g^2 = 1$ (and $g^{-1} = g$). The isometries gf, gf^2 and gf^3 must belong to $\mathcal{S}(X)$, by Proposition 5.17, and it is easy to check that they are the other reflections. So

$$\mathcal{S}(X) = \{ 1, f, f^2, f^3, g, gf, gf^2, gf^3 \},$$

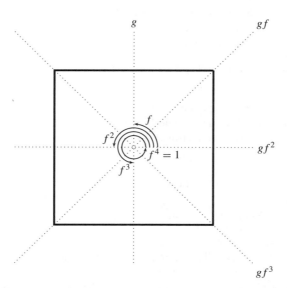

Fig. 5.1 Example 1: The eight symmetries of a square

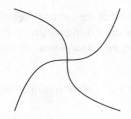

Fig. 5.2   Example 2: $\mathcal{S}(X) = C_4$

where these eight symmetries are all different, and $f^4 = 1 = g^2$. Then $fg \in \mathcal{S}(X)$, so fg must be one of the elements listed above. Either work it out from the definitions of f and g, or note that $(fg)^{-1} = g^{-1}f^{-1} = gf^3$, a reflection, so that $fg = (gf^3)^{-1} = gf^3$. This is now enough information to compute all the products in $\mathcal{S}(X)$; for example,

$$(gf)(gf^2) = g(fg)f^2 = g(gf^3)f^2 = g^2 f^5 = f.$$

The reader is invited to complete the 8×8 multiplication table for $\mathcal{S}(X)$. Here $\mathcal{S}(X)$ is an example of a *dihedral* group, and is denoted D_4. (Some people would call it D_8, because it has eight elements, but we prefer D_4 for the symmetries of a square, and D_8 for the symmetries of a regular octagon. What does D_8 look like?)

The subset $\{1, f, f^2, f^3\}$ is a subgroup of $\mathcal{S}(X)$, consisting of all the rotational symmetries of X (where, of course, we count 1 as a trivial rotation, through the angle zero). This group, which will be defined for general X later, is called the *rotation group* of X, written $\mathcal{S}^+(X)$. It is an example of a *cyclic* group, and is denoted C_4. A cyclic group is a group in which all the elements are powers of a particular element, called a *generator*. (So f is a generator for $\mathcal{S}^+(X)$. Is it the *only* generator?)

- Ex.5.12: D_4 *has a total of eight proper non-trivial subgroups. Find them.*

Example 2.   (Fig. 5.2.) Let $X = \{(x, y) \in \mathbb{R}^2 : (y - x^3)(x + y^3) = 0\}$. So X is the union of two cubic curves, $y = x^3$ and $x = -y^3$, the one being obtainable from the other by the *same* quarter-turn f as in Example 1 above, and it is easy to see that $\mathcal{S}^+(X) = C_4$, again. But this time X has *no* reflectional symmetries (OK?), so that $\mathcal{S}(X) = C_4$ also.

- Ex.5.13:   *Find another set X with $|\mathcal{S}(X)| = 4$, but for which $\mathcal{S}(X)$ is not cyclic.*
- Ex.5.14:   *Find $\mathcal{S}(X)$ when $X = \{(x, y) \in \mathbb{R}^2 : x^3y - xy^3 = 1\}$.*
- Ex.5.15:   *Find a set X with $|\mathcal{S}(X)| = 3$. (A drawing will do; you are not obliged to describe the set by algebra.)*

If you did Exercise 5.14 properly—no, wait: *did* you do it? (Skipping the exercises is called *laziness*, and is a Bad Thing, unless you are a genius.) Once more, then: if you did Exercise 5.14 properly, you will have found that you obtained D_4 again, or rather something that looked like D_4 with (as it happens) the same rotations as in Example 1 above, but four different reflections. If you were to work out the 8×8 multiplication table in this case, you would find it was the 'same' as in Example 1: in some sense, the *groups are the same*. The correct language for this situation is: the groups are *isomorphic*. Here is the full definition:

Definition 5.18. *Let G, H be groups. A map $\alpha : G \to H$ is an **isomorphism** if both (i) α is bijective (that is, it is both one-to-one (injective) and onto (surjective)), and also (ii) for every $a, b \in G$, we have $\alpha(ab) = \alpha(a)\alpha(b)$.*

Because of (i), there is a map $\beta : H \to G$ with $\alpha\beta = 1_G$ and $\beta\alpha = 1_H$, where $1_G : G \to G$ is the identity map, fixing everything, and similarly for 1_H. We write $\beta = \alpha^{-1}$; it is easy to show that β is another isomorphism. Notice that, in (ii), the multiplication on the left takes place in G, whereas the multiplication on the right takes place in H. If our groups were additive, (ii) would become $\alpha(a + b) = \alpha(a) + \alpha(b)$; if G were additive but H were multiplicative, it would become $\alpha(a + b) = \alpha(a)\alpha(b)$; and if G were multiplicative and H were additive, we should require $\alpha(ab) = \alpha(a) + \alpha(b)$. (There is a more general notion, that of a *homomorphism* $\alpha : G \to H$. This is required to satisfy (ii) but not necessarily (i); so an isomorphism is just a bijective homomorphism.)

Example. Let $P_4 = \{ z \in \mathbb{C} : z^4 = 1 \} = \{ 1, i, -1, -i \}$. This is easily seen to be a subgroup of $\mathbb{C}^*$, the group of all non-zero complex numbers under complex multiplication. Then let $\mathbb{Z}_4 = \{ 0, 1, 2, 3 \}$ under addition modulo 4 (so that $1 + 1 = 2$, but $2 + 2 = 0$, for instance). We leave it to the reader to show that the map $\alpha : \mathbb{Z}_4 \to P_4$ given by $\alpha(n) = e^{\pi i n/2}$ is an isomorphism.

Definition 5.19. *We say the groups G, H are **isomorphic** if there is an isomorphism $\alpha : G \to H$, and we write $G \cong H$.*

So the last example says $\mathbb{Z}_4 \cong P_4$. It is easy to show $P_4 \cong C_4$ and $\mathbb{Z}_4 \cong C_4$: this just amounts to saying that P_4 and $\mathbb{Z}_4$ are both cyclic (generated by i and by 1, respectively). But it is not necessary to produce all three of these isomorphisms explicitly; two will do:

• Ex.5.16: *Show that isomorphism is an* equivalence relation, *that is, for any groups G, H, K: (i) $G \cong G$; (ii) $G \cong H \Rightarrow H \cong G$; and (iii) $G \cong H \cong K \Rightarrow G \cong K$. (In words, $\cong$ is (i) reflexive, (ii) symmetric, and (iii) transitive.)*

Here is some more notation. Put $T = \{ z \in \mathbb{C} : |z| = 1 \}$, the *circle* group, and $P_n = \{ z \in \mathbb{C} : z^n = 1 \} = \{ e^{2\pi i r/n} : r = 0, 1, \ldots, n - 1 \}$. Clearly P_n is a subgroup of T, which is a subgroup of $\mathbb{C}^*$, and P_n is cyclic, generated by $e^{2\pi i/n}$. If we join the points of P_n in the obvious order ($r = 0, 1, 2, \ldots, n - 1, 0$), we obtain a *regular n-gon*; more generally, any figure X similar to this is called a regular n-gon. It has n vertices, $X = A_1 A_2 \ldots A_n$ say, with $|A_1 A_2| = |A_2 A_3| = \cdots = |A_n A_1|$ and $\angle A_1 A_2 A_3 = \angle A_2 A_3 A_4 = \cdots = \angle A_{n-1} A_n A_1 = \angle A_n A_1 A_2 = \pi(1 - (2/n))$.

How many symmetries does X have? A symmetry of X must send A_1 to A_k say, where $1 \le k \le n$ (n choices) and then it must send A_2, which is adjacent to A_1, to one of the two vertices adjacent A_k (2 choices, so $2n$ choices altogether). It then has to send A_n, which is also adjacent to A_1, to the other vertex adjacent to A_k, and these three images are enough to determine the symmetry. So $|\mathcal{S}(X)| = 2n$. If O is the centre of X, and f is the rotation about O through the angle $2\pi/n$, then

$$\mathcal{S}^+(X) = \{1, f, f^2, \ldots, f^{n-1}\},$$

where the listed elements are all different, and $f^n = 1$. This is a *cyclic* group of (finite) order n, denoted by C_n.

Now let g be the reflection in the line joining some vertex of X to O; so g is a symmetry of X, and indeed

$$\{g, gf, gf^2, \ldots, gf^{n-1}\}$$

are all different, and all symmetries of X. None of them is a rotation, else $gf^r = f^s$ (say), and then $g = f^{s-r}$: but g is a reflection and f^{s-r} is a rotation (or trivial). We deduce that they are all reflections (in lines through O), and that

$$\mathcal{S}(X) = \{1, f, f^2, \ldots, f^{n-1}, g, gf, gf^2, \ldots, gf^{n-1}\}.$$

Here the $2n$ elements listed are all different, $f^n = 1$ (still), and $g^2 = 1$. Also, $(gf)^2 = 1$, or $fg = gf^{n-1}$. The group we have obtained is a *dihedral* group of (finite) order $2n$, denoted by D_n.

Our description of C_n and D_n made the tacit assumption that $n \geq 3$ (else what is an n-gon?), but to make our theory work we must extend the definition to $n = 1, 2$. So we put $C_1 = \{1\}$, the trivial group, and $D_1 = \{1, g\}$, where g is a reflection. In fact D_1 is the symmetry group of an isosceles (but non-equilateral) triangle, and C_1 is its rotation group. And then we put $C_2 = \{1, f\}$, where f is a half-turn, and $D_2 = \{1, f, g, gf\}$, where f is a half-turn and g is a reflection in some line through the centre of f; note that this implies $fg = gf$, so that D_2 is abelian. In fact $D_2 = \mathcal{S}(X)$, where X is a (non-square) rectangle, and then $C_2 = \mathcal{S}^+(X)$. Both C_2 and D_1 are cyclic of order 2, so they are isomorphic, but we retain the separate notations as the non-trivial element of D_1 is a reflection whereas the non-trivial element of C_2 is a half-turn. So although these groups are indistinguishable to an algebraist, they look different to a geometer.

Somewhat similarly, $\mathcal{S}(T)$ consist of *all* rotations about O, and *all* reflections in lines through O. It is not hard to conclude that the subgroup $\mathcal{S}^+(T)$, consisting of all rotations about O, is isomorphic to T. However, this group (which is infinite) is *not* cyclic. (Proof?) There *are* infinite cyclic groups: $\mathbb{Z}$ (the integers), for instance, is an infinite (additive) cyclic group, generated by 1. For a multiplicative version, we just write

$$C_\infty = \{\ldots, f^{-2}, f^{-1}, 1, f, f^2, \ldots\},$$

where the elements are all supposed to be different. We have $\mathbb{Z} \cong C_\infty$, by $n \mapsto f^n$, and indeed it is not hard to show that two cyclic groups are isomorphic iff they have the same order (= number of elements). The infinite dihedral group is then

$$D_\infty = \{\ldots, f^{-2}, f^{-1}, 1, f, f^2, \ldots; \ldots, gf^{-2}, gf^{-1}, g, gf, gf^2, \ldots\},$$

where $g^2 = 1$ and $fg = gf^{-1}$. We can obtain a concrete example of this by letting f be the translation $f : (x, y) \mapsto (x + 1, y)$, and letting g be the reflection $g : (x, y) \mapsto (-x, y)$.

- Ex.5.17: *Find a subset X of* $\mathbb{R}^2$ *such that* $\mathcal{S}(X) = D_\infty$, *with f and g as above.*

Definition 5.20. *Let G be a group, and a* $\in$ *G. The* **cyclic subgroup generated by** *a is the subgroup* $H = \{a^n : n \in \mathbb{Z}\}$, *and the* **order** *of a,* order(a), *is the order of this subgroup, that is, the number of elements it contains.*

So the order of a is $1, 2, 3, \ldots$ or ∞. As a matter of notation, we sometimes write $|X|$ for the number of elements in a set X, and in this notation, order(a) $= |H|$.

- Ex.5.18: *Show that* order(a) $= 1 \iff a = 1$. *(Here 1 has two different meanings.)*
- Ex.5.19: *Show that* order(a) *is the least* $n \geq 1$ *such that* $a^n = 1$, *if such n exists; and if no such n exists, then* order(a) $= \infty$.
- Ex.5.20: *Show that, if a* $\in C_n$, *then* $a^n = 1$. *Does this mean that* order(a) $= n$?
- Ex.5.21: *Show that, if a* $\in C_\infty$, *then* order(a) $= 1$ *or* ∞.
- Ex.5.22: *Show that, if a* $\in D_\infty$, *then* order(a) $= 1$ *or* 2 *or* ∞.
- Ex.5.23: *Show that, if* $\alpha : G \to H$ *is an isomorphism, and a* $\in G$, *then* order(a) $=$ order($\alpha(a)$). *Deduce that* $T \not\cong C_\infty$ *and* $\mathcal{S}(T) \not\cong D_\infty$.

There is a deeper reason why T and C_∞ cannot be isomorphic. Both sets are infinite, but one is bigger than the other, in the following sense. An infinite set is *countable* if it can be put in one-to-one correspondence with the set of natural numbers $\mathbb{N} = \{1, 2, 3, \ldots\} = \{n \in \mathbb{Z}: n \geq 1\}$. So $\mathbb{N}$ itself is countable (obviously), and so is $\mathbb{Z}$ (less obviously), and so is $\mathbb{Q}$, the rational numbers (harder); but $\mathbb{R}$ is uncountable, that is, not countable (harder still). It is easy to find a bijection between $\mathbb{R}$ and a subset of T (OK?), so that T is uncountable; C_∞, on the other hand, is obviously countable. So we conclude that there is not even a bijection between the groups T and C_∞, let alone an isomorphism: T, as a set, is just too big.

- Ex.5.24: *Classify the letters of the alphabet*

  ABCDEFGHIJKLMNOPQRSTUVWXYZ

  according to their symmetry groups. This means lump together all those whose symmetry group is D_1, *all those whose symmetry group is* D_2, *and so on. (To make it more interesting, redesign the letters first, to give them as much symmetry as possible. So, for example, use* L *in place of* L.*)*

- Ex.5.25: *Let X be the pattern obtained by placing a letter* P *at unit intervals along the x-axis:*

  $\cdots$ P   P   P   P   P $\cdots$

  (Here there is a letter P *with its base at* $(n, 0)$ *for each integer n.) (i) Show that* $\mathcal{S}(X) \cong C_\infty$. *(ii) What if* T *were used, in place of* P? *(iii) What if* N *were used, in place of* P?

5.3   COMPLEX FORMULAE FOR PLANE ISOMETRIES

In this section we introduce a way of writing isometries in terms of complex numbers. Thus we identify $(x, y) \in \mathbb{R}^2$ with $z = x + iy \in \mathbb{C}$, the complex numbers (Fig. 5.3, cf. Fig. 1.8). Here $i^2 = -1$, the real part of z is $\Re(z) = x$, and the imaginary part of z is $\Im(z) = y$. (Note that both $\Re(z)$ and $\Im(z)$ are *real*.) If the polar coordinates of (x, y) are (r, θ), then $z = re^{i\theta}$, where $r = |z|$, the modulus of z, and $e^{i\theta} = \cos\theta + i \sin\theta$. Recall also that the conjugate of z is $\bar{z} = x - iy = re^{-i\theta}$, and $z\bar{z} = x^2 + y^2 = |z|^2$. Further, $|\bar{z}| = |z|$.

Example 1.   Let $a \in \mathbb{C}$ with $|a| = 1$, so that $a = e^{i\alpha}$, say. Let $f : z \mapsto az$, for all $z \in \mathbb{C}$. Then $f(re^{i\theta}) = re^{i\theta}e^{i\alpha} = re^{i(\theta+\alpha)}$ (OK?), so f is rotation about O through the angle α.

Example 1. $f : z \mapsto z + b$, for all $z \in \mathbb{C}$. This is translation through $b = b_1 + ib_2$.

Example 2. $f : z \mapsto \bar{z}$, that is, $(x, y) \mapsto (x, -y)$. This is reflection in the x-axis.

Proposition 5.21.   *Given an isometry* $f : \mathbb{C} \to \mathbb{C}$, *then there exist unique* $a, b \in \mathbb{C}$ *with* $|a| = 1$ *such that either (i)* $f(z) = az + b$, *all* $z \in \mathbb{C}$, *or else (ii)* $f(z) = a\bar{z} + b$, *for all* $z \in \mathbb{C}$. *Further, every such* a, b *give rise to isometries by the above formulae.*

Proof.   Last part: suppose $f(z) = az + b$, where $|a| = 1$. Then $|f(z_1) - f(z_2)| = |(az_1 + b) - (az_2 + b)| = |az_1 - az_2| = |a(z_1 - z_2)| = |a||z_1 - z_2| = |z_1 - z_2|$, so f is an isometry. Then if $g(z) = a\bar{z} + b$, we have $|g(z_1) - g(z_2)| = |f(\bar{z}_1) - f(\bar{z}_2)| = |\bar{z}_1 - \bar{z}_2| = |\overline{z_1 - z_2}| = |z_1 - z_2|$, and g is an isometry also.

Now for the first part. Suppose f is any isometry; put $b = f(0)$ and define g by $g(z) = f(z) - b$. Then g, being the composite of f (an isometry) and a translation (also an isometry) is itself an isometry (since isometries form a group), and furthermore $g(0) = 0$. Put $a = g(1)$, and note that $|a| = |a - 0| = |g(1) - g(0)| = |1 - 0| = 1$,

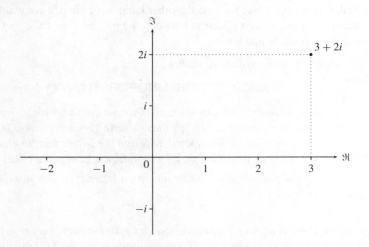

Fig. 5.3   Complex numbers (cf. Fig. 1.8)

so that $a = e^{i\theta}$, say. Define h by $h(z) = e^{-i\theta}g(z)$, so that h is the composite of the isometry g and rotation about O through $-\theta$, another isometry. So h is an isometry, and furthermore $h(0) = e^{-i\theta}g(0) = 0$, and $h(1) = e^{-i\theta}g(1) = e^{-i\theta}e^{i\theta} = 1$. By Corollary 5.8, the real axis is a line of fixed points for h, and by Corollary 5.9, either $h = 1$ or h is reflection in the real axis. In the first case, $h(z) = z$, for all z, so $g(z) = ah(z) = az$, for all z, and $f(z) = g(z) + b = az + b$, for all z, as required. In the second case, $h(z) = \bar{z}$, for all z, so $g(z) = ah(z) = a\bar{z}$, for all z, and $f(z) = g(z) + b = a\bar{z} + b$, for all z, as required.

Finally, in either case $b = f(0)$ and $a = f(1) - f(0)$, whence the uniqueness. $\square$

The following exercise will form a crucial part of the proof of a later result:

- Ex.5.26: *Let $f : \mathbb{C} \to \mathbb{C}$ be an isometry, and let $z_1, z_2, \cdots, z_n \in \mathbb{C}$. Prove that*

$$f\left(\frac{z_1 + z_2 + \cdots + z_n}{n}\right) = \frac{f(z_1) + f(z_2) + \cdots + f(z_n)}{n}.$$

5.3.1 Classification of plane isometries

We are now going to show that plane isometries $f \neq 1$ come in just four types: rotations, translations, reflections, and a fourth mystery type which will creep out of the woodwork in due course; see if you can solve the mystery for yourself before we get to it. We start by classifying isometries into *two* kinds:

Definition 5.22. *A plane isometry f given by $f(z) = az + b$, for all z, is called a* **direct** *isometry, and a plane isometry f given by $f(z) = a\bar{z} + b$, for all z, is called an* **opposite** *isometry. (Here $a, b \in \mathbb{C}$ with $|a| = 1$, as before.)*

This distinction may seem artificial, based as it is merely on the shape of the complex formula for f. There is a geometrical way of making the same distinction, to do with orientation, which will be discussed in Section 5.3.2. In the meantime, the following exercise begins to make the usefulness of the distinction clear:

- Ex.5.27: *Show that if f, g are isometries, then if f, g are both direct or both opposite, then fg is direct, but if one of f, g is direct and the other is opposite, then fg is opposite. Show also that f^{-1} is direct or opposite according as f is direct or opposite.*

There is a neat group-theoretic way of summing up what this exercise is saying. Recall that $P_2 = \{1, -1\}$ is a multiplicative group of two elements. (This is a subgroup of $\mathbb{C}^*$; it is cyclic of order two, generated by -1.) Define $\varphi : \mathcal{I}(\mathbb{R}^2) \to P_2$ by putting

$$\varphi(f) = \begin{cases} 1 & \text{if } f \text{ is direct,} \\ -1 & \text{if } f \text{ is opposite.} \end{cases}$$

Then the exercise says that, for any $f, g \in \mathcal{I}(\mathbb{R}^2)$, we have $\varphi(fg) = \varphi(f)\varphi(g)$; in other words, φ is a homomorphism (see page 89). The exercise also shows that the set $\mathcal{I}^+(\mathbb{R}^2)$ of direct isometries of $\mathbb{R}^2$ (or $\mathbb{C}$) is a subgroup of $\mathcal{I}(\mathbb{R}^2)$. In fact, for any group of isometries G, we can define $G^+ = G \cap \mathcal{I}^+(\mathbb{R}^2)$ to be the set of direct isometries in G, and this is a subgroup of G. In particular, for any set $X \subseteq \mathbb{R}^2$, the *rotation group* of X is $\mathcal{S}^+(X) = \mathcal{S}(X) \cap \mathcal{I}^+(\mathbb{R}^2)$. This agrees with our previous usage of the notation $\mathcal{S}^+(X)$ (page 88).

Example 1.   Every translation is given by $f(z) = z + b$, for suitable b, and this is direct.
Example 2.   Rotation through θ about O is given by $f(z) = az$, where $a = e^{i\theta}$, and this is also direct. More generally, rotation through θ about ω is given by $f(z) - \omega = a(z - \omega)$ (same a), or $f(z) = az + (\omega - a\omega)$, which is direct, so that every rotation is direct.

In fact the above examples are the *only* direct isometries, as we see next:

Proposition 5.23.   *Let f be a non-trivial direct isometry, given by $f(z) = az + b$, where $|a| = 1$. (i) If $a = 1$, then f is a translation, $f(z) = z + b$. (ii) If $a \neq 1$, then f is a rotation about $b/(1 - a)$ through the angle θ, where $a = e^{i\theta}$.*

Proof.   (i) is clear. For (ii), look for a fixed point, ω: solve $f(\omega) = \omega$, that is, $a\omega + b = \omega$, or $\omega(1 - a) = b$, or $\omega = b/(1 - a)$, which is all right since $1 - a \neq 0$. Check that this reverses: $f(\omega) = \omega$. Then, for any z, $f(z) - \omega = (az + b) - (a\omega + b) = a(z - \omega) = e^{i\theta}(z - \omega)$, which is the required rotation.   □

This means, among other things, that the fourth type (mystery) isometries are all *opposite*.

- Ex.5.28:   *Let f_1, f_2 be rotations through angles θ_1, θ_2 respectively, but not necessarily about the same centre. Prove that $f_1 f_2$ is a rotation through the angle $\theta_1 + \theta_2$, provided $\theta_1 + \theta_2$ is not an integer multiple of 2π. What happens if $\theta_1 + \theta_2$ is an integer multiple of 2π?*

- Ex.5.29:   *Let f_1, f_2 be direct isometries, with $f_1 \neq 1$ and $f_2 \neq 1$. Prove that $f_1 f_2 = f_2 f_1$ iff either both f_1, f_2 are translations or both f_1, f_2 are rotations about the same centre.*

Now what about opposite isometries? Since we now know that every direct isometry is a translation or rotation, it follows that every reflection must be an opposite isometry. In fact:

Proposition 5.24.   *Let f be an opposite isometry, given by $f(z) = a\bar{z} + b$, where $|a| = 1$. Then f is a reflection iff $a\bar{b} + b = 0$, that is, iff $f(b) = 0$.*

Proof. $\Rightarrow$: If f is a reflection, then $f^2 = 1$, so in particular $f^2(0) = 0$. Now $f(0) = b$, so we must have $f(b) = 0$—done.
$\Leftarrow$: Suppose $f(b) = 0$. For any z,

$$f^2(z) = f(a\bar{z} + b) = a(\overline{a\bar{z} + b}) + b = a(\bar{a}z + \bar{b}) + b = a\bar{a}z + (a\bar{b} + b) = z.$$

So $f^2 = 1$. Now $f \neq 1$ and f is not a half-turn, since it is opposite. Thus f is a reflection, by Corollary 5.10.   □

Corollary 5.25.   *Let f be an opposite isometry. If f has a fixed point, it is a reflection.*

Proof. Let $f(z) = a\bar{z} + b$, where $|a| = 1$, and suppose $f(z_1) = z_1$, for some z_1. Then also $f^2(z_1) = z_1$. But

$$f^2(z_1) = a(\overline{a\bar{z}_1 + b}) + b = z_1 + (a\bar{b} + b),$$

so $a\bar{b} + b = 0$, and we have a reflection. $\qquad\square$

We now know, then, that the fourth type (mystery) isometries are opposite isometries *without* fixed points. (Can you find an example?)

Let us explore reflections a little further. Suppose $f : z \mapsto a\bar{z} + b$ is a reflection, as above. Then where is its axis, in terms of a and b?

Case 1: $b \neq 0$. Then $f(0) = b$, so the axis is the perpendicular bisector of the line segment joining 0 to b.

Case 2: $b = 0$ and $a \neq 1$. So $f(z) = a\bar{z}$. Then $f(1) = a$, so the axis is the perpendicular bisector of the line segment joining 1 to a. (This passes through 0, of course.)

Case 3: $b = 0$ and $a = 1$. So $f(z) = \bar{z}$, which is reflection in the x-axis.

This process reverses: to write down the equation $f(z) = a\bar{z} + b$ of reflection in a line ℓ, proceed as follows. If ℓ does not pass through 0, then let b be the reflection of 0 in ℓ, and choose a so that $a\bar{b} + b = 0$. (This necessarily entails $|a| = 1$—OK?) If ℓ passes through 0 but is not the x-axis, then let a be the reflection of 1 in ℓ, and put $b = 0$. (Again, check that $|a| = 1$.) If ℓ is the x-axis, put $a = 1$ and $b = 0$.

- Ex.5.30: *Let f be reflection in the line ℓ, where $f(z) = a\bar{z} + b$, for all z. Let z_1, z_2 be points of ℓ. Prove that*

$$(z_1 - z_2)^2 = a|z_1 - z_2|^2$$

and deduce that, if $a = e^{i\theta}$, then ℓ makes an angle $\theta/2$ with the x-axis.

- Ex.5.31: *Let f_1, f_2 be reflections in lines ℓ_1, ℓ_2 respectively, where $f_1(z) = a_1\bar{z} + b_1$ and $f_2(z) = a_2\bar{z} + b_2$, for all z, and suppose also that $f_1 \neq f_2$, that is, $\ell_1 \neq \ell_2$.*

    (i) *Prove that $f_1 f_2 \neq 1$, and that $(f_1 f_2)^{-1} = f_2 f_1$.*
    (ii) *Prove that $\ell_1 \parallel \ell_2 \iff a_1 = a_2$, and that in this case $f_1 f_2$ is a translation, in a direction orthogonal to ℓ_1 and ℓ_2, through twice the distance between them. Show also that here $f_1 f_2 \neq f_2 f_1$.*
    (iii) *Prove that, if ℓ_1, ℓ_2 are not parallel, then $f_1 f_2$ is a rotation about the meet $\ell_1 \ell_2$, through twice the angle between ℓ_1 and ℓ_2.*
    (iv) *Prove that $f_1 f_2 = f_2 f_1 \iff a_1 = -a_2$, and hence or otherwise show that here $\ell_1 \perp \ell_2$, and $f_1 f_2$ is a half-turn about $\ell_1 \ell_2$.*

The time has now come to reveal the fourth type of isometry. Here is an example:

Example. $f(z) = \bar{z} + 1$. This is an opposite isometry, so it is not a translation or rotation, and it is not a reflection either; indeed, $f^2 \neq 1$, since $f^2(z) = \overline{\bar{z} + 1} + 1 = z + 2$, so that f^2 is a non-trivial translation. We can split f up as a composite $f = hg$, where

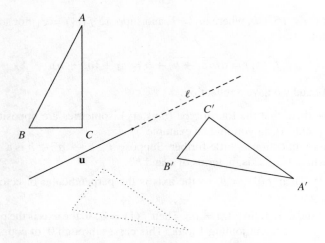

Fig. 5.4   A glide-reflection, the composite of reflection in ℓ and translation by **u**

$g(z) = \bar{z}$ and $h(z) = z + 1$. Thus f is the *composite* of a reflection g and a non-trivial translation h, in a direction parallel to the axis of g. But we stress again, f is *not* a reflection (because $f^2 \neq 1$), and it is *not* a translation (since it is opposite).

Definition 5.26.   *The composite of reflection in a line ℓ and a non-trivial translation in a direction parallel to ℓ is called a* **glide-reflection** *(Fig. 5.4).*

The classification of plane isometries can now be finished:

Proposition 5.27.   *Every (plane) opposite isometry is a reflection or a glide-reflection.*

Proof.   Let f be an opposite isometry which is not a reflection. So $f(z) = a\bar{z} + b$, say, where $|a| = 1$ and $a\bar{b} + b \neq 0$. Note that $f^2(z) = a(\overline{a\bar{z} + b}) + b = z + (a\bar{b} + b)$, so that f^2 is a non-trivial translation. Put $c = \frac{1}{2}(a\bar{b} + b)$, and define isometries g, h by $g(z) = f(z) - c$ and $h(z) = z + c$, for all z. So h is a non-trivial translation, and $f = hg$.

Note that $a\bar{c} = \frac{1}{2}a(\overline{a\bar{b} + b}) = \frac{1}{2}a(\bar{a}b + \bar{b}) = \frac{1}{2}(b + a\bar{b}) = c$. We claim g is a reflection. For $g : z \mapsto a\bar{z} + b - c$ is an opposite isometry, and

$$g^2(z) = a(\overline{a\bar{z} + b - c}) + b - c = z + a\bar{b} + b - a\bar{c} - c = z + a\bar{b} + b - 2c = z$$

for any z, so g is a reflection, by Corollary 5.10.

It remains to show that the direction of the translation h is parallel to the axis of g, and it is sufficient to show that if z lies on the axis of g, then so does $h(z) = z + c$. So suppose $g(z) = z$, that is, $a\bar{z} + b - c = z$. We must show that $g(z + c) = z + c$. We have

$$g(z + c) = a(\overline{z + c}) + b - c = (a\bar{z} + b - c) + a\bar{c} = z + a\bar{c} = z + c,$$

and the proof is thus complete. $\square$

Remark 1. With the above notation, for any z we have $hg(z) = f(z) = a\bar{z} + b$, whereas $gh(z) = g(z + c) = a(\overline{z + c}) + b - c = (a\bar{z} + b) + (a\bar{c} - c)$. So the equation $a\bar{c} = c$, obtained above, amounts to showing that $hg = gh$. We could have predicted this, since by Exercise 5.31(ii) we can write $h = h_1 h_2$ where h_1, h_2 are reflections with parallel axes, orthogonal to the axis of g; and then by Exercise 5.31(iv) we have $hg = h_1 h_2 g = h_1 g h_2 = g h_1 h_2 = gh$.

Remark 2. Again with the same notation, the mid-point of PP' (where $P' = f(P)$) lies on the axis of g. The painful way to prove this is to calculate $g(\frac{1}{2}(z + f(z)))$ and show that it is always the same as $\frac{1}{2}(z + f(z))$. More subtly, since f is an isometry, it must send the mid-point of PP' to the mid-point of $P'P''$. Recalling that $f^2(z) = z + 2c$ (i.e., $f^2 = h^2$), we have

$$g(\tfrac{1}{2}(z + f(z))) = h^{-1} f(\tfrac{1}{2}(z + f(z))) = \tfrac{1}{2}(f(z) + f^2(z)) - c = \tfrac{1}{2}(f(z) + z),$$

as required. Even more easily, if $f(P) = P'$ and $g(P) = Q$, then $h(Q) = P'$. If ℓ is the axis of g, then provided P is not on ℓ, then ℓ is the perpendicular bisector of PQ, and is parallel to QP'. It follows immediately that ℓ meets PP' at its mid-point. (What happens if P *is* on ℓ?) This goes to show that sometimes the best way to solve a geometrical problem is by geometry, rather than algebra!

Definition 5.28. *Let f be an opposite isometry, $f : P \mapsto P'$. The **axis** of f is the unique line containing the mid-point of PP', for every P.*

(Are you sure about the uniqueness?) So if f is a reflection, the terminology is as before, and if f is a glide-reflection with $f = hg$ as above, then the axis of f is defined to be the axis of the reflection g.

- Ex.5.32: *Generalize Exercise 5.30 to arbitrary opposite isometries: prove that if $f(z) = a\bar{z} + b$ and $a = e^{i\theta}$, then the axis of f makes an angle $\theta/2$ with the x-axis.*

We summarize our classification of plane isometries:

Theorem 5.29. *Every non-trivial plane isometry is a translation, a rotation, a reflection, or a glide-reflection.* ☐

It is possible to distinguish the cases by looking at fixed points and invariant lines. A translation has no fixed points, a rotation has exactly one, its centre, and a reflection has a line of fixed points, its axis; a glide-reflection, like a translation, has no fixed points. A translation has many invariant lines, all parallel, and a rotation has none except in the case of a half-turn, when every line through the centre is invariant. A reflection has many invariant lines: its axis, and every line orthogonal to its axis. Finally,

- Ex.5.33: *A glide-reflection has exactly one invariant line, its axis.*

- Ex.5.34: *The inverse of a translation is a translation; the inverse of a rotation is a rotation; the inverse of a reflection is a reflection; and the inverse of a glide-reflection is a glide-reflection.*

• Ex.5.35:   *An opposite isometry and its inverse have the same axis.*

Recall that every plane isometry is the composite of n reflections, where $n \leq 3$. Case $n = 0$: the empty product is 1, by convention. Case $n = 1$: a reflection is a reflection is a reflection. Case $n = 2$: the composite of two reflections is 1 if they coincide, a translation if their axes are parallel, and a rotation about the meet of their axes otherwise. Case $n = 3$: since every reflection is an opposite isometry, the composite of three reflections is also an opposite isometry, by Exercise 5.27, so it is a reflection or a glide-reflection. How do we decide which?

Proposition 5.30.   *Let* f_i *be reflection in* ℓ_i, *for* $1 \leq i \leq 3$, *and let* $f = f_1 f_2 f_3$. *Then* f *is a reflection if* ℓ_1, ℓ_2, ℓ_3 *are either concurrent or all parallel, and it is a glide-reflection otherwise.*

Proof.   First suppose ℓ_1, ℓ_2, ℓ_3 are concurrent at P. Then P is a fixed point of f_1, of f_2, and of f_3, and hence of f, which must therefore be a reflection.

Next suppose ℓ_1, ℓ_2, ℓ_3 are parallel. Then every line orthogonal to (all of) them is an invariant line of f_1, of f_2, and of f_3, and hence of f, which must therefore be a reflection.

Conversely, suppose f is a reflection, with axis ℓ. Remember that $f_3^2 = 1$, so $f_1 f_2 = f f_3 = g$, say, a direct isometry, and the argument splits into three cases, according as g is trivial, a translation, or a rotation.

Firstly, if $g = 1$, then $\ell_1 = \ell_2$ and $\ell = \ell_3$. If ℓ_1 and ℓ_3 are parallel, then ℓ_1, ℓ_2, ℓ_3 and ℓ are all parallel; and if ℓ_1 and ℓ_3 are not parallel, then ℓ_1, ℓ_2, ℓ_3 and ℓ are concurrent at the meet $\ell_1 \ell_3$.

Secondly, if g is a translation, then ℓ_1 and ℓ_2 are parallel, and orthogonal to the direction of translation; and also ℓ_3 and ℓ are parallel, and orthogonal to the (same) direction of translation. So ℓ_1, ℓ_2, ℓ_3, and ℓ, are all parallel.

Finally, if g is a rotation, then ℓ_1 and ℓ_2 meet at the centre of rotation, and ℓ_3 and ℓ meet at the (same) centre of rotation. So ℓ_1, ℓ_2, ℓ_3, and ℓ are concurrent. $\square$

Notice that the proof gives a little more than the statement of the Proposition, namely that when f is a reflection, its axis is parallel to the other axes when they are parallel, and concurrent with the other axes when they are concurrent.

• Ex.5.36:   *With the same notation as Proposition 5.30, show that if* ℓ_2, ℓ_3 *are not parallel, and not concurrent with* ℓ_1, *then the axis of the glide-reflection* f *passes through the foot of the perpendicular from* $\ell_2 \ell_3$ *onto* ℓ_1.

• Ex.5.37:   *Let* AP, BQ, CR *be the altitudes of* $\triangle ABC$, *and write* f_a, f_b, f_c *for the reflections in* BC, CA, AB *respectively. Prove that the glide-reflection* $f_a f_b f_c$ *has axis* PR. *Deduce that the line* $f_c(PR)$ *is invariant under* $f_c f_a f_b$, *and hence show that* $\angle BRP = \angle QRA$.

We finish this section by finding all pairs of commuting isometries, that is, isometries f, g with $fg = gf$. We already know (Exercise 5.29) that two direct isometries commute iff both are translations or else both are rotations about the same centre. We also know

(Exercise 5.31(iv)) that two distinct reflections commute iff their axes are perpendicular; so two arbitrary reflections commute iff their axes are perpendicular or coincident. The next proposition gives a similar answer for two arbitrary opposite isometries:

Proposition 5.31. *Two opposite isometries commute iff either they have the same axis or else both are reflections, with orthogonal axes.*

Proof. Let $f_1(z) = a_1\bar{z} + b_1$ and $f_2(z) = a_2\bar{z} + b_2$, for all z. Then

$$f_1 f_2(z) = a_1\bar{a}_2 z + (a_1\bar{b}_2 + b_1), \quad \text{and} \quad f_2 f_1(z) = a_2\bar{a}_1 z + (a_2\bar{b}_1 + b_2).$$

Thus if f_1, f_2 commute, we must have $a_1\bar{a}_2 = a_2\bar{a}_1$, and also $a_1\bar{b}_2 + b_1 = a_2\bar{b}_1 + b_2$. Since $|a_1| = 1 = |a_2|$, the first equation yields $a_1^2 = a_2^2$, or $a_1 = \pm a_2$.

Case 1: $a_1 = a_2 = a$, say. Then $f_1(z) = a\bar{z} + b_1$, and $f_2(z) = a\bar{z} + b_2$, and also $a\bar{b}_2 + b_1 = a\bar{b}_1 + b_2$. Since $f_1(0) = b_1$, we deduce that $\frac{1}{2}b_1$ lies on the axis of f_1; we claim that $\frac{1}{2}b_2$ also lies on the axis of f_1. Now exactly as in the proof of Proposition 5.27, we can write $f_1 = hg$, where $g(z) = f_1(z) - c$ (a reflection) and $h(z) = z + c$ (a translation), and $c = \frac{1}{2}(a\bar{b}_1 + b_1)$; and we must show that $\frac{1}{2}b_2$ is on the axis of g, or (equivalently) that $f_1(\frac{1}{2}b_2) = \frac{1}{2}b_2 + c$. But

$$f_1(\tfrac{1}{2}b_2) = \tfrac{1}{2}a\bar{b}_2 + b_1 = \tfrac{1}{2}(a\bar{b}_2 + b_1) + \tfrac{1}{2}b_1 = \tfrac{1}{2}(a\bar{b}_1 + b_2) + \tfrac{1}{2}b_1 = \tfrac{1}{2}b_2 + c,$$

as required. So provided $b_1 \neq b_2$, the axis of f_1 is the line joining $\frac{1}{2}b_1$ to $\frac{1}{2}b_2$; and similarly so is the axis of f_2. On the other hand, if $b_1 = b_2$, then $f_1 = f_2$, and again the two axes coincide.

Case 2: $a_1 = -a_2 = a$, say. Then $f_1(z) = a\bar{z} + b_1$, and $f_2(z) = -a\bar{z} + b_2$, and also $a\bar{b}_2 + b_1 = -a\bar{b}_1 + b_2$, or $a\bar{b}_1 + b_1 = -a\bar{b}_2 + b_2$. Multiplying this last equation by $\bar{a}$ and then conjugating, we have $b_1 + a\bar{b}_1 = -b_2 + a\bar{b}_2$. Comparing, we see that $a\bar{b}_1 + b_1 = 0 = -a\bar{b}_2 + b_2$, so that both f_1 and f_2 are reflections. The fact that their axes are orthogonal follows as in Exercise 5.31(iv).

The converse is easy, and is left as an exercise for the reader. □

Finally, we find when a direct isometry commutes with an opposite isometry:

Proposition 5.32. *A direct and an opposite isometry commute iff either the one is a translation and the other has its axis parallel to the direction of the translation, or else the one is a half-turn and the other is a reflection in a line through the centre of the rotation.*

Proof. Let $f_1(z) = a_1 z + b_1$ and $f_2(z) = a_2\bar{z} + b_2$, for all z. The condition $f_1 f_2 = f_2 f_1$ yields $a_1 a_2 = a_2\bar{a}_1$ so that $a_1 = \bar{a}_1 = \pm 1$; and also $a_1 b_2 + b_1 = a_2\bar{b}_1 + b_2$.

Case 1: $a_1 = 1$. Then $f_1(z) = z + b_1$, a translation. Also $b_1 + b_2 = a_2\bar{b}_1 + b_2$, so $b_1 = a_2\bar{b}_1$, and $b_1^2 = |b_1|^2 a_2$. If $a_2 = e^{i\theta}$, then $b_1 = \pm|b_1|e^{i\theta/2}$, and the result follows by Exercise 5.32.

Case 2: $a_1 = -1$. Then $f_1(z) = -z + b_1$, a half-turn about $\frac{1}{2}b_1$. Also $-b_2 + b_1 = a_2\bar{b}_1 + b_2$, so that $\frac{1}{2}a_2\bar{b}_1 + b_2 = \frac{1}{2}b_1$. But this equation just says $\frac{1}{2}b_1$ is a fixed point of f_2; so f_2 is a reflection, and $\frac{1}{2}b_1$ lies on its axis.

The converse is easy, and is once again left as an exercise for the reader.   □

- Ex.5.38: *Let* $f \in \mathcal{I}(\mathbb{R}^2)$, *and put* $C(f) = \{ g \in \mathcal{I}(\mathbb{R}^2) : fg = gf \}$, *the* centralizer *of* f *in* $\mathcal{I}(\mathbb{R}^2)$. *Prove that* $C(f)$ *is a subgroup of* $\mathcal{I}(\mathbb{R}^2)$. *Describe (in terms of types, axes, centres etc.) what isometries belong to* $C(f)$ *when* f *is (i) trivial; (ii) a translation; (iii) a rotation, not a half-turn; (iv) a half-turn; (v) a reflection; and (vi) a glide-reflection.*

- Ex.5.39: *Let* $A_1 A_2 \ldots A_n$ *be a regular n-gon. Let* $f_1, f_2, \ldots, f_n$ *be the reflections in* $A_1 A_2, A_2 A_3, \ldots, A_n A_1$ *respectively. Find the composite* $f_n f_{n-1} \cdots f_2 f_1$. *(This means decide whether it is trivial, a translation, rotation, reflection or glide-reflection; if it is opposite find its axis; if it is a translation or glide-reflection find the translation involved; and if it is a rotation find the centre and angle involved.)*

  Hint: for $n = 3$ *we know part of the answer already, by Exercise 5.37. For* $n = 3, 4,$ *or* 6 *the plane can be tiled with regular n-gons, and the exercise is rather easily done by chasing successive reflections of the original n-gon around the tiling. But for other values of* n, *no such tiling exists, and you will have to think of another method.*

5.3.2   Orientation

When names A, B, C are assigned to the three vertices of a triangle, there is more than one way to do this: indeed, A can first be assigned to any of the three vertices, then B to either of the other two, and finally C to the remaining vertex. This is a total of $3 \times 2 = 6$ ways. In three of these ways, the labels A, B, C (in that order) go around the triangle in one direction, and in the other three ways the labels go around the triangle in the opposite direction. We need to distinguish between the two choices of direction.

Having labelled the triangle, we can refer to it in six possible ways: $\triangle ABC$, $\triangle BCA$, $\triangle CAB$, $\triangle ACB$, $\triangle CBA$, or $\triangle BAC$. In the first three ways, the labels are named going around the triangle in one direction, and in the last three ways, the labels are named going around the triangle the other direction.

We can now give an informal definition of the orientation, or handedness, of a triangle. Recall that in polar coordinates, a point $P = (r, \theta)$, moving with $r = $ constant and θ increasing, is moving *counter-clockwise* on a circle around the origin. In the same spirit, if the origin is inside $\triangle ABC$ and P moves around the perimeter of the triangle such that θ is increasing, then P is moving counter-clockwise around the triangle. If, as this happens, P passes through the vertices A, B, C in that order, we say $\triangle ABC$ is *labelled positively*, or is *positively oriented* (Fig. 5.5). If P has to move clockwise (θ decreasing) to pass through A, B, C in that order, then $\triangle ABC$ is *labelled negatively* or *negatively oriented*. We shall refer to the *orientation* of $\triangle ABC$ as being positive or negative, accordingly. So if $\triangle ABC$ is positively oriented, then so is $\triangle BCA$ and $\triangle CAB$; but $\triangle ACB$, $\triangle CBA$

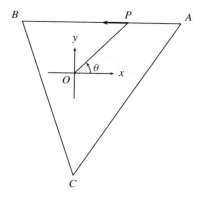

Fig. 5.5 △*ABC* is positively oriented

and △*BAC* are negatively oriented. We say that △*ABC* and △*ACB* (for example) have *opposite* orientations.

Another way of describing what is meant by the orientation of △*ABC* is to imagine the effect of walking around the triangle. If, having walked from *A* to *B*, you need to turn *left* to proceed toward *C*, then △*ABC* is positively oriented; and if you need to turn *right*, it is negatively oriented. Again, the triangle is positively or negatively oriented according as *C* is on the left or right of *AB* as you proceed from *A* to *B*. This works whether or not *O* is inside the triangle.

Concepts involving left and right are often tricky. Pooh, you will remember,

looked at his two paws. He knew that one of them was the right, and he knew that when you had decided which one of them was the right, then the other one was the left, but he never could remember how to begin. [21]

Pooh was quite entitled to be confused: it does not matter which is right and which is left. The only thing that matters is that you decide, and stick to your decision, and preferably agree your decision with all your friends-and-relations (unless you want some fearful arguments on car journeys). Our definition of orientation depends on the three-dimensional nature of our world. We place the paper on the desk, draw a triangle, and look at it from above, and use this view of the triangle to determine its handedness, or orientation. If instead we were to draw the triangle on the inside of a window, note its orientation, and then go outside to look at it, we should notice that the orientation has changed, although the triangle has not. (It is better to label the triangle AHM, rather than *ABC*, for this experiment!) So orientation depends on where you are standing at the time, which is unsatisfactory.

We need a way of determining orientation that refers only to the plane $\mathbb{R}^2$ in which we are doing our geometry, and not on the way this plane is put inside $\mathbb{R}^3$, or where in $\mathbb{R}^3$ we are standing to look at it. The clue as to how to do this is that we have not yet used the fact that $\mathbb{R}^2$ comes equipped with a pair of axes, giving rise to a representation of its points by coordinates (x, y); indeed, it was *defined* in Chapter 3 to be nothing more nor less than the set of all such ordered pairs of real numbers.

Definition 5.33.   *Given* $\triangle A_1 A_2 A_3$, *let* $A_k = (x_k, y_k)$; *or, in complex form,* $z_k = x_k + iy_k$, *for each k. Then the quantity* $\check{\Delta}$ *is defined by*

$$\check{\Delta} = \tfrac{1}{2}\Im(\overline{z}_1 z_2 + \overline{z}_2 z_3 + \overline{z}_3 z_1).$$

The reason for the notation $\check{\Delta}$, and for the presence of the fraction $\frac{1}{2}$ in its formula, is explained by Exercise 5.40 below. We shall regard $\check{\Delta}$ as a function of $A_1 A_2 A_3$ or of z_1, z_2, z_3, as appropriate. Notice that

$$\check{\Delta}(A_1 A_2 A_3) = \check{\Delta}(A_2 A_3 A_1) = \check{\Delta}(A_3 A_1 A_2)$$
$$= -\check{\Delta}(A_1 A_3 A_2) = -\check{\Delta}(A_3 A_2 A_1) = -\check{\Delta}(A_2 A_1 A_3).$$

Proposition 5.34. $\triangle A_1 A_2 A_3$ *is positively oriented iff* $\check{\Delta} > 0$, *and it is negatively oriented iff* $\check{\Delta} < 0$.

Proof.   The condition that A_3 is on the left of $A_1 A_2$ (in that order) becomes, in terms of complex numbers, $z_3 - z_1 = (z_2 - z_1)re^{i\theta}$ for some $r > 0$ and $0 < \theta < \pi$. This condition on θ is equivalent to demanding $\Im(e^{i\theta}) > 0$, so our condition is

$$\Im\left(\frac{z_3 - z_1}{z_2 - z_1}\right) > 0,$$

or $\Im((z_3 - z_1)(\overline{z}_2 - \overline{z}_1)) > 0$, and the result follows on multiplying out the brackets, remembering that $z_1 \overline{z}_1$ is real and $\Im(\overline{w}) = -\Im(w)$. The argument reverses; and the argument for the other orientation is similar. $\square$

This gives a way of deciding orientation that is not dependent on where the observer is standing: we simply turn the last proposition into the *definition* of orientation:

Definition 5.35. $\triangle A_1 A_2 A_3$ *is* **positively oriented** *iff* $\check{\Delta} > 0$, *and it is* **negatively oriented** *iff* $\check{\Delta} < 0$.

As an example, if $A_1 = (1, 0)$, $A_2 = (0, 1)$ and $A_3 = (0, 0)$, we have $\check{\Delta} = \Im(i + 0 + 0) = 1$, so $\triangle A_1 A_2 A_3$ is positively oriented, which agrees with our informal definition. If we were to begin again with the x-axis and y-axis interchanged (see Fig. 5.7), then all our notions of left and right would be interchanged also; everything would work, except that we are now outside (in the snow, probably) watching someone draw geometrical diagrams on the inside of a window. Axes as we normally draw them (Fig. 5.6) are called *right-handed*, and axes as in Fig. 5.7 are called *left-handed*. Because computer monitor screens are scanned electronically from left to right and from top to bottom, it is common for computer software to have its screen coordinates set up in a left-handed system, like Fig. 5.8. This is convenient for text, where we usually count lines downwards from the top of the page, but mathematicians using such software for diagrams need to beware, as using standard formulae involving inequalities can lead to surprises. The best plan is

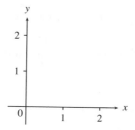

Fig. 5.6   Right-handed axes

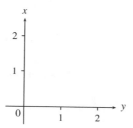

Fig. 5.7   Left-handed axes

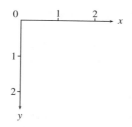

Fig. 5.8   Screen coordinates

immediately to add subroutines to convert from screen coordinates to the conventional right-handed system, and back.

- Ex.5.40: *Use Proposition 4.6 to show that, if the area of $\triangle A_1 A_2 A_3$ is Δ, then, with the notation of Definition 5.33, we have $\Delta = |\check{\Delta}|$.*

- Ex.5.41: *Show also that* $\begin{vmatrix} x_1 & x_2 & x_3 \\ y_1 & y_2 & y_3 \\ 1 & 1 & 1 \end{vmatrix} = 2\check{\Delta}.$

We have met this determinant before, in Proposition 4.24. Its vanishing is the condition for A_1, A_2, A_3 to be collinear; we now know that when it does not vanish, its sign tells us on which side of $A_1 A_2$ the point A_3 lies, and its absolute value is twice the area of $\triangle A_1 A_2 A_3$.

Definition 5.36. $\check{\Delta} = \check{\Delta}(A_1 A_2 A_3)$ *is the* **signed area** *of* $\triangle A_1 A_2 A_3$.

The reader should have met signed areas in calculus, where the area 'under' a curve $y = f(x)$ is given by an integral, but the answer may turn out to be negative if it happens that the curve lies below the x-axis:

- Ex.5.42: *Let* $A_1 = (0, 0)$, $A_2 = (1, 0)$, *and* $A_3 = (a, b)$, *where* $0 < a < 1$. *Write down the equation* $y = f_1(x)$ *of the line* $A_1 A_3$, *and the equation* $y = f_2(x)$ *of the line* $A_2 A_3$, *and define* f *by*

$$f(x) = \begin{cases} f_1(x) & \text{if } x \leq a, \\ f_2(x) & \text{if } x > a. \end{cases}$$

  Show that

$$\int_0^1 f(x)\, dx = \tfrac{1}{2}b = \check{\Delta}(A_1 A_2 A_3).$$

- Ex.5.43: *Given* $\triangle A_1 A_2 A_3$ *and a point* P, *let the barycentric coordinates of* P *with respect to* $\triangle A_1 A_2 A_3$ *be* (ξ, η, ζ). *Prove that* $\xi : \eta : \zeta = \check{\Delta}(PA_2 A_3) :$ $\check{\Delta}(A_1 P A_3) : \check{\Delta}(A_1 A_2 P)$. *If we choose* ξ, η, ζ *so that* $\xi + \eta + \zeta = \check{\Delta}(A_1 A_2 A_3)$ *(as we may), then* $\xi = \check{\Delta}(PA_2 A_3)$, $\eta = \check{\Delta}(A_1 P A_3)$, *and* $\zeta = \check{\Delta}(A_1 A_2 P)$; (ξ, η, ζ) *are then called* areal coordinates *of* P *with respect to* $\triangle A_1 A_2 A_3$.

- Ex.5.44: *Given* $\triangle A_1 A_2 A_3$, *of area* Δ, *let* D_1, D_2, D_3 *lie on* $A_2 A_3$, $A_3 A_1$, $A_1 A_2$ *respectively, with* $A_2 D_1 / D_1 A_3 = A_3 D_2 / D_2 A_1 = A_1 D_3 / D_3 A_2 = 2$. *(So* D_1, D_2, D_3 *trisect the sides of* $\triangle A_1 A_2 A_3$.) *Find the barycentric coordinates of the points* $P_1 = A_2 D_2 \cdot A_3 D_3$, $P_2 = A_3 D_3 \cdot A_1 D_1$ *and* $P_3 = A_1 D_1 \cdot A_2 D_2$, *with respect to* $\triangle A_1 A_2 A_3$, *and deduce that the area of* $\triangle P_1 P_2 P_3$ *is* $\tfrac{1}{7}\Delta$.

We now come to the effect of isometries on signed areas. Given $\triangle A_1 A_2 A_2$, of area Δ, and an isometry $f : \mathbb{R}^2 \to \mathbb{R}^2$, write $A'_k = f(A_k)$, for each k, and let Δ' be the area of $\triangle A'_1 A'_2 A'_3$. Then $\triangle A_1 A_2 A_2 \equiv \triangle A'_1 A'_2 A'_3$, so that $\Delta = \Delta'$. If we write $\check{\Delta}$, $\check{\Delta}'$ for the corresponding signed areas, then it follows that $\check{\Delta}' = \pm\check{\Delta}$, the correct choice of sign being determined by f:

Proposition 5.37.   *With the above notation,* $\check{\Delta}' = \check{\Delta}$ *if* f *is direct, and* $\check{\Delta}' = -\check{\Delta}$ *if* f *is opposite.*

Proof.   Let $z' = f(z) = az + b$, for all z. As in the proof of Proposition 5.34, we have

$$2\check{\Delta} = \Im((z_3 - z_1)(\bar{z}_2 - \bar{z}_1)),$$

and similarly

$$2\check{\Delta}' = \Im((z'_3 - z'_1)(\bar{z}'_2 - \bar{z}'_1))$$
$$= \Im((az_3 - az_1)(\overline{az}_2 - \overline{az}_1)),$$

the b's cancelling, and the result $\check{\Delta}' = \check{\Delta}$ follows immediately, since $a\bar{a} = 1$. The calculation for the opposite case is done similarly, using $f(z) = a\bar{z} + b$. (Check it!) $\square$

So we now have a geometric interpretation of direct/opposite: direct isometries *preserve* orientation, and opposite isometries *reverse* orientation. Indeed, if the function $\varphi : \mathcal{I}(\mathbb{R}^2) \to \{\pm 1\}$ is defined as on page 93, we have

$$\check{\Delta}' = \varphi(f)\check{\Delta}.$$

- Ex.5.45: *Let X, X' be squares of equal size. How many isometries f are there with $f(X) = X'$?*

- Ex.5.46: *Let $ABCD$ and $A'B'C'D'$ be squares of equal size. Prove that, if AA', BB', CC' and DD' are not all parallel, then either the perpendicular bisectors of AA', BB', CC' and DD' are concurrent or the mid-points of AA', BB', CC' and DD' are collinear.*

- Ex.5.47: *Let $A_1 A_2 \cdots A_n$ be a regular n-gon, and let f_k be the rotation about A_k through the angle $\pi(1 - \frac{2}{n})$, in the positive (counter-clockwise) direction. Find the composite $f_n f_{n-1} \cdots f_2 f_1$ (cf. Exercise 5.39). Does it make any difference whether the polygon is oriented positively or negatively? And what is strange about the case $n = 4$?*

5.4 FINITE GROUPS OF PLANE ISOMETRIES

The number of elements in a group G is called its *order*, $|G|$. This can be finite or infinite; and a group is said to be finite or infinite accordingly. In this section we are going to find *all* the finite groups of plane isometries. It would be nice to be able to find all groups of isometries, that is, all the subgroups of $\mathcal{I}(\mathbb{R}^2)$, but this is much too hard. We content ourselves with finding the finite subgroups, which is fairly easy; we shall look at some of the infinite subgroups later.

Proposition 5.38 (Lagrange's theorem). *Let H be a subgroup of the finite group G. Then $|H|$ is a factor of $|G|$.*

Proof. Of course, if G is finite, then so is H. For $g \in G$, write $gH = \{gh: h \in H\}$, the *right coset* of H determined by g. (Some authors call this a *left* coset; there's no accounting for tastes.) Now $h \mapsto gh$ is a bijection $H \to gH$ (OK?), so that the various right cosets of H each have exactly $|H|$ elements. Next, $g \in gH$ (because $1 \in H$), so every element of G lies in some right coset of H. In fact, if $g \in g'H$, it is easy to see that $gH \subseteq g'H$ (OK?), and by comparing sizes it follows that $gH = g'H$; thus gH is the *only* right coset that contains g. So G splits up as a disjoint union of the various right cosets of H, each of which has $|H|$ elements, and the result follows. □

The number of distinct right cosets of H in G is called the *index* of H in G, written $[G : H]$. So Lagrange's theorem can be expressed as the equation: $|G| = [G : H]|H|$.

Let G be a group, and $g \in G$. Recall that the cyclic subgroup H generated by g is the collection of all powers of g, that is,

$$H = \{g^n: n \in \mathbb{Z}\},$$

and then $\operatorname{order}(g) = |H|$.

• Ex.5.48: *If* $g^m = 1$, *then* order(g) *is a factor of* m.

Note that $g^m = 1$ does *not* say that order(g) = m. For example, if we just know that $g^6 = 1$, we can only deduce that order(g) = 1, 2, 3 or 6.

Corollary 5.39. *Let G be a finite group, with* $|G| = n$, *and let* $g \in G$. *Then* $g^n = 1$.

Proof. Let H be the cyclic subgroup generated by g, let $|H| = m$ and $[G : H] = r$. Then $n = rm$, by Lagrange's theorem, and $g^m = 1$, by Exercise 5.19. So $g^n = (g^m)^r = 1$. □

• Ex.5.49: *Suppose* $g \in G$ *and* $g^m = 1$. *Does it follow that* m *is a factor of* $|G|$?

Recall that the group T is the subgroup of $\mathbb{C}^*$ consisting of all complex numbers of modulus 1, and the group P_n is the subgroup of T consisting of all the complex nth roots of unity; and $P_n \cong C_n$.

Corollary 5.40. *Every finite subgroup of T is cyclic; indeed, if G is a finite subgroup of T with* $|G| = n$, *then* $G = P_n$.

Proof. Every element of G is an nth root of unity, by Corollary 5.39, so that $G \subseteq P_n$. But $|G| = n = |P_n|$, so $G = P_n$. □

Corollary 5.41. *Every subgroup of* P_n *is cyclic; indeed, the only subgroups of* P_n *are* P_m, *where m divides n*. □

Said another way, the subgroups of a finite cyclic group are all cyclic, and there is exactly one subgroup of each order dividing the order of the group.

• Ex.5.50: *Show that, in Corollary 5.40,* T *can be replaced by* $\mathbb{C}^*$ *throughout.*

For those who know about fields, an even more general result is true: every finite subgroup of the multiplicative group of non-zero elements of a field is cyclic. (So, for example, $\mathbb{Z}_{23}^* \cong C_{22}$.) For details, see [7].

The groups C_n and D_n, which were constructed from the symmetries of a regular n-gon, had the property that all their members had the centre of the n-gon as a fixed point. We now show that something similar holds for every finite group of isometries:

Lemma 5.42. *Let G be a finite group of plane isometries. Then the elements of G have a common fixed point.*

Proof. Let $G = \{ f_1, f_2, \ldots, f_n \}$. Pick any $z \in \mathbb{C}$, and put $z_r = f_r(z)$, and $X = \{ z_1, z_2, \ldots, z_n \}$. ($X$ is the *orbit* of z under the action of G. Note that we are not assuming that $z_1, z_2, \ldots, z_n$ are distinct; indeed, if we happened to choose z to be the fixed point we are looking for, then we should have $X = \{ z \}$.)

We claim that $G \subseteq \mathcal{S}(X)$. If $f \in G$, then for each r we have $f f_r = f_s$ for some s, and the map $r \mapsto s$ thus obtained is a permutation of $\{1, 2, \ldots, n\}$. (OK?) So $f(z_r) = f f_r(z) = f_s(z) = z_s$, and $f \in \mathcal{S}(X)$.

But now it follows that

$$z_1 + z_2 + \ldots + z_n = f(z_1) + f(z_2) + \ldots + f(z_n)$$

for all $f \in G$, the terms on the right being simply a permutation of the terms on the left. So, dividing by n,

$$\frac{z_1 + z_2 + \ldots + z_n}{n} = \frac{f(z_1) + f(z_2) + \ldots + f(z_n)}{n} = f\left(\frac{z_1 + z_2 + \ldots + z_n}{n}\right),$$

by Exercise 5.26, and thus $\omega = (z_1 + z_2 + \ldots + z_n)/n$ is a fixed point of every $f \in G$.
$\square$

We can now find all the finite groups of *direct* isometries—they are cyclic groups of rotations about some centre:

Proposition 5.43. *Let G be a finite group of direct plane isometries. Then G is cyclic.*

Proof. Let $G = \{f_1, f_2, \ldots, f_n\}$. By Lemma 5.42, there exists $\omega \in \mathbb{C}$ such that $f_r(\omega) = \omega$, for all r, and since translations have no fixed points, the non-trivial elements of G are all rotations about ω. For each r, then, there is an element $a_r \in T$ such that

$$f_r(z) - \omega = a_r(z - \omega)$$

for all z. Thus we have a map $\alpha : G \to T$ given by $f_r \mapsto a_r$. Now, for any r, s,

$$(f_r f_s)(z) - \omega = f_r(f_s(z)) - \omega = a_r(f_s(z) - \omega)$$
$$= a_r(a_s(z - \omega)) = (a_r a_s)(z - \omega).$$

It follows that $\alpha(f_r f_s) = \alpha(f_r)\alpha(f_s)$, that is, α is a group homomorphism. (More geometrically, but more wordily: if the angle of rotation of f_r is θ_r, then $a_r = e^{i\theta_r}$, and if $f_r f_s = f_t$ then $\theta_r + \theta_s \equiv \theta_t \pmod{2\pi}$, since the composite rotation adds the two angles of rotation. But this means $a_r a_s = a_t$, or $\alpha(f_r)\alpha(f_s) = \alpha(f_t)$.)

Now α is clearly injective (OK?), so $G \cong \alpha(G)$, the image of α. By Corollary 5.40, $\alpha(G) = P_n$, so $G \cong P_n \cong C_n$, as required.
$\square$

We can now find all the finite groups of plane isometries:

Theorem 5.44 (Leonardo da Vinci's theorem). *Let G be a finite group of plane isometries. Then G is cyclic or dihedral.*

Proof. By Lemma 5.42, there exists $\omega \in \mathbb{C}$ fixed by all $f \in G$. The subgroup G^+ of direct isometries in G is necessarily finite, so if $|G^+| = n$, then $G^+ = C_n$, by Proposition 5.43. If G contains no opposite isometries, then $G = C_n$, and we are done.

So now suppose $G^+ = \{1, f, f^2, \ldots, f^{n-1}\}$ (all distinct, and all rotations about ω, with $f^n = 1$), and suppose G contains an opposite isometry g. Every opposite isometry in G (including g) fixes ω, and so is a reflection in some line through ω. The coset

$$gG^+ = \{g, gf, gf^2, \ldots, gf^{n-1}\}$$

thus consists of n distinct reflections in lines through ω, and $G^+ \cap gG^+ = \emptyset$. We claim $G = G^+ \cup gG^+$, so that $|G| = 2n$. This is easy: if $h \in G$ then either h is direct, when $h \in G^+$, or h is opposite. But then gh (which certainly belongs to G) is direct, so $gh \in G^+$, or $gh = f^r$ for some r; and now $h = g^{-1}f^r = gf^r \in gG^+$.

So we have

$$G = \{1, f, f^2, \ldots, f^{n-1}, g, gf, gf^2, \ldots, gf^{n-1}\},$$

where the elements are all distinct, $f^n = 1$, $g^2 = 1$, and indeed $(gf^r)^2 = 1$ for all r. In particular $(gf)^2 = 1$, so $fg = gf^{-1} = gf^{n-1}$, and we have enough information to calculate the whole of the multiplication table of G (OK?), and $G = D_n$. $\qquad\square$

The reader may be forgiven for having thought for some while now that all geometrical questions are to be answered by turning them into rather tedious algebraic calculations, and indeed that algebra is merely a tool for doing geometry. So we turn the tables, and give a stunning example of the use of geometry to answer a purely algebraic question:

Corollary 5.45. *Every subgroup of D_n is cyclic or dihedral.*

Proof. D_n is (or can be thought of as) a finite group of plane isometries, so every subgroup is a finite group of plane isometries, which is cyclic or dihedral by Leonardo da Vinci's theorem. $\qquad\square$

Anyone who doubts the power of the above geometrical argument should try to prove the same result by algebra alone. For example, take a large clean sheet of paper and draw up the 24×24 multiplication table of D_{12}, and see if, by staring at it (and not thinking at all about rotations or reflections), you can find *all* the subgroups of D_{12}.

To put you out of your misery, let us find all the subgroups, by geometry, and count them. We have $D_{12} = \{1, f, f^2, \ldots, f^{11}, g, gf, \ldots, gf^{11}\}$, where f is a rotation through $2\pi/12 = \pi/6$ about some point O, and $g, gf, \ldots$ are reflections in lines through O. The direct isometries form a subgroup C_{12}, and this has one (cyclic) subgroup for each divisor of 12, that is, one each of C_1, C_2, C_3, C_4, C_6, and C_{12}: six subgroups.

Since $g.gf = f$, the angle between the axis of g and the axis of gf is $(1/2)(\pi/6) = \pi/12$, by Exercise 5.31(iii), and the same goes for the angle between the axes of gf^r and gf^{r+1}, each r. So the 12 axes are at equal angular intervals, and similarly this will be true for each dihedral subgroup. The 12 axes meet at a point, and a circle centred at this point meets the 12 axes in 24 equally-spaced points. Label these points $A_1, B_1, A_2, B_2, \ldots$, A_{12}, B_{12} (Fig. 5.9). Our group D_{12} is the symmetry group of the dodecagon (12-gon) $A_1 A_2 \ldots A_{12}$, and equally it is the symmetry group of the 12-gon $B_1 B_2 \ldots B_{12}$. (For example, reflection in the diagonal $B_1 B_7$ of the second 12-gon is the same as reflection in the perpendicular bisector of the two opposite sides $A_1 A_2$ and $A_7 A_8$ of the first 12-gon.)

If D_m is a subgroup, then, by Lagrange's theorem, $2m$ is a divisor of 24, so m is a divisor of 12. There is only one subgroup D_{12}, the whole group. What about D_6? Its direct

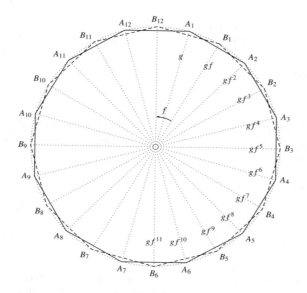

Fig. 5.9 D_{12}

isometries will be C_6, and the only possibility here is $C_6 = \{\,1,\, f^2,\, f^4,\, f^6,\, f^8,\, f^{10}\,\}$. For the reflections, we need to pick out, from our 12 reflections, six with axes at equal angular intervals of $\pi/6$. We get two D_6's, one including $g,\, gf^2,\, \ldots$ and the other including $gf,\, gf^3,\, \ldots$: explicitly, the two subgroups are

$$\{\,1,\, f^2,\, f^4,\, f^6,\, f^8,\, f^{10},\, g,\, gf^2,\, gf^4,\, gf^6,\, gf^8,\, gf^{10}\,\}, \quad \text{and}$$
$$\{\,1,\, f^2,\, f^4,\, f^6,\, f^8,\, f^{10},\, gf,\, gf^3,\, gf^5,\, gf^7,\, gf^9,\, gf^{11}\,\}.$$

If g is reflection in A_1A_7, then the first of these groups is the symmetry group of the hexagon $A_1A_3A_5A_7A_9A_{11}$, and equally of $A_2A_4A_6A_8A_{10}A_{12}$; and the second group is the symmetry group of the hexagon $B_1B_3B_5B_7B_9B_{11}$, and equally of $B_2B_4B_6B_8B_{10}B_{12}$.

Next, D_4, the symmetries of a square, will contain $C_4 = \{\,1,\, f^3,\, f^6,\, f^9\,\}$ as its direct isometries, and the opposite isometries come from selecting every third line of reflection. So we get three D_4's, namely $\{\,1,\, f^3,\, f^6,\, f^9,\, g,\, gf^3,\, gf^6,\, gf^9\,\}$ (symmetries of the square $A_1A_4A_7A_{10}$, or of the square $B_2B_5B_8B_{11}$); $\{\,1,\, f^3,\, f^6,\, f^9,\, gf,\, gf^4,\, gf^7,\, gf^{10}\,\}$ (symmetries of $A_3A_6A_9A_{12}$, or of $B_1B_4B_7B_{10}$); and $\{\,1,\, f^3,\, f^6,\, f^9,\, gf^2,\, gf^5,\, gf^8,\, gf^{11}\,\}$ (symmetries of $A_2A_5A_8A_{11}$, or of $B_3B_6B_9B_{12}$).

Similarly, four D_3's can be found by inscibing equilateral triangles in one or other of the two 12-gons: symmetries of $A_1A_5A_9$ or of $A_3A_7A_{11}$ give $\{\,1,\, f^4,\, f^8,\, g,\, gf^4,\, gf^8\,\}$; symmetries of $B_1B_5B_9$ or of $B_3B_7B_{11}$ give $\{\,1,\, f^4,\, f^8,\, gf,\, gf^5,\, gf^9\,\}$; symmetries of $A_2A_6A_{10}$ or of $A_4A_8A_{12}$ give $\{\,1,\, f^4,\, f^8,\, gf^2,\, gf^6,\, gf^{10}\,\}$; and symmetries of $B_2B_6B_{10}$ or of $B_4B_8B_{12}$ give $\{\,1,\, f^4,\, f^8,\, gf^3,\, gf^7,\, gf^{11}\,\}$.

Finally, we get six copies of D_2 by choosing two lines at right angles all possible ways, giving $\{1, f^6, gf^r, gf^{r+6}\}$ for $0 \leq r \leq 5$; and 12 copies of D_1, each consisting of 1 together with one reflection, that is, $\{1, gf^r\}$ for $0 \leq r \leq 11$. The total number of subgroups of D_{12} is thus

$$6 + (1 + 2 + 3 + 4 + 6 + 12) = 34.$$

For general n, a similar argument shows that the number of subgroups of D_n is $d(n) + \sigma(n)$, where $d(n)$ is the *number* of (positive) divisors of n, and $\sigma(n)$ is the *sum* of the (positive) divisors of n.

ANSWERS TO EXERCISES

5.1: Let f be an isometry of $\mathbb{R}^3$ with $f^2 = 1$. If $f \neq 1$, then f is either a reflection or a half-turn or a *central inversion* (see below). Proof: if the fixed points of f are not collinear, then f is a reflection, by Corollaries 5.8 and 5.9. So now suppose f has at least two fixed points, and its fixed points lie in a line ℓ. For any $A \notin \ell$, the mid-point of AA' is on ℓ (as in the proof of Corollary 5.10), and indeed $AA' \perp \ell$. (OK?) So the half-turn about ℓ sends A to A', and so coincides with f. Finally, if f has at most one fixed point, then as in the proof of Corollary 5.10, the mid-points of AA', for all A, coincide at some point P, and such an f is called a *central inversion* about P.

5.2: For any P, we have $(f(gh))(P) = f((gh)(P)) = f(g(h(P))) = (fg)(h(P)) = ((fg)h)(P)$, and we are done.

5.3: For example, let $f(x, y) = (2x, y)$ and $g(x, y) = (y, x)$, for all x, y. Then f, g are transformations (OK?), but for example $fg(1, 0) = f(0, 1) = (0, 1)$, whereas $gf(1, 0) = g(2, 0) = (0, 2)$.

5.4: If $1, 1'$ are *both* neutral elements, then $x = x1'$, for all $x \in G$, and $1y = y$, for all $y \in G$. Put $x = 1$ and $y = 1'$ and we have $1 = 11' = 1'$, as required. If x^{-1} and x' are *both* inverses for x, then $1 = xx^{-1}$, so that $x' = x'(xx^{-1})$; and also $x'x = 1$, so that $(x'x)x^{-1} = x^{-1}$. But $x'(xx^{-1}) = (x'x)x^{-1}$, so $x' = x^{-1}$.

5.5: $(y^{-1}x^{-1})(xy) = y^{-1}(x^{-1}x)y = y^{-1}1y = y^{-1}y = 1$, so $y^{-1}x^{-1} = (xy)^{-1}$ by Exercise 5.4.

5.6: $xy = xz \Rightarrow x^{-1}(xy) = x^{-1}(xz) \Rightarrow (x^{-1}x)y = (x^{-1}x)z \Rightarrow 1y = 1z \Rightarrow y = z$. Then $xz = yz \Rightarrow (xz)z^{-1} = (yz)z^{-1} = \ldots$, similarly. Next, $xy = x \Rightarrow xy = x1 \Rightarrow y = 1$; and similarly for the other case. (For the part in parentheses, take f, g as in the solution to Exercise 5.3, and let $h(x, y) = (x, 2y)$, all x, y. Then it is easy to check that $fg = gh : (x, y) \mapsto (2y, x)$; but $f \neq h$.)

5.7: $(x^{-1})^2 = x^{-1}x^{-1} = (xx)^{-1} = (x^2)^{-1}$, so either can be taken as x^{-2}. Put $x^0 = 1$, and then inductively for $n > 0$ define $x^n = x^{n-1}x$, and for $n < 0$ define $x^n = x^{n+1}x^{-1}$. The formula $x^n x^m = x^{n+m}$ is now clearly true if $n = 0$ or $m = 0$. For $m = 1$ and $n > 0$ we have $x^n x^1 = x^n x = x^{n+1}$, and for $m = 1$ and $n < 0$ we have $x^n x^1 = x^n x = x^{n+1}x^{-1}x = x^{n+1}$; so the formula is true when $m = 1$, for all n. Then for $m > 0$ we have $x^n x^m = x^n x^{m-1}x = x^{n+m-1}x^1$ (by induction), and this equals x^{n+m} by the case $m = 1$. Finally, for $m < 0$ we have $x^n x^m = x^n x^{m+1}x^{-1} = x^{n+m+1}x^{-1}$ (by induction), and since $x^{n+m+1} = x^{n+m}x^1$ by the case $m = 1$, we get $x^{n+m}xx^{-1}$, or x^{n+m} once more.

5.8: Suppose G is a group. Then $1 \in G$ shows $G \neq \emptyset$; and if $f, g \in G$, then $f, g^{-1} \in G$, so $fg^{-1} \in G$. For the converse, if $G \neq \emptyset$ then pick $f \in G$, and note that the given conditions imply that $ff^{-1} \in G$, or $1 \in G$. Next, if $f \in G$, then since $1 \in G$ also, we have $1f^{-1} \in G$, or $f^{-1} \in G$. Finally, if $f, g \in G$, then $f, g^{-1} \in G$, so $f(g^{-1})^{-1} \in G$, or $fg \in G$, and we are done.

5.9: Clearly $1 \in \mathcal{C}(\mathbb{R}^2)$. If $f, g \in \mathcal{C}(\mathbb{R}^2)$ and ℓ is a line, then $g(\ell)$ is a line, and hence so is $f(g(\ell))$, or $(fg)(\ell)$, whence $fg \in \mathcal{C}(\mathbb{R}^2)$. It remains to show that if $f \in \mathcal{C}(\mathbb{R}^2)$ then $f^{-1} \in \mathcal{C}(\mathbb{R}^2)$, and this is surprisingly hard. So let f be a collineation and let ℓ be a line. We must show that $f^{-1}(\ell)$ is a line. To this end, pick A, B, $C \in f^{-1}(\ell)$, so that $f(A)$, $f(B)$, $f(C) \in \ell$. It is sufficient to show that A, B, C are collinear, so we suppose not, and look for a contradiction. We shall obtain this by showing that $f(\mathbb{R}^2) \subseteq \ell$, which contradicts the fact that f is a transformation, and hence surjective.

So take X on AB: since X, A, B are collinear, so are $f(X)$, $f(A)$, and $f(B)$, which means that $f(X) \in \ell$. (Note that $f(A) \neq f(B)$, since $A \neq B$ and f is a transformation.) Similarly, if X is on BC or CA then we deduce $f(X) \in \ell$. Finally, take $X \notin AB \cup BC \cup CA$. Let m be a line through X not parallel to AB or BC or CA—there are plenty of such lines!—so that m meets each of AB, BC, CA. Since these three lines are not concurrent (OK?), m meets $AB \cup BC \cup CA$ in at least two distinct points, say in Y and Z. By the previous argument, this means $f(Y) \in \ell$ and $f(Z) \in \ell$. But X, Y, Z are collinear, so $f(X)$, $f(Y)$, $f(Z)$ are collinear, whence $f(X) \in \ell$, and we have our contradiction.

5.10: If $P = (x', y')$ lies on the line $ax+by+c = 0$, then $ax'+by'+c = 0$, or $(\frac{1}{2}a)(2x')+by'+c = 0$, so that $f(P) = (2x', y')$ lies on the line $\frac{1}{2}ax + by + c = 0$, whence $f \in \mathcal{C}(\mathbb{R}^2)$. However, whereas $(1, 0)$ and $(0, 1)$ are both distance 1 from $(0, 0)$, their images $f(1, 0) = (2, 0)$ and $f(0, 1) = (0, 1)$ are of distance 2 and 1 respectively from $f(0, 0) = (0, 0)$, so that $f \notin \mathcal{E}(\mathbb{R}^2)$.

5.11: Write $f_{\mathbf{v}}(\mathbf{u}) = \mathbf{u} + \mathbf{v}$ (translation by $\mathbf{v}$), so that $\mathcal{T}(\mathbb{R}^2) = \{f_{\mathbf{v}} : \mathbf{v} \in \mathbb{R}^2\}$. Then $1 = f_0 \in \mathcal{T}(\mathbb{R}^2)$; $f_{\mathbf{v}}f_{\mathbf{w}} = f_{\mathbf{v+w}} \in \mathcal{T}(\mathbb{R}^2)$; $f_{\mathbf{v}}^{-1} = f_{-\mathbf{v}} \in \mathcal{T}(\mathbb{R}^2)$; and $f_{\mathbf{v}}f_{\mathbf{w}} = f_{\mathbf{v+w}} = f_{\mathbf{w}}f_{\mathbf{v}}$, for all $\mathbf{v}$, $\mathbf{w} \in \mathbb{R}^2$.

5.12: $\{1, f^2\}$, $\{1, f, f^2, f^3\}$, $\{1, g\}$, $\{1, gf\}$, $\{1, gf^2\}$, $\{1, gf^3\}$, $\{1, f^2, g, gf^2\}$, and $\{1, f^2, gf, gf^3\}$.

5.13: Take X to be a (non-square) rectangle, or maybe the curve $y = \frac{1}{x}$.

5.14: As a polar curve, this is $r^4 \sin 4\theta = 4$, which has rotational symmetries about O through multiples of $\pi/2$, and reflectional symmetries in the lines $\theta = \pi/8, 3\pi/8, 5\pi/8$, and $7\pi/8$.

5.15: Take X to be a three-legged swastika, or an Isle of Man symbol.

5.16: (i) $1_G : G \to G$ is an isomorphism. (It is bijective, and $1_G(ab) = ab = 1_G(a)1_G(b)$, for all $a, b \in G$.) (ii) If $\alpha : G \to H$ is an isomorphism, it is bijective, and hence so is $\alpha^{-1} : H \to G$. Then, for $a', b' \in H$, put $a = \alpha^{-1}(a')$ and $b = \alpha^{-1}(b')$. We have $\alpha(ab) = \alpha(a)\alpha(b) = a'b'$, so $\alpha^{-1}(a'b') = ab = \alpha^{-1}(a')\alpha^{-1}(b')$. (iii) Let $\alpha : G \to H$ and $\beta : H \to K$ be isomorphisms. Then α and β are bijections, and hence so is $\beta\alpha : G \to K$. Then, for $a, b \in G$, we have $(\beta\alpha)(ab) = \beta(\alpha(ab)) = \beta(\alpha(a)\alpha(b)) = (\beta(\alpha(a)))(\beta(\alpha(b))) = ((\beta\alpha)(a))((\beta\alpha)(b))$.

5.17: Take X to be the graph of $y = |\cos \pi x|$. (See also Exercise 5.25 below.)

5.18: If $a = 1$ then $a^n = 1$, all n, and so $H = \{1\}$, and $|H| = 1$. Conversely, if $|H| = 1$ then $H = \{1\}$, so that $a^n = 1$, for all n, and in particular $a = a^1 = 1$.

5.19: If $a^n \neq 1$, for all $n > 0$, then $a^{n+m} \neq a^m$, for all $n, m > 0$, so that $1, a, a^2, a^3, \ldots$ are all distinct, and thus $|H| = \infty$. Otherwise, if $n > 0$ is least so that $a^n = 1$, then $1, a, a^2, \ldots, a^{n-1}$ are distinct (else $a^i = a^j$ for some i, j with $n > j > i \geq 0$, and now $a^{j-i} = 1$ with $0 < j - i < n$, a contradiction); and further, for any $m \in \mathbb{Z}$ we have $m = nq + r$ with $q, r \in \mathbb{Z}$ and $0 \leq r < n$, and now $a^m = (a^n)^q a^r = 1^q a^r = a^r$, so $H = \{1, a, a^2, \ldots, a^{n-1}\}$ and $|H| = n$.

5.20: $C_n = \{1, f, f^2, \ldots, f^{n-1}\}$, where $f^n = 1$. If $a \in C_n$, then $a = f^r$, say, and now $a^n = (f^r)^n = f^{rn} = (f^n)^r = 1^r = 1$. This does not mean that order$(a) = n$: for example, if $n = 4$ and $a = f^2$, then $a \neq 1$ but $a^2 = f^4 = 1$, so order$(a) = 2$, not 4. (In general, if $a^n = 1$, this just means that the order of a divides n. Proof?)

5.21: $C_\infty = \{\ldots, f^{-2}, f^{-1}, 1, f, f^2, \ldots\}$ where order$(f) = \infty$, so that $f^n = 1$ iff $n = 0$. Let $a \in C_\infty$, so that $a = f^n$ for some $n \in \mathbb{Z}$. If $n = 0$ then $a = 1$ and order$(a) = 1$. If $n \neq 0$ then, for any $m \neq 0$ we have $a^m = f^{nm} \neq 1$, and thus order$(a) = \infty$.

5.22: Let $a \in D_\infty = \{\ldots, f^{-2}, f^{-1}, 1, f, f^2, \ldots; \ldots, gf^{-2}, gf^{-1}, g, gf, gf^2, \ldots\}$. If $a = f^r$, for some $r \in \mathbb{Z}$, then order$(a) = 1$ or ∞, as in the previous exercise. Otherwise $a = gf^r$ for some $r \in \mathbb{Z}$, so $a \neq 1$, and $a^2 = (gf^r)(gf^r) = g(f^r g)f^r = g(gf^{-r})f^r = g^2 = 1$, and thus order$(a) = 2$.

5.23: We have $\alpha(a^n) = (\alpha(a))^n$, for all $n \in \mathbb{Z}$ (OK?), so $a^n = 1$ in G iff $(\alpha(a))^n = 1$ in H; result follows by Exercise 5.19. Then both T and $\mathcal{S}(T)$ contain elements of order 3 (and, indeed, of every finite order), so cannot be isomorphic to C_∞ or D_∞, by Exercises 5.21 and 5.22.

5.24: Trivial symmetry group $1 = \{1\} = C_1$: FGJPR. C_2: NSZ. D_1: ABCDEK†L†MQ†TUVW. D_2: HI. D_3: Y†. D_4: X†. Infinite symmetry group (but *not* D_∞): O†. (†Suitably redesigned.)

5.25: (i) The symmetries are all translations in the x-direction through integer distances: $\mathcal{S}(X) = \{f_n : n \in \mathbb{Z}\}$ where $f_n(x, y) = (x + n, y)$. It is immediate that $f_0 = 1$ and $f_n = f_1^n$, so that $\mathcal{S}(X) = \{\ldots, f_1^{-2}, f_1^{-1}, 1, f_1, f_1^2, \ldots\} \cong C_\infty$. (ii) Here we have the same symmetries as in (1), but also reflections in the lines $y = n/2$ ($n \in \mathbb{Z}$) (through the middle of a T, or halfway between two T's), that is, g_n where $g_n(x, y) = (n - x, y)$. It is easy to see that $g_{-n} = g_0 f_1^n = f_1^{-n} g_0$, and $g_0^2 = 1$, so we have D_∞. (iii) It is easiest to situate the N's with their *centres* at $(n, 0)$, $n \in \mathbb{Z}$. We then have the f_n as symmetries, as before, but this time we have half-turn symmetries about the points $(n/2, 0)$ ($n \in \mathbb{Z}$), that is, h_n where $h_n(x, y) = (n - x, -y)$. It is easy to see that $h_{-n} = h_0 f_1^n = f_1^{-n} h_0$, and $h_0^2 = 1$, so we have D_∞ again (as an abstract group), though this time the symmetry group consists of translations and rotations, not translations and reflections.

5.26: Suppose $f(z) = az + b$, for all z. Then $f(z_1) + \ldots + f(z_n) = (az_1 + b) + \ldots + (az_n + b) = a(z_1 + \ldots + z_n) + nb$; divide through by n. If $f(z) = a\bar{z} + b$, a similar calculation works.

5.27: Let $f(z) = a_1 z + b_1$ and $g(z) = a_2 z + b_2$; then $fg(z) = f(a_2 z + b_2) = a_1(a_2 z + b_2) + b_1 = a_1 a_2 z + (a_1 b_2 + b_1)$, direct. Now let $f(z) = a_1 \bar{z} + b_1$ and $g(z) = a_2 \bar{z} + b_2$; then $fg(z) = f(a_2 \bar{z} + b_2) = a_1 \overline{(a_2 \bar{z} + b_2)} + b_1 = a_1 \bar{a}_2 z + (a_1 \bar{b}_2 + b_1)$, direct again. Next, let $f(z) = a_1 \bar{z} + b_1$ and $g(z) = a_2 z + b_2$; so $fg(z) = f(a_2 z + b_2) = a_1 \overline{(a_2 z + b_2)} + b_1 = a_1 \bar{a}_2 \bar{z} + (a_1 \bar{b}_2 + b_1)$, opposite. Finally, if $f(z) = a_1 z + b_1$ and $g(z) = a_2 \bar{z} + b_2$; so $fg(z) = f(a_2 \bar{z} + b_2) = a_1(a_2 \bar{z} + b_2) + b_1 = a_1 a_2 \bar{z} + (a_1 b_2 + b_1)$, opposite again. For the inverses: if $w = az + b$, then $w - b = az$, so $z = \bar{a}w - \bar{a}b$; and if $w = a\bar{z} + b$ then $\bar{z} = \bar{a}w - \bar{a}b$, and $z = a\bar{w} - \bar{a}b$.

5.28: Let $f_1(z) = a_1 z + b_1$ and $f_2(z) = a_2 z + b_2$, where $a_1 = e^{i\theta_1}$ and $a_2 = e^{i\theta_2}$. Then $f_1 f_2(z) = a_1(a_2 z + b_2) + b_1 = (a_1 a_2)z + (a_1 b_2 + b_1)$. Also $a_1 a_2 = e^{i(\theta_1 + \theta_2)}$, so we have a rotation through $\theta_1 + \theta_2$ if $a_1 a_2 \neq 1$, that is, if $\theta_1 + \theta_2$ is not an integer multiple of 2π.

When $\theta_1 + \theta_2$ *is* an integer multiple of 2π, then $a_1 a_2 = 1$ and $f_1 f_2(z) = z + (a_1 b_2 + b_1)$. This is a translation *or* it is 1; the latter occurs iff $b_1 = -a_1 b_2$, or

$$\frac{b_1}{1 - a_1} = \frac{-a_1 b_2}{1 - a_1} = \frac{a_1 b_2}{a_1 - 1} = \frac{a_2 a_1 b_2}{a_2(a_1 - 1)} = \frac{b_2}{1 - a_2},$$

which is the condition that the two centres of rotation coincide.

5.29: Let $f_1(z) = a_1 z + b_1$ and $f_2(z) = a_2 z + b_2$, so that $f_1 f_2(z) = (a_1 a_2)z + (a_1 b_2 + b_1)$ and $f_2 f_1(z) = (a_1 a_2)z + (a_2 b_1 + b_2)$. Thus $f_1 f_2 = f_2 f_1 \Leftrightarrow a_1 b_2 + b_1 = a_2 b_1 + b_2 \Leftrightarrow b_1(1 - a_2) = b_2(1 - a_1)$. If $a_1 = 1$, then $b_1 \neq 0$ (else $f_1 = 1$), so $a_2 = 1$, and both f_1, f_2 are translations. Similarly if $a_2 = 1$. If both $a_1 \neq 1$ and $a_2 \neq 1$, then both f_1, f_2 are rotations, and $f_1 f_2 = f_2 f_1 \Leftrightarrow b_1/(1 - a_1) = b_2/(1 - a_2)$, which is the condition that the two centres of rotation coincide.

5.30: We have $f(z_1) = z_1$ and $f(z_2) = z_2$, or $z_1 = a\bar{z}_1 + b$ and $z_2 = a\bar{z}_2 + b$. Subtracting, $z_1 - z_2 = a(\overline{z_1 - z_2})$, whence $(z_1 - z_2)^2 = a(\overline{z_1 - z_2})(z_1 - z_2) = a|z_1 - z_2|^2$. Putting $(z_1 - z_2)/|z_1 - z_2| = e^{i\alpha}$, we have $e^{2i\alpha} = e^{i\theta}$, or $\alpha = \frac{\theta}{2} + n\pi$.

5.31: (i) $f_1 f_2 = 1 \Leftrightarrow f_1 = f_2^{-1} = f_2$; and $(f_1 f_2)^{-1} = f_2^{-1} f_1^{-1} = f_2 f_1$. (ii) $\ell_1 \parallel \ell_2 \Leftrightarrow a_1 = a_2$ by Exercise 5.30. Then $f_i(0) = b_i$, and $\frac{b_i}{2}$ lies on ℓ_i, for $i = 1$, 2. So the line through 0, b_1, b_2 is orthogonal to ℓ_1 and ℓ_2, and further, $f_1 f_2(z) = z + b_1 - b_2$ (check). (iii) $f_1 f_2$ fixes $\ell_1 \ell_2$; then $f_1 f_2(z) = a_1 \bar{a}_2 z + \dots$, and $a_1 \bar{a}_2 = e^{i(\theta_1 - \theta_2)}$, so use Exercise 5.30 again. (iv) $f_1 f_2 = f_2 f_1 \Leftrightarrow (f_1 f_2)^2 = 1$, by (i), and this means $f_1 f_2$ is a half turn, by Corollary 5.10. Then $\ell_1 \perp \ell_2$ by (iii), and $a_1 = -a_2$ by Exercise 5.30, once more. The converse is easy.

5.32: Apply Exercise 5.30 to g, defined as in Proposition 5.27.

5.33: If P is not on the axis, then P and its image P' are on opposite sides of the axis. So PP' meets the axis, in Q, say. An invariant line through P must contain P', hence Q; but now Q lies on two invariant lines, so is fixed, which is impossible. So there is no invariant line through P.

5.34: This is immediate from the preceding remarks, noting that an isometry and its inverse have the same fixed points and invariant lines.

5.35: A reflection is equal to its inverse. A glide-reflection and its inverse have the same invariant line(s), so use Exercise 5.33.

5.36: Let $\ell_2 \ell_3 = P$, and note that $f_3(P) = P$ and $f_2(P) = P$, so that $f(P) = f_1(P) = P'$, the reflection of P in ℓ_1. Thus the axis of f passes through the mid-point of PP', which is the foot of the perpendicular from P onto ℓ_1.

5.37: By Exercise 5.36, the axis of $f_a f_b f_c$ passes through P, and the axis of $(f_a f_b f_c)^{-1} = f_c f_b f_a$ passes through R; but the two axes coincide, by Exercise 5.35, so the axis of $f_a f_b f_c$ is PR. Thus $f_a f_b f_c(PR) = PR$, whence $f_c f_a f_b f_c(PR) = f_c(PR)$; but the axis of $f_c f_a f_b$ is RQ, so, by Exercise 5.33, $f_c(PR) = RQ$. Thus the lines PR, RQ are equally inclined to BA, or $\angle BRP = \angle QRA$.

5.38: $1f = f = f1$ shows $1 \in C(f)$. Then if $g, h \in C(f)$, we have $f(gh) = (fg)h = (gf)h = g(fh) = g(hf) = (gh)f$, so $gh \in C(f)$. Finally, if $g \in C(f)$, then $fg = gf$, so $g^{-1}fg = f$, and $g^{-1}f = fg^{-1}$, so $g^{-1} \in C(f)$, and we have a subgroup.

   (i) $C(1) = \mathcal{I}(\mathbb{R}^2)$. (ii) $C(f)$ consists of 1, and every translation, and every opposite isometry whose axis is parallel to the direction of the translation f. (iii) $C(f)$ consists of 1, and all rotations with the same centre as f. (iv) $C(f)$ consists of 1, and all rotations with the same centre as f, and all reflections in lines through the centre of f. (v) $C(f)$ consists of 1, all translations parallel to the axis of f, and all half-turns about points on the axis of f, and f itself, and all reflections with axis orthogonal to the axis of f, and all glide-reflections with the same axis as f. (vi) $C(f)$ consists of 1, and all translations parallel to the axis of f, and reflection in the axis of f, and all glide-reflections with the same axis as f.

5.39: Let B_i be the mid-point of $A_i A_{i+1}$, for all i. Here we are interpreting subscripts of the letters A and B modulo n, so $A_0 = A_n$, $B_0 = B_n$, and so on. Put $g = f_n f_{n-1} \cdots f_2 f_1$. If n is even, g is direct, and if n is odd, g is opposite. Note that $f_i(B_{i-1} B_i) = B_i B_{i+1}$, for all i, so that $B_0 B_1$

is an invariant line of g. Next, put $C_0 = B_0$, and $C_i = f_i(C_{i-1})$, for all i; these subscripts are *not* modulo n, and indeed $C_n = g(C_0) = g(B_n)$. We shall show that $C_n \neq B_n$, so that g is either a translation by $\overrightarrow{B_n C_n}$ or a glide-reflection with axis $B_0 B_1$ and translational component $\overrightarrow{B_n C_n}$. Put $|B_0 B_1| = a$. We have $|B_1 C_1| = |B_1 C_0| = a$, so $\overrightarrow{B_1 C_1} = \overrightarrow{B_2 B_1}$, and $|B_2 C_1| = 2a$. Thus $\overrightarrow{B_2 C_2} = 2\overrightarrow{B_3 B_2}$, so that $|B_3 C_2| = 3a$, and so on; and finally, $\overrightarrow{B_n C_n} = n\overrightarrow{B_1 B_n}$. There are thus two cases: if n is even, g is a translation through $n\overrightarrow{B_1 B_n}$; and if n is odd, g is a glide-reflection with axis $B_n B_1$ and translational component $n\overrightarrow{B_1 B_n}$.

5.40: We have $(z_3 - z_1)/(z_2 - z_1) = (b/c)e^{\pm iA}$, according to the orientation, so that $(z_3 - z_1)(\bar{z}_2 - \bar{z}_1) = bce^{\pm iA}$, and $\Im((z_3 - z_1)(\bar{z}_2 - \bar{z}_1)) = \pm bc \sin A = \pm 2\Delta$. The result follows on multiplying out the brackets.

5.41:
$$\Im((x_1 - iy_1)(x_2 + iy_2) + (x_2 - iy_2)(x_3 + iy_3) + (x_3 - iy_3)(x_1 + iy_1))$$
$$= x_1 y_2 - x_2 y_1 + x_2 y_3 - x_3 y_2 + x_3 y_1 - x_1 y_3$$
$$= \begin{vmatrix} x_1 & x_2 & x_3 \\ y_1 & y_2 & y_3 \\ 1 & 1 & 1 \end{vmatrix}.$$

5.42: We have $f_1(x) = bx/a$ and $f_2(x) = b(x-1)/(a-1)$, so

$$\int_0^1 f(x)\, dx = \int_0^a \frac{bx}{a}\, dx + \int_a^1 \frac{b(x-1)}{a-1}\, dx = \left[\frac{bx^2}{2a}\right]_0^a + \left[\frac{b}{a-1}\left(\frac{x^2}{2} - x\right)\right]_a^1 = \cdots = \frac{1}{2}b.$$

Also, putting $z_1 = 0$, $z_2 = 1$, and $z_3 = a + ib$, we have $\bar{z}_1 z_2 + \bar{z}_2 z_3 + \bar{z}_3 z_1 = a + ib$, whence $\check{\Delta} = \frac{1}{2}b$ also.

5.43: If A_i has cartesian coordinates (a_i, b_i), $i = 1, 2, 3$, and P has cartesian coordinates (x, y), and if $\xi + \eta + \zeta = 1$, then equations (4.2) give

$$\begin{pmatrix} a_1 & a_2 & a_3 \\ b_1 & b_2 & b_3 \\ 1 & 1 & 1 \end{pmatrix} \begin{pmatrix} \xi \\ \eta \\ \zeta \end{pmatrix} = \begin{pmatrix} x \\ y \\ 1 \end{pmatrix},$$

or

$$\begin{pmatrix} a_1 & a_2 & a_3 \\ b_1 & b_2 & b_3 \\ 1 & 1 & 1 \end{pmatrix} \begin{pmatrix} \xi & 0 & 0 \\ \eta & 1 & 0 \\ \zeta & 0 & 1 \end{pmatrix} = \begin{pmatrix} x & a_2 & a_3 \\ y & b_2 & b_3 \\ 1 & 1 & 1 \end{pmatrix}.$$

Taking determinants gives $\xi = \check{\Delta}(PA_2 A_3)/\check{\Delta}(A_1 A_2 A_3)$, and similarly for η and ζ. (In effect, we have proved *Cramer's rule* for solving simultaneous linear equations: see [6].)

5.44: Taking $\triangle A_1 A_2 A_3$ as triangle of reference, we have $D_1 = 2(0, 0, 1) + 1(0, 1, 0) = (0, 1, 2)$; so $A_1 D_1$ is $\zeta = 2\eta$, and similarly $A_2 D_2$ is $\xi = 2\zeta$, and $A_3 D_3$ is $\eta = 2\xi$. Thus $P_1 = (2/7, 4/7, 1/7)$, $P_2 = (1/7, 2/7, 4/7)$ and $P_3 = (4/7, 1/7, 2/7)$, where we have divided the 'obvious' solutions through by 7 to ensure $\xi + \eta + \zeta = 1$ in each case. Then, if in cartesian coordinates $A_i = (a_i, b_i)$ and $P_i = (x_i, y_i)$, for each i, we have

$$\begin{pmatrix} a_1 & a_2 & a_3 \\ b_1 & b_2 & b_3 \\ 1 & 1 & 1 \end{pmatrix} \begin{pmatrix} 2/7 & 1/7 & 4/7 \\ 4/7 & 2/7 & 1/7 \\ 1/7 & 4/7 & 2/7 \end{pmatrix} = \begin{pmatrix} x_1 & x_2 & x_3 \\ y_1 & y_2 & y_3 \\ 1 & 1 & 1 \end{pmatrix}.$$

Taking determinants, the one on the left is twice the area of $\triangle A_1 A_2 A_3$; the one on the right is twice the area of $\triangle P_1 P_2 P_3$; and the one in the middle is

$$7^{-3} \begin{vmatrix} 2 & 1 & 4 \\ 4 & 2 & 1 \\ 1 & 4 & 2 \end{vmatrix} = 7^{-3}.49 = \frac{1}{7}.$$

5.45: Let f, g be isometries with $f(X) = X'$ and $g(X) = X'$. Then $h = fg^{-1}$ is an isometry with $h(X') = fg^{-1}(X') = f(X) = X'$, that is, $h \in S(X')$; and $f = hg$. Conversely, given isometries g with $g(X) = X'$ and $h \in S(X')$, we can define the isometry f by $f = hg$, and then we have $f(X) = hg(X) = h(X') = X'$. So the number of such f is $|S(X')|$, or 8.

5.46: There is a (unique) isometry f with $f(A) = A'$, $f(B) = B'$, $f(C) = C'$, and $f(D) = D'$. If f is a translation, then AA', BB', CC', DD' are all parallel, which the question excludes. Likewise, if f is a reflection, then AA', BB', CC', DD' are all perpendicular to the axis of f, and so are parallel. So f must be a rotation or a glide-reflection. If it is a rotation, then the perpendicular bisectors of AA', BB', CC', DD' pass through the centre of rotation. If it is a glide-reflection, then the mid-points of AA', BB', CC', DD' lie on its axis.

5.47: $g = f_n f_{n-1} \cdots f_2 f_1$ is direct, and $n\pi(1 - 2/n) = \pi(n - 2)$, so that, if n is odd, g is a half-turn, and if n is even, g is either a translation or trivial. The easier case to deal with is when $A_1 A_2 \ldots A_n$ is labelled negatively, for then $f_1(A_1) = A_1$, $f_2(A_1) = A_3$, $f_3(A_3) = A_3$, $f_4(A_3) = A_5, \ldots$, and putting this all together, $g(A_1) = A_1$ if n is even, and $g(A_1) = A_n$ if n is odd. So $g = 1$ if n is even, and g is a half-turn about the mid-point of $A_1 A_n$ if n is odd.

Now suppose $A_1 A_2 \ldots A_n$ is labelled positively. With subscripts modulo n (so that $A_0 = A_n$, etc.), define $B_i = A_{i-2} A_{i-1} \cdot A_i A_{i+1}$, for $1 \leq i \leq n$. So $B_1 B_2 \ldots B_n$ is a regular n-gon, and $\angle B_i A_i B_{i+1} = \angle A_{i+1} A_i A_{i-1} = \pi(1 - \frac{2}{n})$, for all i. Further, $|A_i B_i| = |A_i B_{i+1}|$, for all i, so that $f_i(B_i) = B_{i+1}$, for all i. Thus $g(B_1) = B_1$, so that, when n is even, $g = 1$, and when n is odd, g is a half-turn about B_1.

Two cases need special mention. When $n = 3$, $B_1 = A_{-1} A_0 \cdot A_1 A_2 = A_2 A_3 \cdot A_1 A_2 = A_2$, so in this case g is a half-turn about A_2. When $n = 4$, $A_{-1} A_0 = A_3 A_4 \parallel A_1 A_2$, so B_1 is undefined, and we need a separate argument. We have $f_1(A_1) = A_1$; $f_2(A_1) = C_2$, where $\overrightarrow{A_2 C_2} = \overrightarrow{A_3 A_2}$; $f_3(C_2) = C_3$, where $\overrightarrow{A_3 C_3} = 2\overrightarrow{A_4 A_3}$; and $f_4(C_3) = C_4$, where $\overrightarrow{A_4 C_4} = 3\overrightarrow{A_1 A_4}$, or $\overrightarrow{A_1 C_4} = 4\overrightarrow{A_1 A_4}$. Since $g(A_1) = C_4$, we see that in this case alone, g is a translation, through $4\overrightarrow{A_1 A_4}$.

5.48: Let order$(g) = k$. So $1, g, g^2, \ldots, g^{k-1}$ are distinct, and $g^k = 1$. Then $m = kq + r$ for some $q, r \in \mathbb{Z}$ with $0 \leq r < k$, and $1 = g^m = g^{kq+r} = (g^k)^q g^r = 1^q g^r = g^r$, so that $r = 0$ and k divides m.

5.49: No; for example, if $|G| = n$ and $g \in G$, then $g^n = 1$, so if $m = 2n$ then $g^m = 1$ also; but clearly m does not divide n. For an example where $m < n$, take G cyclic of order 6, generated by h, and put $g = h^3$, so that order$(g) = 2$. If we now put $m = 4$, then $g^4 = (g^2)^2 = 1$, but 4 does not divide 6.

5.50: The proof is exactly the same: if $z \in G$, with $|G| = n$, then $z^n = 1$, so $z \in P_n$ and thus $G \subseteq P_n$. But $|P_n| = n$, so $G = P_n$, which is cyclic.

6

Isometries of $\mathbb{R}^n$

In this chapter we shall make a brief excursion into n-dimensional geometry, in order to see how isometries can be written down, and to see a few examples of what can happen in dimensions higher than 2. The methods will be via matrices, so we first rewrite our isometries of $\mathbb{R}^2$ in matrix form.

6.1   MATRIX FORMS FOR ISOMETRIES OF $\mathbb{R}^2$

As usual, we think of $\mathbb{R}^2$ as being the 'same' as $\mathbb{C}$. Let $f : \mathbb{C} \to \mathbb{C}$ be an isometry, given by $f(z) = a\hat{z} + b$, for all $z \in \mathbb{C}$. Here $a, b \in \mathbb{C}$ with $|a| = 1$, and $\hat{z}$ means z or $\bar{z}$, so that if $z = x + iy$, then $\hat{z} = x \pm iy$, where of course $x, y \in \mathbb{R}$. Let $a = \cos\theta + i\sin\theta$, and let $b = b_1 + ib_2$, where $b_1, b_2 \in \mathbb{R}$. We thus have

$$f(x + iy) = (\cos\theta + i\sin\theta)(x \pm iy) + (b_1 + ib_2),$$

or, in vector form,

$$f : (x, y) \mapsto ((\cos\theta)x \mp (\sin\theta)y + b_1, (\sin\theta)x \pm (\cos\theta)y + b_2).$$

Where there is a choice of sign, the convention is that the top sign is chosen throughout, or else the bottom sign is chosen throughout. In terms of matrices, our equation can now be written

$$f\begin{pmatrix} x \\ y \end{pmatrix} = \begin{pmatrix} \cos\theta & \mp\sin\theta \\ \sin\theta & \pm\cos\theta \end{pmatrix} \begin{pmatrix} x \\ y \end{pmatrix} + \begin{pmatrix} b_1 \\ b_2 \end{pmatrix},$$

that is, $f(\mathbf{v}) = \mathbf{A}\mathbf{v} + \mathbf{b}$, where

$$\mathbf{v} = \begin{pmatrix} x \\ y \end{pmatrix}, \quad \mathbf{A} = \begin{pmatrix} \cos\theta & \mp\sin\theta \\ \sin\theta & \pm\cos\theta \end{pmatrix}, \quad \text{and } \mathbf{b} = \begin{pmatrix} b_1 \\ b_2 \end{pmatrix}.$$

Note that, if f is direct, then $\mathbf{A} = \begin{pmatrix} \cos\theta & -\sin\theta \\ \sin\theta & \cos\theta \end{pmatrix}$ and so $\det \mathbf{A} = 1$; whereas if f is opposite, then $\mathbf{A} = \begin{pmatrix} \cos\theta & \sin\theta \\ \sin\theta & -\cos\theta \end{pmatrix}$, and then $\det \mathbf{A} = -1$. Notice also that the rows of $\mathbf{A}$, namely $(\cos\theta, \mp\sin\theta)$ and $(\sin\theta, \pm\cos\theta)$ are unit vectors which are orthogonal to each other: $(\cos\theta, \mp\sin\theta) \cdot (\sin\theta, \pm\cos\theta) = 0$, for either choice of signs; and similar remarks apply to the columns. Such a matrix $\mathbf{A}$ is called an *orthogonal matrix*, and satisfies the conditions $\mathbf{A}\mathbf{A}^T = \mathbf{I} = \mathbf{A}^T\mathbf{A}$, or $\mathbf{A}^T = \mathbf{A}^{-1}$. Here $\mathbf{A}^T$ is the *transpose* of $\mathbf{A}$: $(\mathbf{A}^T)_{ij} = (\mathbf{A})_{ji}$, for all i, j. There will be more about orthogonal matrices in the next section.

The map f is not in general linear, but is the composite of a linear map $\mathbf{v} \mapsto \mathbf{w} = \mathbf{A}\mathbf{v}$ and a translation $\mathbf{w} \mapsto \mathbf{w} + \mathbf{b}$. It *is* possible nonetheless, by low cunning, to obtain f as a linear map, by the device of placing $\mathbb{R}^2$ inside $\mathbb{R}^3$ as the plane $z = 1$, that is, by identifying $(x, y) \in \mathbb{R}^2$ with $(x, y, 1) \in \mathbb{R}^3$. The formula for f can then be written $f(\mathbf{u}) = \mathbf{M}\mathbf{u}$, where

$$\mathbf{u} = \begin{pmatrix} x \\ y \\ 1 \end{pmatrix} \quad \text{and} \quad \mathbf{M} = \begin{pmatrix} \cos\theta & \mp\sin\theta & b_1 \\ \sin\theta & \pm\cos\theta & b_2 \\ 0 & 0 & 1 \end{pmatrix},$$

so that f is the restriction to the plane $z = 1$ of a linear map $\mathbb{R}^3 \to \mathbb{R}^3$. The 3×3 matrix $\mathbf{M}$ of this map is *not* orthogonal (unless $\mathbf{b} = \mathbf{0}$), but (as is readily seen) it does inherit the property of $\mathbf{A}$ that its determinant is $+1$ or -1 according as f is direct or opposite.

- Ex.6.1: *Prove that the map* $\psi : f \mapsto \mathbf{M}$ *is an isomorphism between* $\mathcal{I}(\mathbb{R}^2)$ *and a certain subgroup of the* general linear *group* $GL_3(\mathbb{R})$ *of all* 3×3 *invertible real matrices, and hence use the composite map* $(\det)(\psi) : \mathcal{I}(\mathbb{R}^2) \to P_2$ *to give a second proof of Exercise 5.27.*

We have met this way of putting $\mathbb{R}^2$ inside $\mathbb{R}^3$ before, in Section 4.8; see, for example, the proof of Proposition 4.25.

6.2   MATRIX FORMS FOR ISOMETRIES OF $\mathbb{R}^n$

We shall now try to make the previous section work in $\mathbb{R}^n$; timid readers may stick to $n \leq 3$ if they wish, but the methods work for any n.

Definition 6.1.   *Let* $(u_1, u_2, \ldots, u_n)$, $(v_1, v_2, \ldots, v_n) \in \mathbb{R}^n$. *The* **dot product** *of these vectors is*

$$(u_1, u_2, \ldots, u_n) \cdot (v_1, v_2, \ldots, v_n) = u_1 v_1 + u_2 v_2 + \cdots + u_n v_n.$$

We leave it to the reader to check that the dot product is *bilinear* (linear in each argument):

$$(\lambda\mathbf{u} + \mu\mathbf{v}) \cdot \mathbf{w} = \lambda(\mathbf{u} \cdot \mathbf{w}) + \mu(\mathbf{v} \cdot \mathbf{w})$$
$$\mathbf{u} \cdot (\lambda\mathbf{v} + \mu\mathbf{w}) = \lambda(\mathbf{u} \cdot \mathbf{v}) + \mu(\mathbf{u} \cdot \mathbf{w}),$$

for all $\mathbf{u}$, $\mathbf{v}$, $\mathbf{w} \in \mathbb{R}^n$ and λ, $\mu \in \mathbb{R}$; and also *symmetric*:

$$\mathbf{u} \cdot \mathbf{v} = \mathbf{v} \cdot \mathbf{u}$$

for all $\mathbf{u}, \mathbf{v} \in \mathbb{R}^n$; and it is *positive definite*:

$$\mathbf{u} \cdot \mathbf{u} > 0$$

whenever $\mathbf{u} \neq \mathbf{0}$. Length $|\mathbf{u}|$ is now defined by $|\mathbf{u}|^2 = \mathbf{u} \cdot \mathbf{u} = u_1^2 + u_2^2 + \ldots + u_n^2$, and the distance from the point $\mathbf{u}$ to the point $\mathbf{v}$ is just $|\mathbf{u} - \mathbf{v}|$. What we are saying is that we want translation to be an isometry, so the distance from $\mathbf{u}$ to $\mathbf{v}$ must be the same as the distance from $\mathbf{u} + \mathbf{w}$ to $\mathbf{v} + \mathbf{w}$, for all $\mathbf{w}$; now put $\mathbf{w} = -\mathbf{v}$.

There are many things to check before one can be sure that all this makes sense. For example, given three points $\mathbf{a}, \mathbf{b}, \mathbf{c}$, put $\mathbf{u} = \mathbf{b} - \mathbf{a}$ and $\mathbf{v} = \mathbf{c} - \mathbf{a}$. To say that $\mathbf{a}, \mathbf{b}, \mathbf{c}$ are collinear is the same as saying that $\mathbf{u}, \mathbf{v}$ are linearly dependent, that is, one is a scalar multiple of the other. If this is *not* the case, then $\mathbf{a}, \mathbf{b}, \mathbf{c}$ determine a *plane*, consisting of all $\mathbf{p} = \mathbf{a} + x\mathbf{u} + y\mathbf{v}$ with $x, y \in \mathbb{R}$. It would be very sad, not to say awkward, if the geometry of such a plane were not the 'same' as the geometry of $\mathbb{R}^2$ that we have been studying. To see that it *is* the same, we need to set up coordinates, essentially by taking $\mathbf{a}$ as origin and letting (x, y) be the coordinates of $\mathbf{p}$, as above. The trouble with this is that the axes are probably at some strange angle to each other, and we want them to be orthogonal. But this can be fixed:

- Ex.6.2: *Show that* $(\lambda_1, \mu_1), (\lambda_2, \mu_2) \in \mathbb{R}^2$ *can be chosen so that* $\lambda_i\mathbf{u} + \mu_i\mathbf{v}$, $i = 1, 2$, *are mutually orthogonal (that is, their dot product vanishes) and of unit length.*

If we now replace $\mathbf{b}$ by $\mathbf{a} + \lambda_1\mathbf{u} + \mu_1\mathbf{v}$ and $\mathbf{c}$ by $\mathbf{a} + \lambda_2\mathbf{b} + \mu_2\mathbf{c}$, then we have the same plane but the axes are now orthogonal. In this plane, the distance from $\mathbf{p} = (x, y)$ to the origin $\mathbf{a}$ is given by

$$\begin{aligned}
|\mathbf{p} - \mathbf{a}|^2 = |x\mathbf{u} + y\mathbf{v}|^2 &= (x\mathbf{u} + y\mathbf{v}) \cdot (x\mathbf{u} + y\mathbf{v}) \\
&= x^2\mathbf{u} \cdot \mathbf{u} + 2xy\mathbf{u} \cdot \mathbf{v} + y^2\mathbf{v} \cdot \mathbf{v} = x^2 + y^2,
\end{aligned}$$

since we have fixed it that $|\mathbf{u}| = 1 = |\mathbf{v}|$ and $\mathbf{u}.\mathbf{v} = 0$. This is exactly what we wanted: the distance function in this plane is the familiar distance function we have used in $\mathbb{R}^2$. A similar calculation shows that, if $m < n$, then $m + 1$ points in general position determine a copy of $\mathbb{R}^m$ inside $\mathbb{R}^n$, with the same geometry as $\mathbb{R}^m$ itself. Here, 'in general position' means that if the first vector is subtracted from the others, we obtain a set of m linearly independent vectors.

An *isometry* of $\mathbb{R}^n$ is just a length-preserving transformation $f : \mathbb{R}^n \to \mathbb{R}^n$, that is, $|f(\mathbf{u}) - f(\mathbf{v})| = |\mathbf{u} - \mathbf{v}|$ for all $\mathbf{u}, \mathbf{v} \in \mathbb{R}^n$. The definitions here are the same as Definitions 3.17 and 3.18, except that $\mathbb{R}^2$ has been replaced by $\mathbb{R}^n$ throughout.

Example. The map $\mathbf{u} \mapsto \mathbf{u} + \mathbf{b}$, where $\mathbf{b}$ is a constant vector, is a *translation*, and is an isometry.

Lemma 6.2. *Every isometry of* $\mathbb{R}^n$ *is of the form* $f : \mathbf{u} \mapsto \mathbf{Au} + \mathbf{b}$ *for some* $n \times n$ *real matrix* $\mathbf{A}$ *and some* $\mathbf{b} \in \mathbb{R}^n$. *That is,* f *is the composite of a linear transformation* $\mathbf{u} \mapsto \mathbf{Au}$ *and a translation* $\mathbf{v} \mapsto \mathbf{v} + \mathbf{b}$.

Proof. Put $\mathbf{b} = f(\mathbf{0})$, and define $g(\mathbf{u}) = f(\mathbf{u}) - \mathbf{b}$. So g is an isometry, being the composite of f and a translation, and also $g(\mathbf{0}) = \mathbf{0}$. We must show that g is linear.

Let $\mathbf{u} \in \mathbb{R}^n$ and $\lambda \in \mathbb{R}$. The point $\lambda \mathbf{u}$ divides the line joining $\mathbf{0}$ to $\mathbf{u}$ in the ratio $\lambda : 1 - \lambda$, and so, since g is an isometry, the point $g(\lambda \mathbf{u})$ divides the line joining $g(\mathbf{0}) = \mathbf{0}$ to $g(\mathbf{u})$ in the ratio $\lambda : 1 - \lambda$. But also $\lambda g(\mathbf{u})$ divides the line joining $\mathbf{0}$ to $g(\mathbf{u})$ in the ratio $\lambda : 1 - \lambda$, and we are forced to conclude that $g(\lambda \mathbf{u}) = \lambda g(\mathbf{u})$.

Again, let $\mathbf{u}, \mathbf{v} \in \mathbb{R}^n$. The mid-point of the line from $2\mathbf{u}$ to $2\mathbf{v}$ is $\mathbf{u} + \mathbf{v}$, and so, since g is an isometry, the mid-point of the line from $g(2\mathbf{u})$ to $g(2\mathbf{v})$ is $g(\mathbf{u}+\mathbf{v})$. But, as we have just seen, $g(2\mathbf{u}) = 2g(\mathbf{u})$ and $g(2\mathbf{v}) = 2g(\mathbf{v})$. The mid-point of the line joining $2g(\mathbf{u})$ to $2g(\mathbf{v})$ is $g(\mathbf{u}) + g(\mathbf{v})$, and so we are forced to conclude that $g(\mathbf{u} + \mathbf{v}) = g(\mathbf{u}) + g(\mathbf{v})$.

We have thus shown that $g : \mathbb{R}^n \to \mathbb{R}^n$ is a linear map, and it follows by a standard argument from linear algebra that $g(\mathbf{u}) = \mathbf{A}\mathbf{u}$ for some $n \times n$ matrix $\mathbf{A}$. Explicitly, we agree to write our vectors as columns, or $n \times 1$ matrices, and we put

$$\mathbf{e}_1 = \begin{pmatrix} 1 \\ 0 \\ 0 \\ \vdots \\ 0 \end{pmatrix}, \quad \mathbf{e}_2 = \begin{pmatrix} 0 \\ 1 \\ 0 \\ \vdots \\ 0 \end{pmatrix}, \quad \ldots, \quad \mathbf{e}_n = \begin{pmatrix} 0 \\ 0 \\ \vdots \\ 0 \\ 1 \end{pmatrix},$$

the *standard basis* of $\mathbb{R}^n$. (Note that $\mathbf{e}_i$ is just the ith column of the identity matrix $\mathbf{I}_n$.) The matrix $\mathbf{A}$ is now defined to be the matrix whose columns are $g(\mathbf{e}_1), g(\mathbf{e}_2), \ldots, g(\mathbf{e}_n)$, and it is easy to check that $g(\mathbf{u}) = \mathbf{A}\mathbf{u}$ for all $\mathbf{u} \in \mathbb{R}^n$.   □

We now ask: for which matrices $\mathbf{A}$ is the map $\mathbf{u} \mapsto \mathbf{A}\mathbf{u}$ an isometry?

Proposition 6.3. *Let $\mathbf{A}$ be an $n \times n$ real matrix. The following conditions are equivalent:*

(i) $\mathbf{u} \mapsto \mathbf{A}\mathbf{u}$ *is an isometry of $\mathbb{R}^n$.*
(ii) $|\mathbf{u}| = |\mathbf{A}\mathbf{u}|$, *for all $\mathbf{u} \in \mathbb{R}^n$.*
(iii) $\mathbf{u} \cdot \mathbf{v} = (\mathbf{A}\mathbf{u}) \cdot (\mathbf{A}\mathbf{v})$, *for all $\mathbf{u}, \mathbf{v} \in \mathbb{R}^n$.*
(iv) $\mathbf{A}^T \mathbf{A} = \mathbf{I}_n$.
(v) $\mathbf{A}$ *is invertible, with $\mathbf{A}^{-1} = \mathbf{A}^T$.*
(vi) $\mathbf{A}\mathbf{A}^T = \mathbf{I}_n$.

Proof. (v) $\Longrightarrow$ (iv): $\mathbf{A}^T \mathbf{A} = \mathbf{A}^{-1}\mathbf{A} = \mathbf{I}_n$.

(iv) $\Longrightarrow$ (v): If $\mathbf{A}^T \mathbf{A} = \mathbf{I}_n$, then $(\det \mathbf{A}^T)(\det \mathbf{A}) = 1$, or $(\det \mathbf{A})^2 = 1$, and $\det \mathbf{A} = \pm 1$. So $\mathbf{A}$ is invertible, and $\mathbf{A}^{-1} = \mathbf{I}_n \mathbf{A}^{-1} = (\mathbf{A}^T \mathbf{A})\mathbf{A}^{-1} = \mathbf{A}^T(\mathbf{A}\mathbf{A}^{-1}) = \mathbf{A}^T \mathbf{I}_n = \mathbf{A}^T$.

(v) $\Longleftrightarrow$ (vi) is done similarly.

(i) $\Longrightarrow$ (ii): $\mathbf{u} \mapsto \mathbf{A}\mathbf{u}$ is an isometry, so $|\mathbf{A}\mathbf{u} - \mathbf{A}\mathbf{0}| = |\mathbf{u} - \mathbf{0}|$, or $|\mathbf{A}\mathbf{u}| = |\mathbf{u}|$, for all $\mathbf{u}$.

(ii) $\Longrightarrow$ (i): $|\mathbf{A}\mathbf{u}| = |\mathbf{u}|$, for all $\mathbf{u}$, so for any $\mathbf{u}, \mathbf{v}$, we have $|\mathbf{u} - \mathbf{v}| = |\mathbf{A}(\mathbf{u} - \mathbf{v})| = |\mathbf{A}\mathbf{u} - \mathbf{A}\mathbf{v}|$, and we have an isometry.

(ii) $\Longrightarrow$ (iii): Note that $|\mathbf{u} + \mathbf{v}|^2 = (\mathbf{u} + \mathbf{v}) \cdot (\mathbf{u} + \mathbf{v}) = \mathbf{u} \cdot \mathbf{u} + 2\mathbf{u} \cdot \mathbf{v} + \mathbf{v} \cdot \mathbf{v} = |\mathbf{u}|^2 + 2\mathbf{u} \cdot \mathbf{v} + |\mathbf{v}|^2$, so that $\mathbf{u} \cdot \mathbf{v} = \frac{1}{2}(|\mathbf{u} + \mathbf{v}|^2 - |\mathbf{u}|^2 - |\mathbf{v}|^2)$. Thus

$$(\mathbf{A}\mathbf{u}) \cdot (\mathbf{A}\mathbf{v}) = \tfrac{1}{2}(|\mathbf{A}\mathbf{u} + \mathbf{A}\mathbf{v}|^2 - |\mathbf{A}\mathbf{u}|^2 - |\mathbf{A}\mathbf{v}|^2)$$

$$= \tfrac{1}{2}(|\mathbf{A}(\mathbf{u} + \mathbf{v})|^2 - |\mathbf{A}\mathbf{u}|^2 - |\mathbf{A}\mathbf{v}|^2)$$
$$= \tfrac{1}{2}(|\mathbf{u} + \mathbf{v}|^2 - |\mathbf{u}|^2 - |\mathbf{v}|^2)$$
$$= \mathbf{u} \cdot \mathbf{v}.$$

(iii) $\implies$ (ii): $|\mathbf{u}|^2 = \mathbf{u} \cdot \mathbf{u} = (\mathbf{A}\mathbf{u}) \cdot (\mathbf{A}\mathbf{u}) = |\mathbf{A}\mathbf{u}|^2$. Take the non-negative square root of each side.

(iii) $\implies$ (iv): The standard basis $\mathbf{e}_1, \mathbf{e}_2, \dots, \mathbf{e}_n$ of $\mathbb{R}^n$ (introduced in Lemma 6.2 above) is *orthonormal*:

$$\mathbf{e}_i \cdot \mathbf{e}_j = \delta_{ij} = \begin{cases} 1 & \text{if } i = j, \\ 0 & \text{if } i \neq j. \end{cases}$$

(δ_{ij} is the *Kronecker delta*, and is the (i, j)th entry of the identity matrix, $\mathbf{I}_n$.) Put $\mathbf{a}_i = \mathbf{A}\mathbf{e}_i$, the ith column of $\mathbf{A}$; so $\mathbf{a}_i^T = \mathbf{e}_i^T \mathbf{A}^T$, the ith row of $\mathbf{A}^T$. Since $\mathbf{e}_i \cdot \mathbf{e}_j = \delta_{ij}$, then also $\mathbf{a}_i \cdot \mathbf{a}_j = \delta_{ij}$, by (iii). This just says that, as matrices, $\mathbf{a}_i^T \mathbf{a}_j = \delta_{ij}$, or $\mathbf{A}^T \mathbf{A} = \mathbf{I}_n$, as required.

(iv) $\implies$ (iii): For any $\mathbf{u}$, $\mathbf{v}$,

$$\begin{aligned} (\mathbf{A}\mathbf{u}) \cdot (\mathbf{A}\mathbf{v}) &= (\mathbf{A}\mathbf{u})^T (\mathbf{A}\mathbf{v}) \quad \text{(the matrix product)} \\ &= (\mathbf{u}^T \mathbf{A}^T)(\mathbf{A}\mathbf{v}) = \mathbf{u}^T (\mathbf{A}^T \mathbf{A})\mathbf{v} = \mathbf{u}^T \mathbf{I}_n \mathbf{v} \quad \text{(by (iv))} \\ &= \mathbf{u}^T \mathbf{v} = \mathbf{u} \cdot \mathbf{v}, \quad \text{as required.} \end{aligned}$$

This completes the proof of the proposition. $\qquad\square$

Definition 6.4.   *The invertible matrix* $\mathbf{A}$ *is an* **orthogonal matrix** *if* $\mathbf{A}^{-1} = \mathbf{A}^T$.

So Proposition 6.3 gives six ways of recognizing an orthogonal matrix. In particular, (iv) and (vi) tell us that $\mathbf{A}$ is orthogonal iff the columns of $\mathbf{A}$ are an orthonormal set (that is, each of unit length, and mutually orthogonal); and similarly for the rows of $\mathbf{A}$.

● Ex.6.3:   *Let* $x_1, x_2, x_3 \in \mathbb{R}$, *not all zero, and let* $y_1, y_2, y_3 \in \mathbb{R}$, *not all zero, and suppose* $x_1 + x_2 + x_3 = 0$, $y_1 + y_2 + y_3 = 0$, *and* $x_1 y_1 + x_2 y_2 + x_3 y_3 = 0$. *Prove that*

$$\frac{x_1^2}{x_1^2 + x_2^2 + x_3^2} + \frac{y_1^2}{y_1^2 + y_2^2 + y_3^2} = \frac{2}{3}, \quad and$$

$$\frac{x_1 x_2}{x_1^2 + x_2^2 + x_3^2} + \frac{y_1 y_2}{y_1^2 + y_2^2 + y_3^2} = -\frac{1}{3}.$$

(There is a trick that makes this easy. If you try to do it by simply flogging the given equations around, you might take some time.)

Recall that, if G, H are (multiplicative) groups, then a map $\alpha : G \to H$ is a *homomorphism* if $\alpha(ab) = \alpha(a)\alpha(b)$ for all $a, b \in G$. From this, $\alpha(a) = \alpha(1a) = \alpha(1)\alpha(a)$, so that $\alpha(1) = 1$. Here the 1 on the left is in G and the 1 on the right is in H; it would be less confusing (but tedious) to write $\alpha(1_G) = 1_H$.

Definition 6.5.   *With G, H, α as above, the* **kernel** *of α is* $\ker \alpha = \{a \in G : \alpha(a) = 1\}$.

- Ex.6.4:   *The subset $\ker \alpha$ is a subgroup of G; and α is injective iff $\ker \alpha = \{1\}$.*

Lemma 6.6.   *The set of all $n \times n$ orthogonal matrices form a group, the $n \times n$* **orthogonal group**, *O_n. The subset of all $\mathbf{A} \in O_n$ with $\det \mathbf{A} = 1$ form a subgroup, the $n \times n$* **special orthogonal group**, *SO_n.*

Proof.   Clearly $\mathbf{I}_n \in O_n$. If $\mathbf{A}, \mathbf{B} \in O_n$, then $(\mathbf{AB})^T(\mathbf{AB}) = \mathbf{B}^T \mathbf{A}^T \mathbf{AB} = \mathbf{B}^T \mathbf{B} = \mathbf{I}_n$, so $\mathbf{AB} \in O_n$. Also $(\mathbf{A}^{-1})^T \mathbf{A}^{-1} = (\mathbf{A}^T)^T \mathbf{A}^{-1} = \mathbf{AA}^{-1} = \mathbf{I}_n$, so $\mathbf{A}^{-1} \in O_n$. Thus O_n is a group, and $\det : O_n \to P_2$ is a homomorphism (OK?); and SO_n is its kernel, so the last part follows by Exercise 6.4. $\square$

We can now combine Lemma 6.2 and Proposition 6.3:

Corollary 6.7.   *Every isometry of $\mathbb{R}^n$ is of the form $\mathbf{u} \mapsto \mathbf{Au} + \mathbf{b}$ for some $\mathbf{A} \in O_n$ and $\mathbf{b} \in \mathbb{R}^n$.* $\square$

Definition 6.8.   *The isometry $\mathbf{u} \mapsto \mathbf{Au} + \mathbf{b}$ is* **direct** *if $\mathbf{A} \in SO_n$, and* **opposite** *otherwise.*

This agrees with previous usage when $n = 2$. If we are given a non-empty subset $X \subseteq \mathbb{R}^n$, then we can define its symmetries (cf. Definition 5.16) to be those isometries f of $\mathbb{R}^n$ such that $f(X) = X$. The symmetries of X form a group (cf. Proposition 5.17), its *symmetry group*, $\mathcal{S}(X)$, and the subgroup of direct isometries form its *rotation group*, $\mathcal{S}^+(X)$. If we take $X = \{\mathbf{0}\}$, then the isometry $\mathbf{u} \mapsto \mathbf{Au} + \mathbf{b}$ is a symmetry of X iff $\mathbf{b} = \mathbf{0}$, or in other words, $\mathcal{S}(X) \cong O_n$; and hence also $\mathcal{S}^+(X) \cong SO_n$. The reader may now protest that surely $O_1, O_2, O_3, \ldots$ are all very different groups—this is true—and surely a one-point set ought to look the 'same' whatever dimension we are working in—also true—so how can $\mathcal{S}(X)$ vary with n? The answer is that the variation is intentional, but the notation is bad: the definition of $\mathcal{S}(X)$ starts by insisting that $X \subseteq \mathbb{R}^n$, so that n is chosen along with X: the symmetries of X depend on the space in which we are working. To avoid ambiguities, we perhaps ought to complicate the notation by writing something like $\mathcal{S}(X, \mathbb{R}^n)$ instead of $\mathcal{S}(X)$; but most of the time the value of n will be clear, and so we shall prefer the simpler notation $\mathcal{S}(X)$.

- Ex.6.5:   *Show that $SO_2 \cong T$, the circle group.*
- Ex.6.6:   *Put $S^n = \{\mathbf{u} \in \mathbb{R}^{n+1} : |\mathbf{u}| = 1\}$, the n-sphere. Prove that $\mathcal{S}(S^n) \cong O_{n+1}$ and $\mathcal{S}^+(S^n) \cong SO_{n+1}$.*

The $n + 1$ in the definition of S^n is not a mistake. One might have expected the circle $x^2 + y^2 = 1$ to be called a 2-sphere, since it needs $\mathbb{R}^2$ for its description, but the topologists got there ahead of us and decided that the (same) circle, $x = \cos \theta$, $y = \sin \theta$, needs only *one* parameter (θ) to describe it, so ought to be called the 1-sphere; and so on.

- Ex.6.7:   *Let* $X \subseteq \mathbb{R}$ *be the interval* $[-1, 1]$. *Prove that* $|\mathcal{S}(X)| = 2$. *Now put* $\mathbb{R}$
    inside $\mathbb{R}^2$ *as the x-axis, so that* $X \subseteq \mathbb{R}^2$ *is the line segment joining* $(-1, 0)$
    to $(1, 0)$. *Prove that we then have* $|\mathcal{S}(X)| = 4$. *What happens to* $|\mathcal{S}(X)|$ *if*
    we now put $\mathbb{R}^2$ *inside* $\mathbb{R}^3$, *so that* $X \subseteq \mathbb{R}^3$?

- Ex.6.8:   *Let* $X \subseteq \mathbb{R}^2$ *be the square with vertices* $(\pm 1, \pm 1)$. *We know that* $\mathcal{S}(X) = D_4$,
    so that $|\mathcal{S}(X)| = 8$. *Now suppose* $\mathbb{R}^2$ *is put inside* $\mathbb{R}^3$ *as the* (x, y)-*plane*.
    Prove that $|\mathcal{S}(X)| = 16$. *(Notice that the reflection of* $\mathbb{R}^2$ *in the y-axis,*
    $(x, y) \mapsto (-x, y)$, *which is a symmetry of X, gives rise to two symmetries*
    in $\mathbb{R}^3$, *namely reflection in the* (y, z)-*plane,* $(x, y, z) \mapsto (-x, y, z)$, *and*
    the half-turn about the y-axis, $(x, y, z) \mapsto (-x, y, -z)$. *What do the two*
    symmetries in $\mathbb{R}^3$ *that arise from (say) a quarter-turn symmetry in* $\mathbb{R}^2$ *look*
    like?)

- Ex.6.9:   *Let* $X \subseteq \mathbb{R}^3$ *be the cube with vertices* $(\pm 1, \pm 1, \pm 1)$. *Prove that* $|\mathcal{S}(X)| =$
    48. *(Hint: it is fairly easy, by playing with a cube, to see that* $|\mathcal{S}^+(X)| = 24$,
    and you should then look at the proof of Leonardo da Vinci's theorem to
    see how you might show that $|\mathcal{S}(X)| = 2|\mathcal{S}^+(X)|$.*) Now place our cube*
    X inside $\mathbb{R}^4$, *so that its vertices are* $(\pm 1, \pm 1, \pm 1, 0)$. *Is it now true that*
    $|\mathcal{S}(X)| = 96$? *(You are* not *being asked to attempt to make a list of 96*
    isometries of $\mathbb{R}^4$!)

6.3   CLASSIFICATION OF ISOMETRIES

Recall from linear algebra that an *eigenvalue* of an $n \times n$ matrix $\mathbf{A}$ is a scalar λ such that
$\mathbf{Au} = \lambda\mathbf{u}$ for some non-zero vector $\mathbf{u}$, called an *eigenvector* of $\mathbf{A}$. Since the equation
$\mathbf{Au} = \lambda\mathbf{u}$ can be rewritten $(\lambda\mathbf{I}_n - \mathbf{A})\mathbf{u} = \mathbf{0}$, and $\mathbf{u} \neq \mathbf{0}$, we require that the matrix $\lambda\mathbf{I}_n - \mathbf{A}$
should be singular, or non-invertible, and so the eigenvalues of $\mathbf{A}$ are the zeros of the
characteristic polynomial $\det(\lambda\mathbf{I}_n - \mathbf{A})$. If $\mathbf{A}$ is real, so is its characteristic polynomial;
but the zeros of a real polynomial can be real or complex, and in the case of a complex
eigenvalue, a corresponding eigenvector will lie in $\mathbb{C}^n$, rather than $\mathbb{R}^n$.

Proposition 6.9.   *Let* $\mathbf{A} \in O_n$ *and let* λ *be an eigenvalue of* $\mathbf{A}$. *Then* $|\lambda| = 1$.

Proof.   Let $\mathbf{u}$ be a corresponding eigenvector. We have $\mathbf{Au} = \lambda\mathbf{u}$, so since $\mathbf{A}$ is orthog-
onal, $|\mathbf{u}| = |\mathbf{Au}| = |\lambda\mathbf{u}| = |\lambda||\mathbf{u}|$. Since $|\mathbf{u}| \neq 0$, we deduce $|\lambda| = 1$. This is fine if
$\lambda \in \mathbb{R}$ (and then $\lambda = \pm 1$), but it is a swindle if λ is complex. (Why?)

So now suppose $\lambda = \mu + i\nu$, where $\mu, \nu \in \mathbb{R}$, and write $\mathbf{u} = \mathbf{v} + i\mathbf{w}$, where $\mathbf{v}$,
$\mathbf{w} \in \mathbb{R}^n$, not both zero. (So, for future use, $|\mathbf{v}|^2 + |\mathbf{w}|^2 \neq 0$.) Then

$$\mathbf{A}(\mathbf{v} + i\mathbf{w}) = (\mu + i\nu)(\mathbf{v} + i\mathbf{w}) = (\mu\mathbf{v} - \nu\mathbf{w}) + i(\nu\mathbf{v} + \mu\mathbf{w}). \qquad (6.1)$$

Since $\mathbf{A}$ is orthogonal, we have, from the real parts of (6.1),

$$|\mathbf{v}|^2 = |\mathbf{Av}|^2 = |\mu\mathbf{v} - \nu\mathbf{w}|^2 = \mu^2|\mathbf{v}|^2 - 2\mu\nu\mathbf{v} \cdot \mathbf{w} + \nu^2|\mathbf{w}|^2,$$

and from the imaginary parts,

$$|\mathbf{w}|^2 = |\mathbf{Aw}|^2 = |\nu\mathbf{v} + \mu\mathbf{w}|^2 = \nu^2|\mathbf{v}|^2 + 2\mu\nu\mathbf{v} \cdot \mathbf{w} + \mu^2|\mathbf{w}|^2.$$

Adding, $|\mathbf{v}|^2 + |\mathbf{w}|^2 = (\mu^2 + \nu^2)(|\mathbf{v}|^2 + |\mathbf{w}|^2)$, whence $\mu^2 + \nu^2 = 1$, or $|\lambda| = 1$. □

In the complex case we must have $\lambda = e^{i\theta}$ for some real θ, not an integer multiple of π. Since the characteristic polynomial is real, the complex zeros come in conjugate pairs, and so $\overline{\lambda} = e^{-i\theta}$ is also an eigenvalue of $\mathbf{A}$; and if $\mathbf{u} = \mathbf{v} + i\mathbf{w}$ is an eigenvector corresponding to λ, then $\overline{\mathbf{u}} = \mathbf{v} - i\mathbf{w}$ is an eigenvector corresponding to $\overline{\lambda}$.

Proposition 6.10. *Let* $\mathbf{A} \in O_n$, *and let* λ_1, λ_2 *be eigenvalues of* $\mathbf{A}$ *such that* $\lambda_1 \lambda_2 \neq 1$, *and with corresponding eigenvectors* $\mathbf{u}_1$, $\mathbf{u}_2$. *Then* $\mathbf{u}_1 \cdot \mathbf{u}_2 = 0$.

Proof. We have

$$(\lambda_1 \lambda_2)\mathbf{u}_1^T \mathbf{u}_2 = (\lambda_1 \mathbf{u}_1)^T (\lambda_2 \mathbf{u}_2) = (\mathbf{A}\mathbf{u}_1)^T (\mathbf{A}\mathbf{u}_2) = \mathbf{u}_1^T \mathbf{A}^T \mathbf{A}\mathbf{u}_2 = \mathbf{u}_1^T \mathbf{u}_2,$$

and the result is immediate. □

Corollary 6.11. *Let* $f : \mathbf{x} \mapsto \mathbf{A}\mathbf{x}$ *be an isometry of* $\mathbb{R}^n$, *where* $\mathbf{A} \in O_n$ *has complex eigenvalue* λ *and corresponding eigenvector* $\mathbf{u} = \mathbf{v} + i\mathbf{w}$. *Then the plane* π *determined by* $\mathbf{0}$, $\mathbf{v}$, $\mathbf{w}$ *is* f-*invariant, and the restriction of* f *to this plane is a rotation through* θ, *where* $\lambda = e^{i\theta}$.

Proof. We have $\lambda^2 \neq 1$, so $\mathbf{u} \cdot \mathbf{u} = 0$, by Proposition 6.10. This looks like a disaster, because eigenvectors are supposed to be non-zero. However, the rule '$\mathbf{u} \cdot \mathbf{u} > 0$ for $\mathbf{u} \neq \mathbf{0}$' does not hold when we are using the ordinary dot product in $\mathbb{C}^n$.[1] We have

$$0 = \mathbf{u} \cdot \mathbf{u} = (\mathbf{v} + i\mathbf{w}) \cdot (\mathbf{v} + i\mathbf{w}) = |\mathbf{v}|^2 - |\mathbf{w}|^2 + 2i\mathbf{v} \cdot \mathbf{w},$$

from which we deduce that $|\mathbf{v}| = |\mathbf{w}|$ and $\mathbf{v} \cdot \mathbf{w} = 0$. Adjusting by a scalar multiple if necessary, we may as well assume $|\mathbf{v}| = 1 = |\mathbf{w}|$. Noting that $\lambda = \cos\theta + i\sin\theta$, we obtain, as in (6.1),

$$\mathbf{A}\mathbf{v} = (\cos\theta)\mathbf{v} - (\sin\theta)\mathbf{w} \quad \text{and} \quad \mathbf{A}\mathbf{w} = (\sin\theta)\mathbf{v} + (\cos\theta)\mathbf{w},$$

and so if we take the coordinate system in π with axes determined by $\mathbf{v}$, $\mathbf{w}$, then the restriction of f to π has matrix

$$\begin{pmatrix} \cos\theta & \sin\theta \\ -\sin\theta & \cos\theta \end{pmatrix} \tag{6.2}$$

as required. □

- Ex.6.10: *Given* f, λ *and* π, *as above, are the vectors* $\mathbf{v}$, $\mathbf{w}$ *uniquely determined?*

Now suppose we are given an isometry $f : \mathbf{x} \mapsto \mathbf{A}\mathbf{x}$ of $\mathbb{R}^n$, and pick an eigenvalue λ of $\mathbf{A}$. Two cases arise:

1. If λ is real, with corresponding eigenvector $\mathbf{x}_1$, then take X_1 to be the one-dimensional subspace determined by $\mathbf{x}_1$ (that is, the line through $\mathbf{0}$ and $\mathbf{x}_1$), which is f-invariant.

---

[1] If you *want* it to hold, you should be using the *Hermitian* product
$$(u_1, u_2, \ldots, u_n) \cdot (v_1, v_2, \ldots, v_n) = u_1 \overline{v}_1 + u_2 \overline{v}_2 + \ldots + u_n \overline{v}_n$$
instead; see, for example, [6].

We may as well also take $|\mathbf{x}_1| = 1$. If $\lambda = 1$, then X_1 is actually a line of fixed points of f; and if $\lambda = -1$, then f acts as a reflection on X_1.

2. If λ is complex, $\lambda = e^{i\theta}$, then take X_1 to be the two-dimensional subspace invariant under f (that is, the π of Corollary 6.11) so that f acts on X_1 as rotation through θ. Take $\mathbf{x}_1, \mathbf{x}_2$ to be the $\mathbf{v}$, $\mathbf{w}$ of Corollary 6.11.

We claim that $\mathbb{R}^n$ has an orthonormal basis $\mathbf{x}_1, \mathbf{x}_2, \ldots, \mathbf{x}_n$, that is, that starts with the $\mathbf{x}_1$ (or, in case 2, with the $\mathbf{x}_1, \mathbf{x}_2$) just chosen. The proof will be sketched, and the reader is left to fill in or look up the missing details.

Method 1: Put $X_1^{\perp} = \{ \mathbf{y} \in \mathbb{R}^n : \mathbf{x} \cdot \mathbf{y} = 0$, for all $\mathbf{x} \in X_1 \}$, the *orthogonal complement* of X_1. In case 1, we just have to solve one non-trivial homogeneous linear equation, and the solution space $X_1^{\perp}$ therefore has dimension $n - 1$, a *hyperplane* of $\mathbb{R}^n$. In case 2, we have two independent equations, and $X_1^{\perp}$ is the intersection of two distinct hyperplanes Y, Z, so that $Y + Z = \mathbb{R}^n$, and $\dim X_1^{\perp} = \dim(Y \cap Z) = \dim Y + \dim Z - \dim(Y + Z) = (n - 1) + (n - 1) - n = n - 2$. Now, if we believe that every subspace of $\mathbb{R}^n$ has an orthonormal basis, we just need to choose such a basis for $X_1^{\perp}$ and tag it onto the basis for X_1 that we already have.

Method 2: Use the *Gram–Schmidt orthogonalization process*. This means we write down *any* basis $\mathbf{x}_1, \mathbf{x}_2, \ldots, \mathbf{x}_n$ that starts with the $\mathbf{x}_1$ (or, in case 2, with the $\mathbf{x}_1, \mathbf{x}_2$) already chosen—is this OK?—and then successively adjust each $\mathbf{x}_i$ by adding to it a suitable linear combination of the preceding basis elements so as to make it orthogonal to them. For details, see [6].

- Ex.6.11: *Let X be any subspace of $\mathbb{R}^n$, and prove that $\dim X + \dim X^{\perp} = n$. Deduce that $\mathbb{R}^n = X \oplus X^{\perp}$ (that is, $\mathbb{R}^n = X + X^{\perp}$ and $X \cap X^{\perp} = \{\mathbf{0}\}$), and that $(X^{\perp})^{\perp} = X$. Show also that if $f : \mathbf{x} \mapsto \mathbf{A}\mathbf{x}$ is an isometry, and X is f-invariant, then $X^{\perp}$ is f-invariant also.*

So f restricts to an isometry $X_1^{\perp} \rightarrow X_1^{\perp}$; let $\mathbf{A}_1$ be the matrix of this restriction with respect to the basis $(\mathbf{x}_2,)\mathbf{x}_3, \ldots, \mathbf{x}_n$. Then the matrix of f with respect to $\mathbf{x}_1, \mathbf{x}_2, \ldots, \mathbf{x}_n$ is

$$\begin{pmatrix} \pm 1 & 0 & \ldots & 0 \\ 0 & & & \\ \vdots & & \mathbf{A}_1 & \\ 0 & & & \end{pmatrix} \quad \text{or} \quad \begin{pmatrix} \cos\theta & \sin\theta & 0 & \ldots & 0 \\ -\sin\theta & \cos\theta & 0 & \ldots & 0 \\ 0 & 0 & & & \\ \vdots & \vdots & & \mathbf{A}_1 & \\ 0 & 0 & & & \end{pmatrix}$$

in cases 1, 2 respectively, where $\mathbf{A}_1 \in O_{n-1}$ or O_{n-2}, as appropriate. We now put $X_2 = X_1^{\perp}$ and repeat the argument for the restriction of f to X_2, and thus split $\mathbb{R}^n$ up as the orthogonal sum of f-invariant lines and planes, on each of which f acts either trivially, or as a reflection, or as a rotation. If we insist on taking any occurrences of the eigenvalue 1 first, followed by any occurrences of the eigenvalue -1, then the matrix of

f will take the form of the block-diagonal sum

$$\mathbf{I}_r \oplus (-\mathbf{I}_s) \oplus \mathbf{A}_{\theta_1} \oplus \cdots \oplus \mathbf{A}_{\theta_t},$$

where $r \geq 0$, $s \geq 0$, $t \geq 0$, and $r + s + 2t = n$, and $\mathbf{A}_\theta$ stands for the 2×2 matrix (6.2).

It is time to test our theory, and see what it says when n is small. When $n = 2$, then A will take one of the forms

$$\begin{pmatrix} 1 & 0 \\ 0 & 1 \end{pmatrix}, \quad \begin{pmatrix} 1 & 0 \\ 0 & -1 \end{pmatrix}, \quad \begin{pmatrix} -1 & 0 \\ 0 & -1 \end{pmatrix}, \quad \begin{pmatrix} \cos\theta & \sin\theta \\ -\sin\theta & \cos\theta \end{pmatrix},$$

which correspond to f being trivial, a reflection, a half-turn, or some other rotation, respectively. These are the only possibilities for isometries of $\mathbb{R}^2$ that fix the origin. The half-turn is of course a rotation through π, but we keep it separate because it has invariant lines, whereas the general rotation does not.

Now for $n = 3$; we find six possible solutions for r, s, t:

1. If $\mathbf{A} = \mathbf{I}_3$, then $f = 1$, the trivial isometry.
2. If $\mathbf{A} = \mathbf{I}_2 \oplus (-1)$, then $f(x, y, z) = (x, y, -z)$, reflection in the (x, y)-plane.
3. If $\mathbf{A} = (1) \oplus (-\mathbf{I}_2)$, then $f(x, y, z) = (x, -y, -z)$, a half-turn about the x-axis.
4. If $\mathbf{A} = -\mathbf{I}_3$, then $f(x, y, z) = (-x, -y, -z)$, which is called a *central inversion*.
5. If $\mathbf{A} = (1) \oplus \mathbf{A}_\theta$, then f is a rotation through θ about the x-axis.
6. If $\mathbf{A} = (-1) \oplus \mathbf{A}_\theta$, then f is the composite (in either order) of a rotation through θ about the x-axis, and a reflection in its orthogonal complement, the (y, z)-plane. This is called a *rotatory reflection* (Fig. 6.1).

Again, note that (3) is the special case of (5), and likewise (4) is the special case of (6), when $\theta = \pi$.

For $n = 4$, the list just gets longer! We mention explicitly only the case $\mathbf{A} = \mathbf{A}_\theta \oplus \mathbf{A}_\varphi$, which is called a *double rotation*.

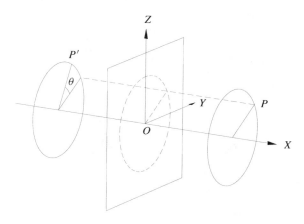

Fig. 6.1 The rotatory reflection $(x, y, z) \mapsto (-x, y\cos\theta - z\sin\theta, y\sin\theta + z\cos\theta)$

● Ex.6.12: *Prove that the double rotation $\mathbf{A}_\theta \oplus \mathbf{A}_\varphi$ is not a rotation. Does it make any difference if $\theta = \varphi$?*

Recall that $T = \{z \in \mathbb{C} : |z| = 1\}$, the unit circle in $\mathbb{C}$. The cartesian product of two copies of T is the set $T^2 = T \times T = \{(z_1, z_2) \in \mathbb{C}^2 : |z_1| = 1 = |z_2|\}$, and is called a *torus*. Warning: we also have $S^1 = \{(x, y) \in \mathbb{R}^2 : x^2 + y^2 = 1\}$, the 1-sphere, so that on identifying $\mathbb{R}^2$ with $\mathbb{C}$ as usual, we have $S^1 = T$. So $T^2 = T \times T = S^1 \times S^1$, but this is *not at all* the same object as S^2, the 2-sphere. Recall that $S^2 = \{(x, y, z) \in \mathbb{R}^3 : x^2 + y^2 + z^2 = 1\}$, and we shall see below (Exercise 6.13) that T^2 can be put inside $\mathbb{R}^3$, in a certain sense, as the more familiar torus or anchor-ring, the surface of revolution formed by rotating a circle about a line in its plane that it does not meet (Fig. 6.2); S^2, on the other hand, is formed by rotating the circle about a diameter (Fig. 6.3). Topologically, these surfaces are distinguished from each other by the fact that T^2 has a hole in it,

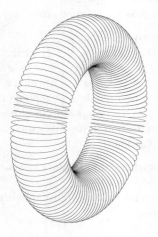

Fig. 6.2 The torus T^2

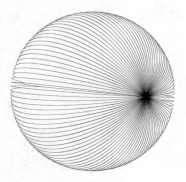

Fig. 6.3 The sphere S^2

whereas S^2 does not; but the technicalities of this distinction are not insignificant, and we do not pursue them here.

It is convenient to change the size of the circle T, and write $T_a = \{z \in \mathbb{C} : |z| = a\}$, so that $T = T_1$. Choose, once and for all, two positive real numbers a and b with $a^2 + b^2 = 1$. Then the torus $T_a \times T_b$ will, by abuse of notation, be referred to as T^2. The point of doing this is to fit T^2 inside S^3, the 3-sphere. For, identifying $\mathbb{C}$ with $\mathbb{R}^2$, and so $\mathbb{C}^2$ with $\mathbb{R}^2 \times \mathbb{R}^2 = \mathbb{R}^4$,

$$T = \{(w, x, y, z) \in \mathbb{R}^4 : w^2 + x^2 = a^2, \ y^2 + z^2 = b^2\}$$
$$\subseteq \{(w, x, y, z) \in \mathbb{R}^4 : w^2 + x^2 + y^2 + z^2 = 1\} = S^3,$$

using the fact that $a^2 + b^2 = 1$. If $w^2 + x^2 = a^2$ we can write $w = a \cos \alpha$ and $x = a \sin \alpha$ for some angle α; and if $y^2 + z^2 = b^2$ we can write $y = b \cos \beta$ and $z = b \sin \beta$ for some angle β; and so we have a pair of parameters α, β to describe the points of T^2.

Now apply the double rotation $\mathbf{A}_\theta \oplus \mathbf{A}_\varphi$ to $\mathbb{R}^4$. We know this is a symmetry of S^3, because it fixes $\mathbf{0}$ and is an isometry; but it is also a symmetry of T^2. Indeed, it is immediate that the point of T^2 with parameters α, β is mapped to the point of T^2 with parameters $\alpha + \theta, \beta + \varphi$ by this isometry. The next exercise is designed to establish what this map $T^2 \to T^2$ looks like when applied to the three-dimensional manifestation of the torus, the anchor-ring. It will *not* give an isometry in this case (except when $\alpha = 0$), but the map it does give helps us visualize what the double rotation is doing to $\mathbb{R}^4$.

- Ex.6.13: *Show that the two versions of the torus (in $\mathbb{R}^4$ and in $\mathbb{R}^3$) are related by* stereographic projection, *thus:*
  (i) First obtain a formula for the familiar stereographic projection, which is a map $\psi_2 : S^2 \to \mathbb{R}^2$ used by cartographers to make flat (plane) maps of the curved (spherical) earth on which we live, and is set up as follows. Take $S^2 \subseteq \mathbb{R}^3$ as $x^2 + y^2 + z^2 = 1$, as usual, with the north pole at $(1, 0, 0)$, so that the equatorial plane is $x = 0$. Show that the line joining the north pole to $(x, y, z) \in S^2$ meets the equatorial plane at $(0, y/(1 - x), z/(1 - x))$. So ψ_2 is given by $\psi_2(x, y, z) = (y/(1 - x), z/(1 - x))$. (See Fig. 7.19.)
  (ii) The corresponding formula for stereographic projection $\psi_3 : S^3 \to \mathbb{R}^3$ is $\psi_3(w, x, y, z) = (x/(1 - w), y/(1 - w), z/(1 - w))$. Show that this sends $T^2 = T_a \times T_b$ (where $a^2 + b^2 = 1$) to the surface of revolution obtained by rotating the circle $x^2 + (y - 1/b)^2 = (a/b)^2$ about the x-axis. Now apply ψ_3 to the points α, β and $\alpha + \theta, \beta + \varphi$ of T^2 and describe what the resulting 'stereographic projection of the double rotation' does to the surface of revolution in $\mathbb{R}^3$.

Our classification of isometries thus far has only looked at those isometries that are linear maps, or in other words that fix the origin. If an isometry f fixes some point $\mathbf{c}$, then by shifting the origin to $\mathbf{c}$ we obtain a linear map again. When can this be done? If $f(\mathbf{x}) = \mathbf{A}\mathbf{x} + \mathbf{b}$, then we must solve $\mathbf{x} = \mathbf{A}\mathbf{x} + \mathbf{b}$, or $(\mathbf{I}_n - \mathbf{A})\mathbf{x} = \mathbf{b}$. This is possible if the matrix $\mathbf{I}_n - \mathbf{A}$ is invertible, that is, if 1 is *not* an eigenvalue of $\mathbf{A}$.

If 1 *is* an eigenvalue of **A**, then f may have no fixed points, and we already know that this can happen, as in the case of translations and glide-reflections in $\mathbb{R}^2$. Now the glide-reflection was a composite of a reflection and a translation, and we saw that as long as there were no fixed points, the translation could be arranged to be parallel to the line of fixed points of the reflection. In symbols, the glide-reflection was of the form $f : \mathbf{x} \mapsto \mathbf{Ax} + \mathbf{b}$, where $\mathbf{x} \mapsto \mathbf{Ax}$ is a reflection, and $\mathbf{Ab} = \mathbf{b}$. (This means that the translation and reflection can be composed in either order: $\mathbf{Ax} + \mathbf{b} = \mathbf{A}(\mathbf{x} + \mathbf{b})$.)

So consider an isometry $f : \mathbf{x} \mapsto \mathbf{Ax} + \mathbf{b}$ of $\mathbb{R}^n$, having no fixed points. Let us try the effect of shifting the origin to some point **c**. The formula for f then becomes $\mathbf{x} \mapsto \mathbf{A}(\mathbf{x} + \mathbf{c}) + \mathbf{b} - \mathbf{c} = \mathbf{Ax} + \mathbf{d}$, where $\mathbf{d} = \mathbf{Ac} + \mathbf{b} - \mathbf{c} = (\mathbf{A} - \mathbf{I}_n)\mathbf{c} + \mathbf{b}$. We should like to fix it so that $\mathbf{Ad} = \mathbf{d}$, or $(\mathbf{A} - \mathbf{I}_n)\mathbf{d} = \mathbf{0}$. Now $\mathbf{A} = \mathbf{I}_r \oplus \mathbf{B}$ for some $r > 0$ and some $\mathbf{B} \in O_{n-r}$, such that 1 is not an eigenvalue of **B**; in other words, $\mathbf{C} = \mathbf{B} - \mathbf{I}_{n-r}$ is invertible. Writing $\mathbb{R}^n = \mathbb{R}^r \oplus \mathbb{R}^{n-r}$, and partitioning all the matrices and vectors accordingly, the equation $(\mathbf{A} - \mathbf{I}_n)\mathbf{d} = \mathbf{0}$ reads

$$\begin{pmatrix} \mathbf{0} & \mathbf{0} \\ \mathbf{0} & \mathbf{C} \end{pmatrix} \left(\begin{pmatrix} \mathbf{0} & \mathbf{0} \\ \mathbf{0} & \mathbf{C} \end{pmatrix} \begin{pmatrix} \mathbf{c}_1 \\ \mathbf{c}_2 \end{pmatrix} + \begin{pmatrix} \mathbf{b}_1 \\ \mathbf{b}_2 \end{pmatrix} \right) = \begin{pmatrix} \mathbf{0} \\ \mathbf{0} \end{pmatrix}$$

or

$$\begin{pmatrix} \mathbf{0} & \mathbf{0} \\ \mathbf{0} & \mathbf{C} \end{pmatrix} \begin{pmatrix} \mathbf{b}_1 \\ \mathbf{Cc}_2 + \mathbf{b}_2 \end{pmatrix} = \begin{pmatrix} \mathbf{0} \\ \mathbf{0} \end{pmatrix},$$

which is satisfied provided $\mathbf{C}(\mathbf{Cc}_2 + \mathbf{b}_2) = \mathbf{0}$. Since **C** is invertible, we only need to put $\mathbf{c}_2 = -\mathbf{C}^{-1}\mathbf{b}_2$, and we have achieved our objective. To sum up:

Theorem 6.12. *Let f be an isometry of $\mathbb{R}^n$. If f has a fixed point **c**, then f has a collection of p invariant lines and q invariant planes through **c**, all mutually orthogonal and with $p + 2q = n$, and such that f acts either trivially or as reflection in **c** on each of the lines, and f acts as a rotation about **c** on each of the planes. If f has no fixed point, then it can be written as a composite $f = gh = hg$ of two commuting isometries g, h, where $g(\mathbf{0}) = \mathbf{0}$ and h is a translation parallel to a line of fixed points of g.* □

When $n = 2$, the only isometries without fixed points we obtain are the translation and the glide-reflection, corresponding to g being trivial or a reflection. So, with our previous remarks about the isometries with fixed points, we have once again our four kinds of non-trivial plane isometry: translation, rotation, reflection, and glide-reflection.

When $n = 3$, we obtained, above, four kinds of isometry with fixed points: trivial, reflection, rotation, rotatory reflection. Of these, only the first three have 1 as an eigenvalue. The trivial map composed with a translation is just a translation. A reflection composed with a translation parallel to the plane of reflection is the three-dimensional version of the glide-reflection, and is still called a glide-reflection (Fig. 6.4). Finally, a rotation composed with a translation parallel to the axis of rotation is called a *screw displacement*, because its effect is like the motion of a screw in a thread, which when turned through an angle in the thread will also move forward (or back) (Fig. 6.5). So:

Corollary 6.13. *Every non-trivial isometry of $\mathbb{R}^3$ is a translation, rotation, screw displacement, reflection, glide-reflection, or rotatory reflection.* □

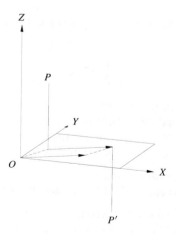

Fig. 6.4   The glide reflection $(x, y, z) \mapsto (x + a, y + b, -z)$

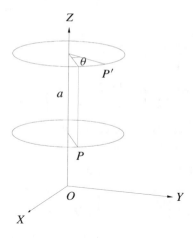

Fig. 6.5   The screw displacement $(x, y, z) \mapsto (x \cos \theta - y \sin \theta, x \sin \theta + y \cos \theta, z + a)$

• Ex.6.14: *Write down the corresponding list for* $\mathbb{R}^4$*, inventing suitably descriptive names for the various types.*

6.4   PERMUTATION GROUPS

We shall use permutation groups as a convenient way of describing certain groups of isometries, and so here is a brief introduction. Let $X_n = \{ 1, 2, 3, \ldots, n \}$. A *permutation* (of X_n) is a bijection $\rho : X_n \to X_n$.

Example.   Let $n = 4$. The permutation ρ with $\rho(1) = 2$, $\rho(2) = 1$, $\rho(3) = 4$, $\rho(4) = 3$ can be written as

$$\rho = \left\{ \begin{matrix} 1 & 2 & 3 & 4 \\ 2 & 1 & 4 & 3 \end{matrix} \right\},$$

where we have simply listed the elements of X_4 in the first row, and their respective images under ρ in the second row. There is no particular reason to list the first row in increasing order, as long as the images are put under the proper elements. The notation is rather fulsome and clumsy, but we shall introduce a slicker one shortly.

Definition 6.14.   *The set of all permutations of* X_n *is denoted* S_n. *It is a group under composition of maps, and is called the* nth **symmetric group.**

Because we are writing maps on the left (i.e. we write $\rho(1)$, not $(1)\rho$), the composite $\rho\sigma$ means do σ first, then ρ. For example, with ρ as in the example above, and

$$\sigma = \left\{ \begin{matrix} 1 & 2 & 3 & 4 \\ 2 & 3 & 4 & 1 \end{matrix} \right\},$$

we have

$$\rho\sigma = \left\{ \begin{matrix} 1 & 2 & 3 & 4 \\ 2 & 1 & 4 & 3 \end{matrix} \right\}\left\{ \begin{matrix} 1 & 2 & 3 & 4 \\ 2 & 3 & 4 & 1 \end{matrix} \right\}$$

$$= \left\{ \begin{matrix} 2 & 3 & 4 & 1 \\ 1 & 4 & 3 & 2 \end{matrix} \right\}\left\{ \begin{matrix} 1 & 2 & 3 & 4 \\ 2 & 3 & 4 & 1 \end{matrix} \right\} = \left\{ \begin{matrix} 1 & 2 & 3 & 4 \\ 1 & 4 & 3 & 2 \end{matrix} \right\},$$

where in the second line, to make things easier, we have simply reordered the columns of ρ so that its first row is the same as the second row of σ.

Definition 6.15.   *Let* $\rho \in S_n$. *The* **sign** *of* ρ *is*

$$\text{sign}(\rho) = \prod_{i>j} \frac{i-j}{\rho(i) - \rho(j)}.$$

For instance, using the same examples as above, we have

$$\text{sign}(\rho) = \frac{2-1}{1-2} \cdot \frac{3-1}{4-2} \cdot \frac{4-1}{3-2} \cdot \frac{3-2}{4-1} \cdot \frac{4-2}{3-1} \cdot \frac{4-3}{3-4} = +1, \text{ and}$$

$$\text{sign}(\sigma) = \frac{2-1}{3-2} \cdot \frac{3-1}{4-2} \cdot \frac{4-1}{1-2} \cdot \frac{3-2}{4-3} \cdot \frac{4-2}{1-3} \cdot \frac{4-3}{1-4} = -1.$$

It should be plain that, up to change of sign, the terms in the numerator and denominator of $\text{sign}(\rho)$ will always cancel, so that the answer is always ± 1. In fact, sign: $S_n \to \{\pm 1\}$ is a homomorphism, since, for any $\rho, \sigma \in S_n$,

$$\text{sign}(\rho\sigma) = \prod_{i>j} \frac{i-j}{\rho\sigma(i) - \rho\sigma(j)} = \prod_{i>j} \frac{\sigma(i) - \sigma(j)}{\rho\sigma(i) - \rho\sigma(j)} \prod_{i>j} \frac{i-j}{\sigma(i) - \sigma(j)}.$$

The presence of the σ's in both numerators and denominators of the first product merely alters (possibly) the order of the terms in the product, and changes (possibly) the sign of both numerator and denominator of some terms, and so has no overall effect. So

$$\prod_{i>j} \frac{\sigma(i) - \sigma(j)}{\rho\sigma(i) - \rho\sigma(j)} = \prod_{i>j} \frac{i - j}{\rho(i) - \rho(j)} = \text{sign}(\rho),$$

whence $\text{sign}(\rho\sigma) = \text{sign}(\rho)\,\text{sign}(\sigma)$.

Definition 6.16. *The permutation ρ is **even** if* $\text{sign}(\rho) = +1$, *and **odd** if* $\text{sign}(\rho) = -1$.

So the product of two even permutations is even, the product of an even and an odd permutation (in either order) is odd, and the product of two odd permutations is even.

Definition 6.17. *The set of all even permutations in S_n (the kernel of the sign homomorphism) is a subgroup of S_n, called the nth **alternating group**, and denoted A_n.*

● Ex.6.15: *Show that, for $n \geq 2$, we have $|S_n| = n!$ and $|A_n| = \frac{1}{2}(n!)$.*

We promised a better notation for permutations, and here it is: it is called *cycle* notation. As an example, the permutation

$$\rho = \left\{ \begin{array}{cccc} 1 & 2 & 3 & 4 \\ 2 & 3 & 4 & 1 \end{array} \right\}$$

sends $1 \mapsto 2 \mapsto 3 \mapsto 4 \mapsto 1$, and is called a 4-*cycle*. We write it as (1234) (or, equivalently, as (2341), or (3412), or (4123)). More generally, the m-*cycle* $(a_1\, a_2\, \cdots\, a_m)$ is the permutation that sends a_1 to a_2, and a_2 to $a_3, \ldots$, and a_{m-1} to a_m, and a_m back to a_1; and leaves everything/anything else alone. In fact, every permutation can be written as the product of disjoint cycles: for instance, if

$$\rho = \left\{ \begin{array}{cccccccc} 1 & 2 & 3 & 4 & 5 & 6 & 7 & 8 \\ 2 & 4 & 3 & 8 & 7 & 5 & 6 & 1 \end{array} \right\}, \tag{6.3}$$

then it is not hard to calculate that $\rho = (1248)(3)(576) = (1248)(576)$ (the latter because a 1-cycle is trivial—OK?—and so can be omitted). We leave the reader to construct the argument for the general case. Not only is cycle notation more compact than the list notation, but it makes certain calculations much easier. For example, a 4-cycle clearly has multiplicative order 4, and a 3-cycle has order 3, and since (1248) and (576), being disjoint, commute with each other, we deduce that the multiplicative order of $\rho = (1248)(576)$ is the least common multiple of 3, and 4, namely 12. We invite the reader to attempt this same calculation using the notation (6.3)!

● Ex.6.16: *Some you win and some you lose: show that every 2-cycle is an odd permutation, every 3-cycle is even, and in general an m-cycle is even if m is odd, and odd if m is even. You just have to learn to live with this.*

- Ex.6.17: *Every cycle can be written as a product of 2-cycles—not disjoint!—and consequently every element of S_n can be written as a product of 2-cycles.*

- Ex.6.18: *Every product of two 2-cycles can be written as a product of 3-cycles, and consequently every element of A_n can be written as a product of 3-cycles.*

Example. S_1 is trivial, and hence so is A_1. Then $S_2 = \{1, (12)\} \cong C_2$, the cyclic group of order 2, and so A_2 is trivial. Then

$$S_3 = \{1, (12), (13), (23), (123), (132)\}.$$

If we take an equilateral triangle X in $\mathbb{R}^2$, and label its vertices 1, 2, 3, then the six symmetries of the triangle induce permutations of the vertices, and we obtain an isomorphism $\mathcal{S}(X) \cong S_3$, so that $S_3 \cong D_3$, the dihedral group. In this isomorphism, the rotation group $\mathcal{S}^+(X)$ corresponds to $\{1, (123), (132)\} = A_3 \cong C_3$. However, there is no reason in general why, in an isomorphism between a group of isometries and a permutation group, there should be any correspondence between direct isometries and even permutations:

- Ex.6.19: *Label the vertices of a square X in $\mathbb{R}^2$ as 1, 2, 3, 4, and show that there is an isomorphism between $\mathcal{S}(X)$ and a subgroup G of S_4, in which the rotation group $\mathcal{S}^+(X)$ does not correspond to $G \cap A_4$.*

- Ex.6.20: *List all the elements of S_4, and of A_4, in cycle notation.*

- Ex.6.21: *Find six matrices in $GL_2(\mathbb{R})$ which form a subgroup isomorphic to S_3.*

6.5 POLYHEDRA

Example 1.   If four points A, B, C, D in $\mathbb{R}^3$ are not coplanar, then they are the vertices of a *tetrahedron*, a figure with (as its name implies) four faces, namely the triangles BCD, ACD, ABD, ABC, and six edges, the line segments AB, AC, AD, BC, BD, CD. If the edges are all the same length, then the faces are all equilateral, and the tetrahedron is called *regular* (Fig. 6.6). (Think of it as a triangular-based pyramid.)

Given a regular tetrahedron X, then, what can we say about $\mathcal{S}(X)$? Every symmetry of X will permute its four vertices, so if we label the vertices 1, 2, 3, 4, we obtain a group homomorphism $\alpha : \mathcal{S}(X) \to S_4$. First note that if $f \in \mathcal{S}(X)$ has trivial image in

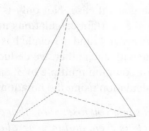

Fig. 6.6   A (regular) tetrahedron

S_4, this means that the four vertices are all fixed points of f, and so $f = 1$ by Corollary 5.6. Thus ker $\alpha = 1$, and α is an injection, and $\mathcal{S}(X)$ is isomorphic to a subgroup of S_4, namely the image of α. We claim this is actually the whole of S_4, so that $\mathcal{S}(X) \cong S_4$. One method: show that the image of α contains all the 2-cycles, and invoke Exercise 6.17. All that is needed here is to produce an isometry that swaps two vertices while leaving the other two alone, and reflection in the perpendicular bisector of the chosen edge will do just what we want:

- Ex.6.22: *Let $A = (1, 1, 1)$, $B = (1, -1, -1)$, $C = (-1, 1, -1)$, and $D = (-1, -1, 1)$. Show that these are the vertices of a regular tetrahedron, and that the plane which is the perpendicular bisector of (say) AB contains both C and D.*

We can also invoke Exercise 6.18 to show that $\mathcal{S}^+(X) \cong A_4$. For if we pick a 3-cycle in S_4, say (123), then the perpendicular from vertex 4 onto the opposite face is the axis of a $\frac{1}{3}$-turn symmetry of the tetrahedron whose corresponding permutation is (123), or its inverse, (132) (Fig. 6.7).

- Ex.6.23: *A second method of proving that α is an isomorphism is to construct its inverse: go through the list you produced for Exercise 6.20 and find the corresponding symmetry of X for each permutation in your list. What is the axis of the half-turn corresponding to (12)(34)? What sort of (opposite) isometry corresponds to (1234), and what is its axis (i.e., its invariant line)?*

- Ex.6.24: *Consider the line segment joining the mid-points of two opposite edges of a regular tetrahedron. The perpendicular bisector of this line segment is a plane, parallel to both the chosen edges. Describe the section of the tetrahedron by this plane, and relate your answer to the preceding exercise.*

- Ex.6.25: *The converse of Lagrange's theorem fails: for example, $|A_4| = 12$, and 6 is a factor of 12, but A_4 has no subgroup of order 6. The proof goes by showing first that, if H were such a subgroup, then $\rho^2 \in H$ for all $\rho \in A_4$. Fill in the details in this sketch: if $\rho^2 \notin H$, then $\rho \notin H$, so $A_4 = H \cup \rho H$ and $\rho^2 \in \rho H$ (the coset); multiply through by ρ^{-1} to get a contradiction. But*

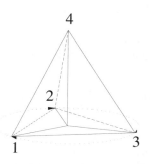

Fig. 6.7 A $\frac{1}{3}$-turn rotational symmetry of a tetrahedron

now H has to contain the square of every element of A_4, *and you should work these out to see why this is impossible.*

A tetrahedron is an example of a *polyhedron*, or many-faced figure. We shall confine attention to *convex* polyhedra, which are certain subsets of $\mathbb{R}^3$ bounded by a finite number of planes, called *faces* of the polyhedron. (A set is *convex* if for every two of its points it contains the line segment they define. It is immediate that any intersection of convex sets is convex. Each plane divides $\mathbb{R}^3$ into two *half-spaces*, which are convex. A convex polyhedron X is a bounded set which is the intersection of a finite number of these half-spaces. The intersection of one of the defining planes with X is a face of X, a convex plane polygon whose edges are edges of X, each lying on two faces, and whose vertices are vertices of X, each lying on at least three faces.)

A polyhedron is *regular* if each of its faces is a regular n-gon, the same n for each face, and m faces meet at each vertex, the same m for each vertex. This is enough to ensure that all faces, and all vertices, and all edges are congruent, in the sense that there will be a symmetry of the polyhedron sending any face to any other face, and likewise for vertices, and for edges. Such a polyhedron is then denoted by its *Schläfli symbol* $\{n, m\}$. The regular tetrahedron, having triangular faces, three at each vertex, is thus $\{3, 3\}$. In fact the values of n and m determine the polyhedron up to similarity, and there are rather a small number of possible values of n and m, as we shall see.

Example 2. A cube X has square faces, three meeting at each of its vertices; so it is the regular polyhedron $\{4, 3\}$. What can we say about $\mathcal{S}(X)$? We could simply mimic what we did for the tetrahedron, number the eight vertices, and get a copy of $\mathcal{S}(X)$ inside S_8. But $|S_8| = 8! = 40\,320$, which is rather big; how big is $\mathcal{S}(X)$? Now a rotation of X sends a chosen face to any one of 6 others, and we can follow this by one of 4 rotations that are symmetries of this face, namely 1 and rotations through $\pi/2, \pi, 3\pi/2$ about the line perpendicular to the face through the centre O of the cube. So $|\mathcal{S}^+(X)| = 6 \times 4 = 24$, and since there *are* reflectional symmetries of a cube, $|\mathcal{S}(X)| = 2|\mathcal{S}^+(X)| = 48$. So $|S_8|$ is extravagantly large; we could instead number the six faces and get a copy of $\mathcal{S}(X)$ inside S_6, but although $6! = 720$ is a lot closer to 48, it is still going to be much too much like hard work to use such a large group.

We need to be a little more ingenious. Now the cube has four main diagonals, the lines joining a vertex through the centre O to the opposite vertex. If we number these diagonals 1, 2, 3, 4 (Fig. 6.8), then we get a map $\alpha : \mathcal{S}(X) \rightarrow S_4$, which is a homomorphism of groups, but is plainly *not* injective, since $|\mathcal{S}(X)| = 48$ but $|S_4| = 24$. However, $|\mathcal{S}^+(X)| = 24$, and we shall show that the restriction of α to $\mathcal{S}^+(X)$ gives an isomorphism $\mathcal{S}^+(X) \cong S_4$.

● Ex.6.26: *Find a non-trivial* $f \in \mathcal{S}(X)$ *with* $\alpha(f) = 1$, *i.e. such that all four of the main diagonals are f-invariant.*

Choose any two adjacent vertices A, B of X, so that AB is an edge. Let the vertex opposite A be A', and let the vertex opposite B be B', so that AA', BB' are main diagonals. Then $A'B'$ is the edge opposite AB, and the half-turn about the line joining the mid-point of AB to the mid-point of $A'B'$ is a symmetry of X. This half-turn swaps

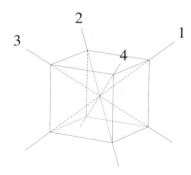

Fig. 6.8 A cube, with its four main diagonals

the diagonals AA', BB', but it leaves the other two diagonals invariant, and so it is mapped by α to a 2-cycle in S_4. Since X has 12 edges, in 6 opposite pairs, we thus obtain all 6 of the 2-cycles of S_4 in the image of α, and so $\mathcal{S}^+(X) \cong S_4$, by Exercise 6.17.

- Ex.6.27: *Instead of the above, it would have been enough to show that if $f \in \mathcal{S}^+(X)$ is such that $\alpha(f) = 1$, then $f = 1$. Would this have been easier? (Compare Exercise 6.26.)*

We have a nice description of $\mathcal{S}^+(X)$; what about $\mathcal{S}(X)$? We claim that $\mathcal{S}(X) \cong S_4 \times C_2$, where C_2 is cyclic of order 2. This is a *direct product* of groups: if G, H are groups, then the ordinary cartesian product of sets $G \times H = \{\,(g, h) : g \in G, h \in H\,\}$ is made into a group by defining $(g, h)(g', h') = (gg', hh')$, where the first product gg' takes place in G, and the second product hh' takes place in H. (Check the axioms!) It should be clear that, for finite groups, $|G \times H| = |G||H|$.

Recall that there is a homomorphism $\phi : \mathcal{S}(X) \to C_2$ given by $\phi(f) = 1$ (f direct) and $\phi(f) = -1$ (f opposite). So define $\beta : \mathcal{S}(X) \to S_4 \times C_2$ by $\beta(f) = (\alpha(f), \phi(f))$. It is simple to check that β is a homomorphism:

$$\beta(fg) = (\alpha(fg), \phi(fg)) = (\alpha(f)\alpha(g), \phi(f)\phi(g))$$
$$= (\alpha(f), \phi(f))(\alpha(g), \phi(g)) = \beta(f)\beta(g).$$

Then $\ker \beta = 1$, for if $\beta(f) = 1$, then $\alpha(f) = 1$ and $\phi(f) = 1$; from the latter, $f \in \mathcal{S}^+(X)$, and since α restricted to $\mathcal{S}^+(X)$ is an isomorphism, then $f = 1$. Since $|\mathcal{S}(X)| = 48$, and $|S_4 \times C_2| = 4! \times 2 = 48$ also, we have our isomorphism.

It is easy to obtain the symmetries of a cube as a matrix group, a subgroup of O_3. Take the cube X with vertices $(\pm 1, \pm 1, \pm 1)$, so that its symmetries are all linear maps given by $\mathbf{u} \mapsto \mathbf{A}\mathbf{u}$ for suitable $\mathbf{A} \in O_3$. The diagonal matrices in O_3 are the eight matrices

$$\begin{pmatrix} \pm 1 & 0 & 0 \\ 0 & \pm 1 & 0 \\ 0 & 0 & \pm 1 \end{pmatrix}$$

which form a subgroup $G \subset O_3$ isomorphic to $C_2^3 = C_2 \times C_2 \times C_2$, and these are all symmetries of X. Then it is clear that the matrices

$$\begin{pmatrix} 1 & 0 & 0 \\ 0 & 1 & 0 \\ 0 & 0 & 1 \end{pmatrix}, \quad \begin{pmatrix} 0 & 1 & 0 \\ 1 & 0 & 0 \\ 0 & 0 & 1 \end{pmatrix}, \quad \begin{pmatrix} 0 & 0 & 1 \\ 0 & 1 & 0 \\ 1 & 0 & 0 \end{pmatrix},$$

$$\begin{pmatrix} 1 & 0 & 0 \\ 0 & 0 & 1 \\ 0 & 1 & 0 \end{pmatrix}, \quad \begin{pmatrix} 0 & 1 & 0 \\ 0 & 0 & 1 \\ 1 & 0 & 0 \end{pmatrix}, \quad \begin{pmatrix} 0 & 0 & 1 \\ 1 & 0 & 0 \\ 0 & 1 & 0 \end{pmatrix}$$

are all orthogonal and permute the vertices of X, and indeed they are called *permutation matrices*. For each one, its rows (or columns) are a permutation of the rows (or columns) of the identity matrix, and they form a subgroup H of O_3 isomorphic to S_3 (cf. Exercise 6.21). We can now choose $\mathbf{D} \in G$ (8 ways) and $\mathbf{P} \in H$ (6 ways) and obtain 48 orthogonal matrices $\mathbf{A} = \mathbf{DP}$, all different and all symmetries of X; and so these matrices,

$$\begin{pmatrix} \pm 1 & 0 & 0 \\ 0 & \pm 1 & 0 \\ 0 & 0 & \pm 1 \end{pmatrix}, \quad \begin{pmatrix} 0 & \pm 1 & 0 \\ \pm 1 & 0 & 0 \\ 0 & 0 & \pm 1 \end{pmatrix}, \quad \begin{pmatrix} 0 & 0 & \pm 1 \\ 0 & \pm 1 & 0 \\ \pm 1 & 0 & 0 \end{pmatrix},$$

$$\begin{pmatrix} \pm 1 & 0 & 0 \\ 0 & 0 & \pm 1 \\ 0 & \pm 1 & 0 \end{pmatrix}, \quad \begin{pmatrix} 0 & \pm 1 & 0 \\ 0 & 0 & \pm 1 \\ \pm 1 & 0 & 0 \end{pmatrix}, \quad \begin{pmatrix} 0 & 0 & \pm 1 \\ \pm 1 & 0 & 0 \\ 0 & \pm 1 & 0 \end{pmatrix}$$

form the required subgroup $\mathcal{S}(X) \subset O_3$. Notice that, as a set, $\mathcal{S}(X)$ can be identified with $G \times H$, or $C_2^3 \times S_3$, by identifying $\mathbf{A} = \mathbf{DP}$ with $(\mathbf{D}, \mathbf{P})$. However, the group structure is *not* that of the direct product, else we should have that (with the obvious notation) $\mathbf{A}_1\mathbf{A}_2 = (\mathbf{D}_1\mathbf{P}_1)(\mathbf{D}_2\mathbf{P}_2) = (\mathbf{D}_1\mathbf{D}_2)(\mathbf{P}_1\mathbf{P}_2)$, which is patently false, unless $\mathbf{P}_1\mathbf{D}_2 = \mathbf{D}_2\mathbf{P}_1$. The *correct* calculation is $(\mathbf{D}_1\mathbf{P}_1)(\mathbf{D}_2\mathbf{P}_2) = (\mathbf{D}_1\mathbf{D}_2')(\mathbf{P}_1\mathbf{P}_2)$ where $\mathbf{D}_2' = \mathbf{P}_1\mathbf{D}_2\mathbf{P}_1^{-1}$ is diagonal, and is obtained from $\mathbf{D}_2$ by permuting the diagonal entries by the permutation corresponding to the permutation matrix $\mathbf{P}_1$. (Check!) This is a particular sort of *semidirect product* of groups, called a *wreath product*, written $S_3 \wr C_2$. For more details of the group theory, see [6].

• Ex.6.28: *Show that the matrix*

$$\begin{pmatrix} 0 & -1 & 0 \\ 0 & 0 & -1 \\ -1 & 0 & 0 \end{pmatrix}$$

has multiplicative order 6. What is the corresponding symmetry of the cube X (as above)? Find an element of $S_4 \times C_2$ of order 6.

• Ex.6.29: *Describe the section of the cube X (as above) by the plane $x_1 + x_2 + x_3 = 0$.*

The regular polyhedron with Schläfli symbol $\{3, 4\}$ has triangular faces, four at each vertex, and looks like two square-based pyramids stuck base-to-base. It has eight triangular faces, and is called a (regular) *octahedron* (Fig. 6.9).

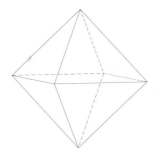

Fig. 6.9 An octahedron

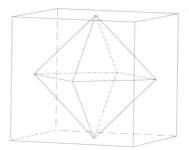

Fig. 6.10 The octahedron dual to a cube

• Ex.6.30: *Show that an octahedron can be placed with its six vertices at $(\pm 1, 0, 0)$, $(0, \pm 1, 0)$, $(0, 0, \pm 1)$, and deduce that the the symmetries of this octahedron are the* same *as the symmetries of the cube X, above.*

The octahedron just obtained had its vertices at the centres of the faces of a cube (Fig. 6.10). It is easy to see that, conversely, the points at the centres of the eight faces of an octahedron form the eight vertices of a cube (Fig. 6.11). The polyhedra $\{4, 3\}$ and $\{3, 4\}$ are thus called *dual* or *reciprocal* polyhedra. It is easy to see that the tetrahedron $\{3, 3\}$ is *self*-reciprocal: the points at the centres of the four faces are vertices of another tetrahedron (Fig. 6.12).

The generalization of the *polyhedron* to higher dimensions is called a *polytope*. In $\mathbb{R}^2$, the regular polygon with n sides is denoted by its Schläfli symbol $\{n\}$; in $\mathbb{R}^3$ the regular polyhedron with m $\{n\}$'s meeting at each vertex has Schläfli symbol $\{n, m\}$; and in $\mathbb{R}^4$ the regular polytope with r $\{n, m\}$'s meeting along each edge has Schläfli symbol $\{n, m, r\}$. We give just one example: $\{4\}$ is a square, $\{4, 3\}$ is a cube, and in $\mathbb{R}^4$ there is a polytope $\{4, 3, 3\}$ with 3 cubes meeting along each edge. It is called a *hypercube*. The points $(\pm 1, \pm 1)$ are the vertices of a square in $\mathbb{R}^2$; the points $(\pm 1, \pm 1, \pm 1)$ are the vertices of a cube in $\mathbb{R}^3$, so naturally the points $(\pm 1, \pm 1, \pm 1, \pm 1)$ are the vertices of a

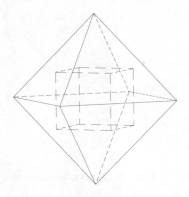

Fig. 6.11   The cube dual to an octahedron

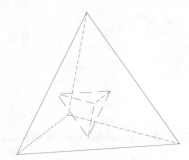

Fig. 6.12   The tetrahedron dual to a tetrahedron

hypercube X in $\mathbb{R}^4$. One face of the cube is the square $(\pm 1, \pm 1, 1)$, and the opposite face is the square $(\pm 1, \pm 1, -1)$, and if we move one of these faces to the other in the direction orthogonal to the 2-space containing it, each of its four sides traces out one of the other (square) faces. In a similar manner, the cube $(\pm 1, \pm 1, \pm 1, 1)$ is a *cell* of the hypercube, and the opposite cell is the cube $(\pm 1, \pm 1, \pm 1, -1)$. If the one cube is moved to the other in the direction orthogonal to the 3-space containing it, then each of its six (square) faces traces out another (cubical) cell of the hypercube. The hypercube thus has eight cells in all, and is therefore also known as the *8-cell* (Fig. 6.13).

- Ex.6.31: *Choose an edge of the hypercube X, and verify that exactly three cubical cells share this edge.*

- Ex.6.32: *Show that the section of X by the hyperplane $x_1 + x_2 + x_3 + x_4 = 0$ is a regular polyhedron. Which one?*

- Ex.6.33: *Show that $\mathcal{S}(X) \cong S_4 \wr C_2$, and deduce that $|\mathcal{S}(X)| = 384$.*

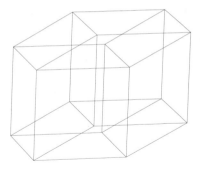

Fig. 6.13   A hypercube

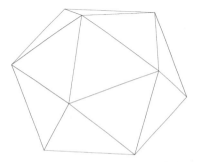

Fig. 6.14   An icosahedron

6.6   THE PLATONIC SOLIDS

The regular polyhedra in $\mathbb{R}^3$ are known as the *Platonic solids*. We have met three of them already: the tetrahedron {3, 3} (Fig. 6.6), the cube {4, 3} (Fig. 6.8), and the octahedron {3, 4} (Fig. 6.9). To find all possible values of {n, m}, note first that $n \geq 3$ and $m \geq 3$. Now we can fit m triangles around a vertex for $m = 3$, 4 or 5, but not 6, because the angles then add up to $6 \times \pi/3 = 2\pi$, which would force the faces to lie in a plane. So {3, 3}, {3, 4} and {3, 5} are the possibilities with $n = 3$, and here only {3, 5} is new. It is the *icosahedron* (Fig. 6.14). For $n \geq 4$ we cannot fit more than 3 n-gons at a vertex, and for $n \geq 6$ we cannot manage even that. So we acquire {4, 3}, the cube, and {5, 3}, the *dodecahedron* (Fig. 6.15), which is the reciprocal of the icosahedron {3, 5}.

- Ex.6.34:   *Formalize the above argument, as follows. Show that the internal angles of a regular n-gon are each $\pi - (2\pi)/n$, and deduce that if {n, m} is a regular polyhedron, then $(\pi - (2\pi)/n)m < 2\pi$. Show that this inequality is equivalent to $(n - 2)(m - 2) < 4$, and hence by factoring the numbers 1, 2, 3 in all possible ways, find the five possible symbols {n, m} again.*

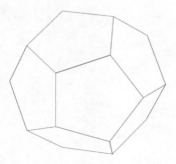

Fig. 6.15   A dodecahedron

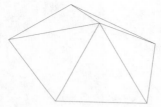

Fig. 6.16   A pentagonal pyramid

• Ex.6.35:   *If n, m are chosen so that* $(\pi - (2\pi/n))m = 2\pi$, *then the n-gons fit together to form a regular* tessellation *or* tiling *of* $\mathbb{R}^2$. *What are the possibilities?*

Finding a Schläfli symbol {3, 5} is no guarantee that there is a corresponding polyhedron, of course. One can start by fixing together five equilateral triangles around a vertex, which gives five more vertices, each having (so far) two triangles meeting there. So now we attempt to add three more triangles at each of these vertices (some of which will be shared between vertices), and so on, hoping against hope that the figure will 'close up' to form a regular polyhedron. This is hard to visualize; one can try it out with card and glue, with success depending on the skill and accuracy deployed, but it is hardly mathematics. (The reader is nonetheless encouraged to have a go.)

Here is an alternative construction of the icosahedron that makes it easier to visualize the result. Pick a vertex A, and let the five triangles adjacent to A be ABC, ACD, ADE, AEF, AFB. Then clearly $BCDEF$ is a regular pentagon, and indeed A, B, C, D, E, F are the vertices of a *pentagonal pyramid*, a solid with a pentagonal base and five other (triangular) faces meeting at the *apex* A (Fig. 6.16). We shall construct the icosahedron from two of these pyramids, one attached to each side of a *pentagonal anti-prism*.

A pentagonal *prism* is easy: place two regular pentagons one directly above the other, in parallel planes, and join corresponding vertices with vertical edges. If the distance between the planes is adjusted correctly, the vertical faces will be squares, and we obtain a *pentagonal prism* (Fig. 6.17). It is an example of a *semi*-regular polyhedron: its faces

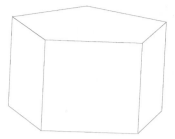

Fig. 6.17   A pentagonal prism

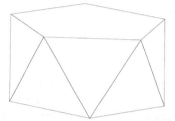

Fig. 6.18   A pentagonal anti-prism

are all regular n-gons, for more than one value of n (here, 5 and 4), but all the vertices are the 'same', having one pentagon and two squares meeting there.

The anti-prism is just a little harder: take the two pentagons as above, but then rotate one of them through $\pi/5$ in its plane, so that its vertices are no longer directly above the vertices in the other pentagon. Now join each vertex in one pentagon to the *two* closest vertices in the other, and we obtain ten triangular faces. If the distance between the two pentagons is adjusted a little, these triangles can be made equilateral, and we obtain a *pentagonal anti-prism* (Fig. 6.18), another semiregular solid with, this time, a pentagon and three triangles meeting at each vertex.

- Ex.6.36:   *A prism is usually taken to mean a triangular prism, which (in its semi-regular manifestation) has two parallel triangular faces, six vertices, and one triangle and two squares meeting at each vertex. What is a triangular anti-prism? (You will find you have met it before!)*

We can now assemble three large 'chunks' to form an icosahedron: take a pentagonal anti-prism, and glue to each of its two pentagonal faces an appropriately-sized pentagonal pyramid (Fig. 6.19). Each of the pyramids has one vertex, its apex, that is surrounded by five triangular faces. Its other vertices each adjoin two triangular faces of the pyramid, and each is now glued to a vertex of the anti-prism, which adjoins three triangular faces of the anti-prism, so these vertices are also surrounded by five triangular faces. So our icosahedron has 12 vertices: $2 \times 5 = 10$ from the anti-prism, and two more from

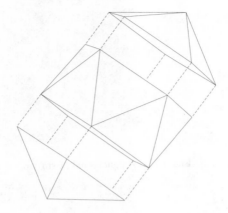

Fig. 6.19   Assembling an icosahedon from an anti-prism and two pyramids

the two pyramids. To count the faces, just count the triangles on the three 'chunks': $10 + 5 + 5 = 20$, which is why it is called an icosahedron. (Alternatively, there are 5 triangles at each of 12 vertices, and $5 \times 12 = 60$. But this has counted all three vertices of every triangle, so divide: $60/3 = 20$ again.)

- Ex.6.37:   *How many edges does the icosahedron have?*
- Ex.6.38:   *The reciprocal figure, the dodecahedron* $\{5, 3\}$*, has a vertex at the centre of each face of the icosahedron, and so has 20 vertices. Show that (as you can guess from its name) it must have 12 faces. How many edges does it have?*

The rest of this section is devoted to the calculation of the rotation and symmetry groups of an icosahedron, X, centre O. (The answers for the dodecahedron are the same, since it is the reciprocal solid.)

First, it is clear from the construction above that the vertex which is the apex of one of the two pentagonal pyramids is opposite the apex of the other, and that X has rotational symmetries about the diagonal joining these vertices through multiples of $2\pi/5$. Any rotation of X will send a chosen vertex to one of 12 others (including itself), followed by one of 5 rotations (including the trivial one) about the diagonal through this vertex. So $|S^+(X)| = 12 \times 5 = 60$, and $|S(X)| = 2 \times 60 = 120$.

- Ex.6.39:   *There is a face argument to produce the same result, finishing with* $20 \times 3 = 60$*. Do it. Is there an edge argument also?*

Since $120 = 5!$, one might hope that $S(X)$ is S_5, but this is false:

- Ex.6.40:   *Show that a pentagonal anti-prism has as one of its symmetries a rotatory reflection* f *with angle* $\pi/5$*, and deduce that* X *also has such a symmetry. What is the least n such that* $f^n = 1$*? By considering cycle structures, show that* S_5 *does not have an element of this order.*

As with the cube, we shall calculate $\mathcal{S}^+(X)$ first, and it will then turn out, as with the cube, that $\mathcal{S}(X) \cong \mathcal{S}^+(X) \times C_2$. Now $|\mathcal{S}^+(X)| = 60 = \frac{1}{2}(5!) = |A_5|$, so is there any hope that $\mathcal{S}^+(X) \cong A_5$? This time the guess *will* turn out to be correct, but will need some effort to verify, because we have to work a little to find five things which are permuted by rotations of X.

Pick any two adjacent vertices A, B of X, and let A', B' be the opposite vertices. Then $AB \parallel A'B'$, and indeed $ABA'B'$ is a rectangle. How many such rectangles are there? By Exercise 6.37, there are 30 edges and so 15 such rectangles. We shall see that these 15 rectangles fall naturally into 5 sets of 3, each such set lying in three mutually perpendicular planes.

- Ex.6.41: *Let $ABCDEF$ be a pentagonal pyramid, with apex A and regular faces. Prove that $AB \perp DE$.*

- Ex.6.42: *Let ABC, ACD be faces of the icosahedron X. Denoting the opposite vertices of A, B, C, D by A', B', C', D' respectively, show that BD' is an edge of X, deduce that $AC \perp BD'$, and hence that the planes of the rectangles $ACA'C'$ and $BD'B'D$ are perpendicular. Where is the third rectangle, perpendicular to both the others?*

- Ex.6.43: *Let $\tau = \frac{1+\sqrt{5}}{2}$, the golden number. Prove that, if $PQRST$ is a regular pentagon, then $|PR|/|PQ| = \tau$, and deduce that the rectangles obtained above are golden rectangles: their sides are in the ratio $\tau : 1$.*

- Ex.6.44: *Prove that the vertices of the three rectangles*

$$(\pm\tau, \pm 1, 0), \quad (0, \pm\tau, \pm 1), \quad (\pm 1, 0, \pm\tau)$$

are the 12 vertices of an icosahedron.

In the last exercise, one of the sets of mutually perpendicular planes is just the three coordinate planes. Label the five *sets* of rectangles 1, 2, 3, 4, 5. Each symmetry of X permutes these sets—the *sets*, not (necessarily) the *elements* of any of the sets—and thus gives a permutation of $\{1, 2, 3, 4, 5\}$. We obtain a homomorphism $\alpha : \mathcal{S}(X) \to S_5$, and we claim that the restriction of α to $\mathcal{S}^+(X)$ gives an isomorphism $\mathcal{S}^+(X) \to A_5$.

- Ex.6.45: *Show that the $\frac{1}{3}$-turn about the line joining the centres of two opposite faces maps to a 3-cycle in S_5, and show that all possible 3-cycles occur thus.*

By Exercise 6.18, the image $\alpha(\mathcal{S}^+(X))$ includes the whole of A_5, and since $|\mathcal{S}^+(X)| = |A_5|$, we have our result: $\mathcal{S}^+(X) \cong A_5$. The fact that $\mathcal{S}(X) \cong A_5 \times C_2$ follows by an argument similar to that for the cube.

There are many alternative ways to obtain the above isomorphisms. First, one could work with the dodecahedron instead. This has 20 vertices, and it is possible to choose a subset of 8 of them which lie at the vertices of a cube, which is thus *inscribed* in the dodecahedron (Fig. 6.20). There are 5 cubes so inscribed, which are permuted by

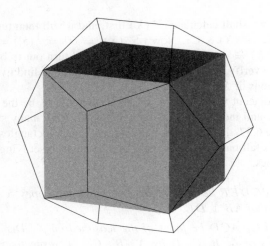

Fig. 6.20   A cube inscribed in a dodecahedron

rotations of the dodecahedron. The geometrical details are probably no harder and no easier than the above, and the reader is invited to have a go.

One alternative argument uses practically no geometry, but quite a lot of group theory, and so is for keen group theorists only. Here is a sketch; you will need to know about group actions and simple groups to follow it. (All of the necessary group theory can be found in [6].) The icosahedron has six (main) diagonals which are permuted by its rotations, and so it is easy to get an isomorphism $\mathcal{S}^+(X) \cong G \subset S_6$. It is a remarkable fact, and all we need to know to finish, that *every* subgroup of S_6 with 60 elements is isomorphic to A_5. Now every group of order 30 contains an element of order 15, and in consequence S_6 does *not* have a subgroup of order 30. But $|G \cap A_6| = 30$ or 60, so it is 60, and thus $G \subset A_6$, and $[A_6 : G] = 6$. Then A_6 acts on the six left cosets of G, and the action is faithful since A_6 is simple. By restriction, we get a faithful action of G on these six cosets where G stabilizes itself, and thus we get a faithful action of G on the five non-trivial cosets, that is, a monomorphism from G to S_5. Since S_5 also has no subgroup of order 30, we obtain $G \cong A_5$, and we are done. (There are some obvious subgroups of S_6 isomorphic to S_5, such as all the permutations that fix 6, or all the permutations that fix 5, or ... ; but there are also some such subgroups which are highly non-obvious, which is why the above proof is so technical.)

● Ex.6.46:   *Mimic the method of Exercise 6.25 to give a direct proof that A_5 has no subgroup of order 30.*

Suppose we count the faces, vertices and edges of a convex polyhedron, and find F faces, V vertices and E edges. There is a famous formula from topology (Euler's formula) relating these numbers: $F + V - E = 2$. Proofs can be found, for instance, in [9] and [10].

• Ex.6.47: *Let X be the regular polyhedron* $\{n, m\}$. *Show that* $nF = mV = 2E = |S^+(X)|$, *and use Euler's formula to deduce that*

$$|S^+(X)| = \frac{4nm}{4 - (n-2)(m-2)}.$$

Note that, by Exercise 6.34, this last expression makes sense.

Euler's formula generalizes to higher dimensions; to see what to do, we need a better notation. So, in some polytope, let N_0 be the number of 0-cells (vertices), let N_1 be the number of 1-cells (edges), and so on. Euler's formula for a 3-polytope (a polyhedron) now reads $N_0 - N_1 + N_2 = 2$, or $N_0 - N_1 + N_2 - N_3 = 1$, since the polyhedron is itself a 3-cell, and thus $N_3 = 1$. Let us check what happens in *lower* dimensions: a 2-polytope is just a convex polygon, a bounded intersection of a finite number of half-planes. If the polygon is an n-gon, then $N_0 = n = N_1$ and $N_2 = 1$, so $N_0 - N_1 + N_2 = 1$. A 1-polytope is an intersection of half-lines in $\mathbb{R}$, so is an interval: $N_0 = 2$ and $N_1 = 1$, whence $N_0 - N_1 = 1$. Finally, if you insist, a 0-polytope can only be a single point: $N_0 = 1$. The pattern seems clear, and so we guess that for a convex n-polytope,

$$N_0 - N_1 + N_2 - \cdots + (-1)^n N_n = 1. \tag{6.4}$$

Actually, topologists are such peculiar people that they cannot bear the sight of that number 1 cluttering up their formula, so they invent a mysterious term N_{-1}, which they secretly put equal to 1, and rewrite (6.4) as $N_{-1} - N_0 + N_1 - \cdots = 0$. For a proof of (6.4), see [10].

• Ex.6.48: *Verify (6.4) for the hypercube.*

6.7   FINITE GROUPS OF ISOMETRIES OF $\mathbb{R}^3$

In Section 5.4 we found all the finite groups of isometries of $\mathbb{R}^2$—Leonardo da Vinci's theorem—and we now do something similar for $\mathbb{R}^3$.

Proposition 6.18.   *Let G be a finite group of isometries of* $\mathbb{R}^3$. *Then the elements of G have a common fixed point.*

Proof.   This is the three-dimensional version of Lemma 5.42, and the proof is exactly the same, except that it needs to be translated into vector and matrix notation. We leave the details to the reader.   □

So we need only concern ourselves with finite subgroups of O_3. Since O_2 can be thought of as a subgroup of O_3 (by identifying $\mathbf{A} \in O_2$ with $\mathbf{A} \oplus \mathbf{I}_1 \in O_3$), we shall obtain all the cyclic and dihedral groups again; and we have been finding a few other examples, such as S_4 and A_5. What else might we expect? Our next proposition essentially says we may confine our search to SO_3:

Proposition 6.19.   *Let G be a subgroup of* O_3. *Then there is a subgroup H of* SO_3 *such that either* $G \cong H$ *or else* $G \cong H \times C_2$.

Proof. We construct a homomorphism $\alpha : O_3 \to SO_3$ as follows: for $\mathbf{A} \in O_3$, define $\alpha(\mathbf{A}) = (\det \mathbf{A})\mathbf{A}$. It is trivial to check that this is a homomorphism, and its kernel is $\{1, \iota\} \cong C_2$, where ι is the central inversion $\iota : \mathbf{u} \mapsto -\mathbf{u}$. Define $H = \alpha(G)$, a subgroup of SO_3.

Case 1: Suppose $\iota \notin G$. Then the kernel of the restriction $\alpha : G \to H$ is trivial, whence $G \cong H$. (If it happens that $G \subseteq SO_3$, then in fact $G = H$.)

Case 2: Suppose $\iota \in G$. Then the map $\beta : G \to H \times C_2$ given by $\beta(\mathbf{A}) = (\alpha(\mathbf{A}), \det \mathbf{A})$ is clearly a homomorphism; to see that it is an isomorphism, check that $\ker \beta$ is trivial (easy), and also, that, for $\mathbf{A} \in SO_3$ and $\lambda = \pm 1$, we have $\beta(\lambda \mathbf{A}) = (\mathbf{A}, \lambda)$.   □

As a special case, note that $O_3 \cong SO_3 \times C_2$.

- Ex.6.49: $O_n \cong SO_n \times C_2$, *for all n: true or false?*

Corollary 6.20.   *In case 1 of the last proposition, if $G \nsubseteq SO_3$, then H has a subgroup K of index 2.*

Proof.   Put $K = G \cap SO_3$, so that $[G : K] = 2$. Applying the isomorphism $\alpha : G \to H$, we have $[H : \alpha(K)] = 2$, or $[H : K] = 2$, since α acts trivially on SO_3.   □

As an example of how to apply this, we have seen that both A_4 and A_5 can occur as groups of isometries of $\mathbb{R}^3$; but since by Exercises 6.25 and 6.46 neither of these has a subgroup of index 2, they can in fact only appear as groups of *direct* isometries of $\mathbb{R}^3$. Contrast this with the behaviour of S_4, which can appear inside O_3 as symmetries of a tetrahedron (including some opposite isometries), or as rotations of a cube (direct isometries only). The former can happen because S_4 has a subgroup of index 2, namely A_4, the *rotations* of a tetrahedron.

We shall concentrate now on finding the finite subgroups of SO_3. Let G be such a subgroup, with $|G| = n > 1$. By Exercise 6.6, $G \subset S^+(S^2)$. The non-trivial elements of G are rotations, by Corollary 6.13, so each one has an axis meeting the 2-sphere S^2 in two diametrically opposite points, which we call *poles* of G. Then, if $\mathbf{x} \in S^2$, we define the *stabilizer* of $\mathbf{x}$ to be

$$G_{\mathbf{x}} = \{ f \in G : f(\mathbf{x}) = \mathbf{x} \},$$

and it is immediate that (i) $G_{\mathbf{x}}$ is a subgroup of G, and (ii) $\mathbf{x}$ is a pole of G iff $|G_{\mathbf{x}}| > 1$.

- Ex.6.50: *Let $\mathbf{x}$ be a pole of G, so that the diameter of S^2 through $\mathbf{x}$ is a line of fixed points for every element of $G_{\mathbf{x}}$. Use Exercise 6.11 and Leonardo da Vinci's theorem to deduce that $G_{\mathbf{x}}$ is cyclic.*

If $G_{\mathbf{x}} \cong C_r$, we say that the pole $\mathbf{x}$ is *of order r*, or is an *r-pole* of G; the corresponding axis is likewise *of order r*, or is an *r-axis* of G.

As in Lemma 5.42, we define the *orbit* of $\mathbf{x} \in S^2$ to be $\mathbf{x}^G = \{ f(\mathbf{x}) : f \in G \}$.

Proposition 6.21.   (Orbit–stabilizer theorem) $|\mathbf{x}^G||G_{\mathbf{x}}| = |G|$.

Proof.   An equivalent way of stating the theorem is (dividing through by $|G_{\mathbf{x}}|$) that $[G : G_{\mathbf{x}}] = |\mathbf{x}^G|$. We have to show that the elements of the orbit of $\mathbf{x}$ correspond to the cosets of its stabilizer. Now, as f runs through the elements of G, then $f(\mathbf{x})$ runs through the elements of the orbit $\mathbf{x}^G$, but (in general) with repetitions. All we have to do is to show that the repetitions come from elements of G lying in the same coset of $G_\mathbf{x}$. So let $f, g \in G$. We have

$$f(\mathbf{x}) = g(\mathbf{x}) \iff g^{-1}f(\mathbf{x}) = \mathbf{x} \iff g^{-1}f \in G_\mathbf{x} \iff f \in gG_\mathbf{x},$$

which completes the proof. $\qquad\qquad\square$

Corollary 6.22. *Let* $\mathbf{x}$ *be an* r-*pole of* G. *Then every element of* $\mathbf{x}^G$ *is an* r-*pole of* G.

Proof. If $\mathbf{y} \in \mathbf{x}^G$, then $\mathbf{y}^G = \mathbf{x}^G$, so $|G_\mathbf{y}| = |G|/|\mathbf{y}^G| = |G|/|\mathbf{x}^G| = |G_\mathbf{x}| = r$. $\qquad\square$

- Ex.6.51: *Do this again, by geometry: if* $f, g \in SO_3$, *then* g *and* fgf^{-1} *are rotations through the same angle; and if* $\mathbf{x}$ *lies on the axis of* g, *then* $f(\mathbf{x})$ *lies on the axis of* fgf^{-1}.

- Ex.6.52: *Deduce that if* $\mathbf{z} \in S^2$ *is very close to the* r-*pole* $\mathbf{x}$ *(closer than any other pole, and in particular not a pole itself), then* $\mathbf{z}^G$ *consists of the vertices of a regular* r-*gon surrounding each of the points of* $\mathbf{x}^G$. *Hence give another proof of the orbit–stabilizer theorem.*

Notice that it may or may not be the case that poles of the same order belong to the same orbit. For example, the rotations of a cube give twelve 2-poles, one for each edge, and eight 3-poles, one for each vertex, and six 4-poles, one for each face, and in this case any two poles of the same order lie in the same orbit. But if we consider the rotations of a tetrahedron, there are six 2-poles, one for each edge, and eight 3-poles, four corresponding to the vertices, and with the diametrically opposite ones corresponding to the faces. This time, no 3-pole lies in the same orbit as the opposite 3-pole.

Let us return to our group $G \subset SO_3$, of order n. Make a (finite!) list of all the poles of G, and from this list select *one* from each orbit, say $\mathbf{x}_1, \mathbf{x}_2, \ldots, \mathbf{x}_s$. So no two of these are in the same orbit, but every pole is in the same orbit as one of them.

Lemma 6.23. *With* n, s, r_i *as above, we have*

$$2\left(1 - \frac{1}{n}\right) = \sum_{i=1}^{s}\left(1 - \frac{1}{r_i}\right). \tag{6.5}$$

Proof. Consider the set of all pairs $(f, \mathbf{x})$ such that f is a non-trivial element of G and $\mathbf{x}$ is a pole of f. Since each rotation has two poles, we have $2(n-1)$ such pairs. On the other hand, given a pole $\mathbf{x}$, there is a value of i such that $\mathbf{x}$ and $\mathbf{x}_i$ are in the same orbit, and thus $G_\mathbf{x} \cong C_{r_i}$. This means there are $r_i - 1$ non-trivial $f \in G$ with pole $\mathbf{x}$, that is $r_i - 1$ pairs $(f, \mathbf{x})$ with *this* $\mathbf{x}$. Since there are n/r_i poles in the orbit of $\mathbf{x}$, each giving rise to the same number of pairs, we can sum over all the (orbits of) poles to obtain

$$2(n - 1) = \sum_{i=1}^{s} \frac{(r_i - 1)n}{r_i},$$

whence (6.5). $\qquad\qquad\square$

Lemma 6.24.   *In (6.5), we must have $s = 2$ or 3.*

Proof.   Since $n \geq 2$, we have $1 > 1 - \frac{1}{n} \geq 1 - \frac{1}{2} = \frac{1}{2}$, so

$$2 > 2\left(1 - \frac{1}{n}\right) \geq 1. \tag{6.6}$$

Similarly, since $r_i \geq 2$, we have $1 > 1 - \frac{1}{r_i} \geq 1 - \frac{1}{2} = \frac{1}{2}$, so

$$s > \sum_{i=1}^{s}\left(1 - \frac{1}{r_i}\right) \geq \frac{s}{2}. \tag{6.7}$$

From (6.5), (6.6) and (6.7), we have $2 > \frac{s}{2}$ and $s > 1$, or $4 > s > 1$, whence $s = 2$ or 3. $\square$

Proposition 6.25.   *The solutions of (6.5) are as follows:*

	s	n	r_1	r_2	r_3	$\frac{n}{r_1}$	$\frac{n}{r_2}$	$\frac{n}{r_3}$
Case 1:	2	n	n	n	$-$	1	1	$-$
Case 2:	3	$2m$	2	2	m	m	m	2
Case 3:	3	12	2	3	3	6	4	4
Case 4:	3	24	2	3	4	12	8	6
Case 5:	3	60	2	3	5	30	20	12

where we are assuming $r_1 \leq r_2 \leq r_3$, and in cases 1 and 2 the values of n and m (respectively) are arbitrary.

Proof.   If $s = 2$, then

$$2\left(1 - \frac{1}{n}\right) = \left(1 - \frac{1}{r_1}\right) + \left(1 - \frac{1}{r_2}\right),$$

whence

$$\frac{1}{r_1} + \frac{1}{r_2} = \frac{2}{n}, \quad \text{or} \quad \frac{n}{r_1} + \frac{n}{r_2} = 2.$$

Since the terms on the left are both positive integers, we deduce

$$\frac{n}{r_1} = 1 = \frac{n}{r_2},$$

or $r_1 = r_2 = n$, which is case 1.

So now assume $s = 3$. We have

$$2\left(1 - \frac{1}{n}\right) = \left(1 - \frac{1}{r_1}\right) + \left(1 - \frac{1}{r_2}\right) + \left(1 - \frac{1}{r_3}\right),$$

or

$$\frac{1}{r_1} + \frac{1}{r_2} + \frac{1}{r_3} = 1 + \frac{2}{n} > 1.$$

Suppose $r_1 \geq 3$. Then

$$\frac{1}{r_1} + \frac{1}{r_2} + \frac{1}{r_3} \leq \frac{1}{3} + \frac{1}{3} + \frac{1}{3} = 1,$$

which is impossible. So we must have $r_1 = 2$ in all remaining cases, and

$$\frac{1}{r_2} + \frac{1}{r_3} = \frac{1}{2} + \frac{2}{n} > \frac{1}{2}.$$

Next, if $r_2 \geq 4$, then

$$\frac{1}{r_2} + \frac{1}{r_3} \leq \frac{1}{4} + \frac{1}{4} = \frac{1}{2},$$

which is impossible. Thus $r_2 = 2$ or 3.
    When $r_2 = 2$, then

$$\frac{1}{r_3} = \frac{2}{n}, \quad \text{or} \quad r_3 = \frac{n}{2};$$

so $n = 2m$ say, and $r_3 = m$. This is case 2.
    When $r_2 = 3$, then

$$\frac{1}{r_3} = \frac{1}{6} + \frac{2}{n} > \frac{1}{6},$$

so that $r_3 < 6$, or in other words $r_3 = 3, 4,$ or 5, and

$$n = \frac{12 r_3}{6 - r_3} = 12, \ 24, \ \text{or} \ 60$$

respectively. This gives the remaining three cases.   □

We shall now find the group G in each of the five cases.

Case 1: We have only one axis, with an n-pole at either end, and so $G \cong C_n$.

Case 2: Assume $m > 2$. There is only one m-axis, ℓ, with both its poles in the same orbit. Therefore the half-turns corresponding to the 2-poles have *their* axes orthogonal to ℓ, since each of them interchanges the two m-poles. We need:

- Ex.6.53: *If* $\rho \in S_m$ *has order* m, *then* ρ *need not be an* m-*cycle.*
- Ex.6.54: *If* $\rho \in S_m$ *has order* m *and has no fixed points, then* ρ *need not be an* m-*cycle.*
- Ex.6.55: *If* $\rho \in S_m$ *has order* m, *and if none of* $\rho, \rho^2, \rho^3, \ldots, \rho^{m-1}$ *has any fixed points, then* ρ *is an* m-*cycle.*

We deduce that the 2-poles in either of the orbits are permuted cyclically by one of the rotations through $2\pi/m$, and so lie at the vertices of a regular m-gon, and $G = D_m$, the rotational symmetries of this m-gon.

- Ex.6.56: *Finish this case: show that when $m = 2$, then $G = D_2$, consisting of 1 and three half-turns about mutually orthogonal axes.*

We are left with cases 3, 4, and 5, with $|G| = 12$, 24, and 60 respectively, and we suspect that $G \cong A_4$, S_4, and A_5 respectively. It is not hard to deduce this from the data in the table:

- Ex.6.57: *Case 3: G permutes the four elements in either orbit of 3-poles, and we get a homomorphism $G \to S_4$ whose image has order 12. Deduce that $G \cong A_4$.*

- Ex.6.58: *Case 4: The eight 3-poles lie in twos on four axes of order 3, which are permuted by G, and we get a homomorphism $G \to S_4$. Deduce that $G \cong S_4$.*

- Ex.6.59: *Case 5: The twelve 5-poles lie in twos on six axes of order 5, which are permuted by G, and we get a homomorphism $G \to S_6$ whose image has order 60. Deduce that $G \cong A_5$.*

However, we want a little more than this. In case 2 above, for example, we did not just show that $G \cong D_m$, as groups: we found a regular m-gon whose rotation group was the given group G. Similarly we want to know in case 3, for instance, not just that $G \cong A_4$, but that there is a tetrahedron whose rotation group is precisely the given group G:

Case 3: Let the four 3-poles in one of the orbits be A, B, C, D. There is a $\frac{1}{3}$-turn f about the axis through A which fixes exactly one other 3-pole, which must therefore be in the *other* orbit, and f must act as a 3-cycle on B, C, D. Since f is an isometry, we have $|AB| = |AC| = |AD|$. Similar arguments with $\frac{1}{3}$-turns fixing B, C show that all six of the lengths $|AB|, |AC|, |AD|, |BC|, |BD|, |CD|$ are equal, so the triangles BCD, ACD, ABD, ABC are equilateral, and $ABCD$ is a regular tetrahedron, of which the 12 elements of G are distinct rotational symmetries. Thus $G = A_4$, the rotation group of this tetrahedron.

Case 4: Let the six 4-poles be A, B, C, D, E, F. Since there is only one orbit of 4-poles, they lie in twos on three axes, say AB, CD and EF. So the $\frac{1}{4}$-turn about AB must act as a 4-cycle on C, D, E, F, whence $|AC| = |AD| = |AE| = |AF|$, and also $|BC| = |BD| = |BE| = |BF|$. Similar arguments for the other axes show that all 12 of the line segments $AC, AD, AE, AF, BC, BD, BE, BF, CE, ED, DF, FC$ are of the same length, the eight triangles $ACE, AED, ADF, AFC, BCE, BED, BDF, BFC$ are thus equilateral, and so A, B, C, D, E, F are the vertices of a (regular) octahedron, of which the 24 elements of G are distinct rotational symmetries. Thus $G = S_4$, the rotation group of this octahedron.

- Ex.6.60: *The eight 3-poles are the vertices of a cube (the reciprocal cube), and G is also the rotation group of this cube.*

Case 5: The twelve 5-poles lie in twos on six axes, $\ell_1, \ldots, \ell_6$ say. Any two of these axes lie in a plane, and the four 5-poles they contain lie at the vertices of a rectangle, of which the axes are diagonals. There are $\binom{6}{2} = 15$ of these rectangles, and we claim they are all congruent. First, given a particular 5-axis, ℓ_1 say, there are five other 5-axes, and so five rectangles having ℓ_1 as a diagonal. These are mapped to each other in a 5-cycle by a $\frac{1}{5}$-turn about ℓ_1, and so the five rectangles sharing a diagonal are congruent. Next, given the rectangle with diagonals ℓ_1, ℓ_2 and the rectangle with diagonals ℓ_3, ℓ_4, then the $\frac{1}{5}$-turn about ℓ_1 acts as a 5-cycle on the other five axes, so that there is a rotation in G which fixes ℓ_1 and sends ℓ_2 to ℓ_3. Thus by the earlier argument, the rectangle with diagonals ℓ_1, ℓ_2 is congruent to the rectangle with diagonals ℓ_1, ℓ_3, and again this is congruent to the rectangle with diagonals ℓ_3, ℓ_4. So all 15 of the rectangles are congruent.

- Ex.6.61: *These rectangles cannot be squares.*

Let the sides of these rectangles be of length a, b, where $a < b$. We join with an edge every two 5-poles whose distance apart is a, which gives $15 \times 2 = 30$ edges. Since each 5-pole belongs to five rectangles, it will lie on five edges. Suppose the 5-pole A is joined to B, C, D, E, F: then there is a $\frac{1}{5}$-turn in G that fixes A, and it must act as a 5-cycle on these five points, say $(BCDEF)$. This means that these five points lie in a plane, and in fact $BCDEF$ is a regular pentagon. Thus $|BC| < |BD|$, and no two of these points lie on the same axis, so BC is an edge, and $\triangle ABC$ is equilateral. In this way we obtain five equilateral triangles meeting at A, and at each of the twelve 5-poles, and so $\frac{5 \times 12}{3} = 20$ such triangles in all. Thus the twelve 5-poles are the vertices of a (regular) icosahedron of which the 60 elements of G are distinct rotational symmetries. Thus $G = A_5$, the rotation group of this icosahedron. To sum up:

Proposition 6.26.   *Every finite group of direct isometries of* $\mathbb{R}^3$ *is one of the following:*

> Case i: C_n, *cyclic of order n;*
> Case ii: D_m, *the rotation group of a regular m-gon;*
> Case iii: A_4, *the rotation group of a tetrahedron;*
> Case iv: S_4, *the rotation group of an octahedron;*
> Case v: A_5, *the rotation group of an icosahedron.*

Note that a square (4-gonal) prism, as we have defined it, is a cube, and a triangular (3-gonal) anti-prism is an octahedron. To make the results below work when m (or n) is 3 or 4, we distort the prisms in these two cases by moving the m-gonal faces a little further apart so that, in the case of the square prism, we have two (opposite) square faces and four rectangular faces (not squares); and in the case of the triangular anti-prism we have two (opposite) equilateral triangular faces and six isosceles triangular faces (not equilateral).

- Ex.6.62: *Show that the rotation group of an m-gonal prism is D_m.*
- Ex.6.63: *What is the rotation group of an m-gonal anti-prism?*
- Ex.6.64: *Find a polyhedron whose rotation group is C_n.*

We can now complete our search for finite subgroups of O_3:

Proposition 6.27.   *Let G be a finite subgroup of O_3, not a subgroup of SO_3, and put $G^+ = G \cap SO_3$. Then the possibilities are:*

		G	G^+
Case 1	(a)	$C_n \times C_2$	C_n
	(b)	C_{2n}	C_n
	(c)	D_n	C_n
Case 2	(a)	$D_m \times C_2$	D_m
	(b)	D_{2m}	D_m
Case 3	(a)	$A_4 \times C_2$	A_4
	(b)	S_4	A_4
Case 4		$S_4 \times C_2$	S_4
Case 5		$A_5 \times C_2$	A_5

Here the C_2, where $G = G^+ \times C_2$, stands for $\{1, \iota\}$, where $\iota(\mathbf{u}) = -\mathbf{u}$, for all $\mathbf{u} \in \mathbb{R}^3$.

Proof.   If $\iota \in G$, then by Proposition 6.19, $G = G^+ \times C_2$, and we obtain cases 1(a), 2(a), 3(a), 4 and 5, corresponding to cases 1–5 of Proposition 6.26.

If $\iota \notin G$, then G is isomorphic to one of the groups of Proposition 6.26, by Proposition 6.19, and G has a subgroup G^+ of index 2, that is, half the size of G. Now C_n has a (unique, cyclic) subgroup of index 2 iff n is even, which gives case 1(b). Next, D_m has a cyclic subgroup C_m of index 2, giving case 1(c); and if m is even it also has a dihedral subgroup of index 2, giving case 2(b). As we have already noted, by Exercises 6.25 and 6.46 neither A_4 nor A_5 has any subgroup of index 2; but S_4 does, to wit A_4, and this gives case 3(b).   □

- Ex.6.65:   *Prove that $C_n \times C_2 \cong C_{2n}$ iff n is odd.*

(This is a special case of the *Chinese remainder theorem*, which, expressed in group-theoretic terms, says that $C_n \times C_m \cong C_{nm}$ iff n, m are coprime. For a proof, see [6].)

So does this exercise imply that cases 1(a) and 1(b) of the last proposition coincide if n is odd? The facile answer is 'no', because in case 1(a) we assumed $\iota \in G$ and in case 1(b) we assumed $\iota \notin G$. So should we perhaps have insisted in case 1(b) that n is even, to avoid this apparent contradiction? The answer again is 'no', as C_{2n} can appear in two quite different geometric guises when n is odd. To see this, let f be a rotatory reflection with angle π/n, so that f has order $2n$ and generates a cyclic group G of order $2n$. Here f^2 is direct and has order n, and is a rotation through $2\pi/n$, and generates G^+. Then the unique element of order 2 in G is f^n, which is opposite, and is a rotatory reflection with angle $n\pi/n = \pi$, that is, $f^n = \iota \in G$, so we are in case 1(a). Now instead let f be a rotatory reflection with angle $2\pi/n$. So f^2 is a rotation through $4\pi/n$, and has order n, and f^n is a rotatory reflection with angle 2π, that is, a reflection. So f has order $2n$ still, but this time the group G which it generates does *not* contain the central inversion

ι, because it contains only one element of order 2, and that is a reflection. So we are now in case 1(b), *not* case 1(a), in spite of the fact that the group G obtained is isomorphic to that just obtained in case 1(a).

• Ex.6.66: *Prove that $D_m \times C_2 \cong D_{2m}$ iff m is odd.*

Rather as before, we must ask whether this means that cases 2(a) and 2(b) coincide if m is odd. The answer is 'no', once more:

• Ex.6.67: *Let X be an m-gonal prism, where m is odd. Show that $\iota \notin S(X)$, and that $S(X) = D_{2m}$, case 2(b).*

• Ex.6.68: *Let X be an m-gonal anti-prism, where m is odd. Show that $\iota \in S(X)$, and that $S(X) = D_m \times C_2$, case 2(a).*

• Ex.6.69: *Find the symmetry group of (i) an m-gonal prism, and (ii) an m-gonal anti-prism, when m is even.*

• Ex.6.70: *Find subsets $X \subset \mathbb{R}^3$ whose symmetry groups are each of the groups in the table in Proposition 6.27.*

• Ex.6.71: *Recall that $D_1 \cong C_2$, and show that $D_2 \cong C_2 \times C_2$. So when $n = 2$ and $m = 1$, cases 1(a), 1(c), 2(a) and 2(b) of Proposition 6.27 all give $G \cong C_2 \times C_2$; but are the four cases distinguishable geometrically?*

ANSWERS TO EXERCISES

6.1: If $f_i(\mathbf{v}) = \mathbf{A}_i \mathbf{v} + \mathbf{b}_i$ and $\psi(f_i) = \mathbf{M}_i$, then

$$\mathbf{M}_i = \begin{pmatrix} \mathbf{A}_i & \mathbf{b}_i \\ \mathbf{0} & 1 \end{pmatrix}.$$

Then $f_1 f_2(\mathbf{v}) = \mathbf{A}_1(\mathbf{A}_2 \mathbf{v} + \mathbf{b}_2) + \mathbf{b}_1 = (\mathbf{A}_1 \mathbf{A}_2)\mathbf{v} + (\mathbf{A}_1 \mathbf{b}_2 + \mathbf{b}_1)$, so that

$$\psi(f_1 f_2) = \begin{pmatrix} \mathbf{A}_1 \mathbf{A}_2 & \mathbf{A}_1 \mathbf{b}_2 + \mathbf{b}_1 \\ \mathbf{0} & 1 \end{pmatrix} = \begin{pmatrix} \mathbf{A}_1 & \mathbf{b}_1 \\ \mathbf{0} & 1 \end{pmatrix} \begin{pmatrix} \mathbf{A}_2 & \mathbf{b}_2 \\ \mathbf{0} & 1 \end{pmatrix} = \psi(f_1)\psi(f_2).$$

For the last part, just note that

$$\begin{vmatrix} \mathbf{A} & \mathbf{b} \\ \mathbf{0} & 1 \end{vmatrix} = |\mathbf{A}|.$$

6.2: Initially, put $\lambda_1 = \mu_2 = 1$ and $\mu_1 = 0$, and choose λ_2 so that $\mathbf{u} \cdot (\lambda_2 \mathbf{u} + \mathbf{v}) = 0$, that is, $\lambda_2 = -(\mathbf{u} \cdot \mathbf{v})/(\mathbf{u} \cdot \mathbf{u})$. Now normalize.

6.3: The vectors (x_1, x_2, x_3), (y_1, y_2, y_3), $(1, 1, 1)$ are mutually orthogonal. So normalize: put $\Delta_x = |(x_1, x_2, x_3)| = \sqrt{x_1^2 + x_2^2 + x_3^2}$ and $\Delta_y = |(y_1, y_2, y_3)| = \sqrt{y_1^2 + y_2^2 + y_3^2}$, and note that $|(1, 1, 1)| = \sqrt{3}$. Thus

$$\begin{pmatrix} x_1/\Delta_x & x_2/\Delta_x & x_3/\Delta_x \\ y_1/\Delta_y & y_2/\Delta_y & y_3/\Delta_y \\ 1/\sqrt{3} & 1/\sqrt{3} & 1/\sqrt{3} \end{pmatrix}$$

is an orthogonal matrix, since its rows are an ortho*normal* set. So its *columns* are also an orthonormal set. So in particular, the first column is a unit vector: $x_1^2/\Delta_x^2 + y_1^2/\Delta_y^2 + 1/3 = 1$; and the first two columns are orthogonal: $(x_1/\Delta_x)(x_2/\Delta_x) + (y_1/\Delta_y)(y_2/\Delta_y) + (1/\sqrt{3})(1/\sqrt{3}) = 0$; and these are the required equations.

6.4: $\alpha(1) = 1$ shows $1 \in \ker \alpha$; and if $a, b \in \ker \alpha$ then $\alpha(ab) = \alpha(a)\alpha(b) = 1 \times 1 = 1$, so $ab \in \ker \alpha$; and also $\alpha(a)\alpha(a^{-1}) = \alpha(aa^{-1}) = \alpha(1) = 1$, so $\alpha(a^{-1}) = (\alpha(a))^{-1} = 1^{-1} = 1$, and $a^{-1} \in \ker \alpha$. So we have a subgroup.

If α is injective and $a \in \ker \alpha$, then $\alpha(a) = 1 = \alpha(1)$, so $a = 1$ and $\ker \alpha = \{1\}$. Conversely, if $\ker \alpha = \{1\}$ and $\alpha(a) = \alpha(b)$, then $\alpha(a^{-1}b) = \alpha(a^{-1})\alpha(b) = (\alpha(a))^{-1}\alpha(b) = 1$, so $a^{-1}b \in \ker \alpha$, or $a^{-1}b = 1$, or $b = a$. So α is injective.

6.5: The isomorphism is

$$\begin{pmatrix} \cos\theta & -\sin\theta \\ \sin\theta & \cos\theta \end{pmatrix} \mapsto \cos\theta + i\sin\theta.$$

6.6: If an isometry f fixes $\mathbf{0}$ then, since S^n consists of points whose distance from $\mathbf{0}$ is 1, their images under f are distant 1 from $f(\mathbf{0}) = \mathbf{0}$; so they are in S^n, and f is a symmetry of S^n. Now we must show that a typical symmetry f of S^n fixes $\mathbf{0}$. Put $A = (1, 0, \ldots, 0)$ and $B = (-1, 0, \ldots, 0)$, so that $A, B \in S^n$ and $|AB| = 2$; further, $\mathbf{0}$ is the mid-point of AB. Let $f(A) = A' = (x_1, x_2, \ldots, x_{n+1})$ and $f(B) = B' = (y_1, y_2, \ldots, y_{n+1})$. Since $f(A')$, $f(B') \in S^n$, we have $x_1^2 + \ldots + x_{n+1}^2 = 1$ and $y_1^2 + \ldots + y_{n+1}^2 = 1$, and since $|A'B'| = |AB| = 2$, we have $(x_1 - y_1)^2 + \ldots + (x_{n+1} - y_{n+1})^2 = 4$. Taking twice the first plus twice the second minus the third of these equations gives $(x_1 + y_1)^2 + \ldots + (x_{n+1} + y_{n+1})^2 = 0$, whence $(x_1, \ldots, x_{n+1}) = -(y_1, \ldots, +y_{n+1})$, so that $\mathbf{0}$ is the mid-point of $A'B'$, whence $f(\mathbf{0}) = \mathbf{0}$. The result now follows from Corollary 6.7, noting that $\mathbf{b} = \mathbf{0}$ if $\mathbf{0}$ is a fixed point.

6.7: If $x, y \in X$ and $|x - y| = 2$, then $x = 1$ and $y = -1$ or vice versa. So symmetries of X permute $\{1, -1\}$. In $\mathbb{R}^1$, the only isometry fixing 1 and -1 is $1 : a \mapsto a$, for all a, and the only isometry swapping 1 and -1 is reflection in 0, or $a \mapsto -a$, for all a. So $|S(X)| = 2$.

In $\mathbb{R}^2$, 1 *and* reflection in the x-axis fix 1 and -1; and reflection in the y-axis *and* the half-turn about O swap 1 and -1. So here $|S(X)| = 4$. In $\mathbb{R}^3$, *every* rotation about the x-axis fixes all the points of X, so here $|S(X)| = \infty$.

6.8: The symmetries of X are given by $\mathbf{v} \mapsto \mathbf{Av}$ where $\mathbf{A} = (a_{ij})$ is one of the eight 2×2 matrices

$$\begin{pmatrix} \pm 1 & 0 \\ 0 & \pm 1 \end{pmatrix}, \quad \begin{pmatrix} 0 & \pm 1 \\ \pm 1 & 0 \end{pmatrix}.$$

Then the 16 possible matrices

$$\begin{pmatrix} a_{11} & a_{12} & 0 \\ a_{21} & a_{22} & 0 \\ 0 & 0 & \pm 1 \end{pmatrix}$$

give the 16 symmetries of X in $\mathbb{R}^3$.

To see that X has no other symmetries in $\mathbb{R}^3$, note that any symmetry must fix the origin, and leave the plane $z = 0$ invariant. So it is given by a 3×3 matrix (a_{ij}) with $a_{31} = a_{32} = 0$. This matrix must be orthogonal, so $a_{33} = \pm 1$ and $a_{13} = a_{23} = 0$, and so now

$$\begin{pmatrix} a_{11} & a_{12} \\ a_{21} & a_{22} \end{pmatrix}$$

must be orthogonal, and gives a symmetry of X in $\mathbb{R}^2$.

6.9: To rotate the cube, choose which of the 6 faces to rotate to the top, and then which of 4 rotations (including the trivial one) to apply, leaving the top face invariant. This is $6 \times 4 = 24$

choices, so $|\mathcal{S}^+(X)| = 24$. (Or choose which of the 8 vertices to bring to a chosen position, and then which of 3 rotations about the corresponding main diagonal to apply, and then $8 \times 3 = 24$. Or again, choose which of the 12 edges to bring to a chosen position, and which of 2 ways round to place it, and then $12 \times 2 = 24$.) The cube *has* symmetries which are opposite isometries (e.g. reflection in $x = 0$), so $\mathcal{S}^+(X)$ has at least two cosets (including itself) in $\mathcal{S}(X)$. If f, g are opposite isometries in $\mathcal{S}(X)$, then $f^{-1}g$ is direct, so $f^{-1}g \in \mathcal{S}^+(X)$, or $g\,\mathcal{S}^+(X) = f\,\mathcal{S}^+(X)$. So $\mathcal{S}^+(X)$ has *exactly* two cosets, and so $|\mathcal{S}(X)| = 2 \times 24 = 48$.

The 48 symmetries of X are given by 3×3 matrices. (These matrices are listed explicitly on page 136.) To get the corresponding 4×4 matrices of the symmetries in $\mathbb{R}^4$, use a similar method to that used in Exercise 6.8: just enlarge each of these 48 matrices by placing ± 1 in the $(4, 4)$ position and filling out the final row and column with zeros. This gives $2 \times 48 = 96$ matrices, all the symmetries of X in $\mathbb{R}^4$.

6.10: No; for example, we could replace $\mathbf{u}$ by $e^{i\alpha}\mathbf{u}$ (still an eigenvector corresponding to λ), which would replace $\mathbf{v}$ by $(\cos \alpha)\mathbf{v} - (\sin \alpha)\mathbf{w}$ and $\mathbf{w}$ by $(\sin \alpha)\mathbf{v} + (\cos \alpha)\mathbf{w}$, a rotation of axes in π.

6.11: Let $\mathbf{x}_1, \ldots, \mathbf{x}_r$ be a basis of X, and extend to a basis $\mathbf{x}_{r+1}, \ldots, \mathbf{x}_n$ of $\mathbb{R}^n$. Apply the Gram–Schmidt process to this basis to obtain an orthonormal basis $\mathbf{y}_1, \ldots, \mathbf{y}_n$ of $\mathbb{R}^n$. Here $\mathbf{y}_1$ is the normalization of $\mathbf{x}_1$; then $\mathbf{y}_2$ is a certain linear combination of $\mathbf{y}_1$ and $\mathbf{x}_2$; $\mathbf{y}_3$ is a certain linear combination of $\mathbf{y}_1$, $\mathbf{y}_2$, and $\mathbf{x}_3$; and so on. It follows that, for each i, the sets $\mathbf{x}_1, \ldots, \mathbf{x}_i$ and $\mathbf{y}_1, \ldots, \mathbf{y}_i$ span the same subspace; and in particular, $\mathbf{y}_1, \ldots, \mathbf{y}_r$ is an orthonormal basis of X. We claim $\mathbf{y}_{r+1}, \ldots, \mathbf{y}_n$ is a basis (orthonormal, of course) of $X^\perp$. For certainly if $\mathbf{u} = \lambda_1\mathbf{y}_1 + \cdots + \lambda_r\mathbf{y}_r$ and $\mathbf{v} = \mu_{r+1}\mathbf{y}_{r+1} + \cdots + \mu_n\mathbf{y}_n$, then the orthogonality shows that $\mathbf{v} \in X^\perp$. Conversely, if $\mathbf{w} \in X^\perp$, then $\mathbf{w} = \nu_1\mathbf{y}_1 + \cdots + \nu_n\mathbf{y}_n$, say; and since $\mathbf{w}.\mathbf{y}_i = 0$ for $1 \le i \le r$, we deduce that $\nu_i = 0$ for $1 \le i \le r$, so that $\mathbf{w}$ is a linear combination of $\mathbf{y}_{r+1}, \ldots, \mathbf{y}_n$. So $\dim X^\perp = n - r = n - \dim X$.

It is clear from the above that every element of $\mathbb{R}^n$ can be written uniquely in the form $\mathbf{u} + \mathbf{v}$ with $\mathbf{u} \in X$ and $\mathbf{v} \in X^\perp$, and this means that $\mathbb{R}^n = X \oplus X^\perp$. To see that $X^{\perp\perp} = X$, just write down the basis $\mathbf{y}_1, \ldots, \mathbf{y}_n$ in reverse order, $\mathbf{y}_n, \ldots, \mathbf{y}_1$, and apply the above argument to show that $\mathbf{y}_r, \ldots, \mathbf{y}_1$ is a basis of $X^{\perp\perp}$; *or* note that (straight from the definitions) $X \subseteq X^{\perp\perp}$ (OK?), and $\dim X^{\perp\perp} = n - \dim X^\perp = n - (n - r) = r = \dim X$.

Now suppose X is f-invariant, and pick $\mathbf{y} \in X^\perp$. Then, for all $\mathbf{x} \in X$, we have $\mathbf{x} \cdot \mathbf{y} = 0$ and thus $f(\mathbf{x}) \cdot f(\mathbf{y}) = 0$, by Proposition 6.3. Now f is bijective, so the restriction $f : X \to X$ is injective, hence surjective also. (OK?) So for any $\mathbf{u} \in X$, we can find $\mathbf{x} \in X$ with $f(\mathbf{x}) = \mathbf{u}$, whence $\mathbf{u} \cdot f(\mathbf{y}) = 0$, and $f(\mathbf{y}) \in X^\perp$, as required.

6.12: A rotation, such as $\mathbf{I}_2 \oplus \mathbf{A}_\alpha$, has a *plane* of fixed points (its axis; in $\mathbb{R}^4$ rotations are about a *plane*). In particular, 1 is an eigenvalue. But 1 is *not* an eigenvalue of $\mathbf{A}_\theta \oplus \mathbf{A}_\varphi$ (whether or not $\theta = \varphi$), so the double rotation has no fixed points other than $\mathbf{0}$, and is *not* a rotation.

6.13: The points $(1, 0)$, (x, y), $(0, t)$ in $\mathbb{R}^2$ are collinear iff

$$\begin{vmatrix} 1 & x & 0 \\ 0 & y & t \\ 1 & 1 & 1 \end{vmatrix} = 0,$$

or $y - t + xt = 0$, or $t = y/(1 - x)$. So stereographic projection $\psi_1 : S^1 \to R^1$ is given by $\psi_1(x, y) = y/(1 - x)$. The formulae for ψ_2 and ψ_3 are immediate. Then $\psi_3(a \cos \alpha, a \sin \alpha, b \cos \beta, b \sin \beta) = (a \sin \alpha/(1 - a \cos \alpha), b \cos \beta/(1 - a \cos \alpha), b \sin \beta/(1 - a \cos \alpha))$. Changing β merely rotates this last point about the x-axis, so the image of T^2 is the rotation about the x-axis of the locus $x = a \sin \alpha/(1 - a \cos \alpha)$, $y = b/(1 - a \cos \alpha)$. This yields $b/y = 1 - a \cos \alpha$, or $a \cos \alpha = 1 - b/y$, and $a \sin \alpha = bx/y$, so $(bx/y)^2 + (1 - b/y)^2 = a^2$, or $b^2x^2 + y^2 - 2by + b^2 = a^2y^2$, or $b^2x^2 + b^2y^2 - 2by + 1 = a^2$ (using $a^2 + b^2 = 1$), and this is the required circle. Adding θ to α moves (x, y) around this circle (though not as a rotation), and adding φ to β gives a rotation through φ about the x-axis. Thinking of the torus as a cycle inner tube mounted on a cylindrical wheel-rim, the angle φ corresponds to rotating the wheel about its axle, whereas the angle θ corresponds to rolling the tyre sideways, as if to remove it from the rim of the wheel. This latter movement is not an isometry, as the parts of the tube initially in contact with the rim have to stretch as the tube is rolled; and other parts that now come into contact with the rim have to shrink. The three-dimensional projection of the double rotation thus involves rotating the wheel while at the same time trying to get the tube off the rim, and is not recommended as a practical experiment!

6.14: First the isometries that fix $\mathbf{0}$. (a) $\mathbf{I}_4$, the identity. (b) $\mathbf{I}_3 \oplus (-1)$, reflection (in a 3-space, since we are in $\mathbb{R}^4$). (c) $\mathbf{I}_2 \oplus (-\mathbf{I}_2)$, half-turn (about a plane). (d) $(1) \oplus (-\mathbf{I}_3)$. A (central) 3-inversion? (e) $-\mathbf{I}_4$. This is $(-\mathbf{I}_2) \oplus (-\mathbf{I}_2)$, so it is a double rotation with both angles equal to π: a double half-turn. (f) $\mathbf{I}_2 \oplus \mathbf{A}_\theta$, rotation (about a plane). (g) $(1) \oplus (-1) \oplus \mathbf{A}_\theta$. This is (plane) reflection $\oplus$ (plane) rotation, so call it a reflective rotation? (It is also $(1)\oplus$ a rotatory reflection, so maybe call this a rotatory reflection also?) (h) $(-\mathbf{I}_2) \oplus \mathbf{A}_\theta$ is a double rotation (with angles π, θ). (i) $\mathbf{A}_\theta \oplus \mathbf{A}_\varphi$ is a double rotation.

Now compose these with translations, as in Theorem 6.12. In cases (e), (h), and (i) there is no line of fixed points. In the other cases, we get: (a) a translation; (b) a glide-reflection; (c) a screw half-turn; (d) a (half) screw reflection? (f) a screw displacement; (g) a screw reflection?

6.15: We have $1 \mapsto$ any of the n numbers, then $2 \mapsto$ any of the remaining $n - 1$ numbers, and so on, so $n(n - 1) \dots (2)(1) = n!$ choices in all, and $|S_n| = n!$. Then A_n is a subgroup, and if ρ is odd, then the coset ρA_n consists entirely of odd permutations; and if σ is odd, then $\rho^{-1}\sigma$ is even, or $\rho^{-1}\sigma \in A_n$, or $\sigma \in \rho A_n$. Thus A_n has index 2, and $|A_n| = \frac{1}{2}(n!)$.

6.16: Take $\rho = (12)$. Then $\text{sign}(\rho) = \frac{2-1}{1-2} = -1$. I worked this out thinking of (12) as being an element of S_2; if we think of it as lying in S_3 then the calculation goes $\text{sign}(\rho) = (\frac{2-1}{1-2})(\frac{3-1}{3-2})(\frac{3-2}{3-1}) = -1$ again. More generally, if $\sigma \in S_n$, and we put S_n inside S_{n+1} as permutations fixing $n+1$, then it doesn't matter whether we compute $\text{sign}(\sigma)$ thinking of σ as belonging to S_n or S_{n+1}: we get the same answer. Prove this. It follows that *every* 2-cycle is odd. For 3-cycles, either work out $\text{sign}(123)$ from the formula, or note that $(123) = (13)(12)$ (remembering that we compose from the *right*), so that $\text{sign}(123) = (-1)^2 = 1$. The general result follows from the formula in the solution to Exercise 6.17 below, which shows that an m-cycle is a product of $m - 1$ 2-cycles, and so has sign $(-1)^{m-1}$.

6.17: $(123 \dots m) = (1m)(1\,m - 1) \dots (13)(12)$, and similarly for other m-cycles.

6.18: Let i, j, k, ℓ be distinct. Then $(ij)(ik) = (ikj)$, and $(ij)(k\ell) = (ij)(ik)(ki)(k\ell) = (ikj)(k\ell i)$.

6.19: Given $f \in S(X)$, the restriction of f to the vertex set $\{1, 2, 3, 4\}$ of X is a permutation $\rho \in S_4$, and it is immediate that this gives a homomorphism $S(X) \to S_4$. Now if $f, g \in S(X)$ map to the *same* $\rho \in S_4$, then $g^{-1}f$ fixes all four vertices of X, so $g^{-1}f = 1$, by Corollary 5.6, and $f = g$. So our homomorphism is injective, and $S(X) \cong$ the image of our homomorphism, G say. But the quarter-turn rotational symmetries of the square, which are direct, give rise to 4-cycles in S_4, which are odd, so our homomorphism does *not* map $S^+(X)$ to $G \cap A_4$.

6.20: $A_4 = \{1, (123), (132), (124), (142), (134), (143), (234), (243), (12)(34), (13)(24), (14)(23)\}$, and $S_4 = A_4 \cup \{(12), (13), (14), (23), (24), (34), (1234), (1243), (1324), (1342), (1423), (1432)\}$.

6.21: Since $S_3 \cong D_3$, we take the symmetries of a suitable equilateral triangle; to get 2×2 (rather than 3×3) matrices, we take the centroid of our triangle at O. For example, if the vertices are

$(1, 0)$ and $(-\frac{1}{2}, \pm\frac{\sqrt{3}}{2})$, then the six matrices are

$$\begin{pmatrix} 1 & 0 \\ 0 & 1 \end{pmatrix}, \quad \begin{pmatrix} -1/2 & -\sqrt{3}/2 \\ \sqrt{3}/2 & -1/2 \end{pmatrix}, \quad \begin{pmatrix} -1/2 & \sqrt{3}/2 \\ -\sqrt{3}/2 & -1/2 \end{pmatrix}, \quad \begin{pmatrix} 1 & 0 \\ 0 & -1 \end{pmatrix},$$

$$\begin{pmatrix} -1/2 & \sqrt{3}/2 \\ \sqrt{3}/2 & 1/2 \end{pmatrix}, \quad \begin{pmatrix} -1/2 & -\sqrt{3}/2 \\ -\sqrt{3}/2 & 1/2 \end{pmatrix}.$$

6.22: $|AB| = |(0, 2, 2)| = 2\sqrt{2}$, and similarly AC, AD, BC, BD, CD all have length $2\sqrt{2}$. So the triangles ABC, ABD, ACD, BCD are congruent equilateral triangles, three meeting at each of A, B, C, D, and we have a regular tetrahedron. The perpendicular bisector of AB consists of all points equidistant from A and B, so it contains C and D. (Its equation is $(x - 1)^2 + (y - 1)^2 + (z - 1)^2 = (x - 1)^2 + (y + 1)^2 + (z + 1)^2$, or $y + z = 0$, and C, D satisfy this equation.)

6.23: The half-turn corresponding to $(12)(34)$ fixes the mid-point of the edge 12 and the mid-point of the edge 34, so the axis is the line joining these mid-points (of two opposite edges). We now have all of $S^+(X)$: 1, and the eight $\frac{1}{3}$-turns and the three half-turns. So (1234) must correspond to an opposite isometry, and its order is 4, so it is not a reflection or central inversion. (We already have the six reflections in $S(X)$, corresponding to the six 2-cycles in S_4.) So (1234) must be a rotatory reflection, with angle $\pi/2$; and since $(1234)^2 = (13)(24)$, its axis is the line joining the mid-points of edges 13 and 24.

6.24: Let the vertices 1, 2, 3, 4 be $(1, 1, 1)$, $(1, -1, -1)$, $(-1, 1, -1)$, and $(-1, -1, 1)$ respectively. The mid-points of edges 13 and 24 are $(0, 1, 0)$ and $(0, -1, 0)$, so the axis of the rotatory reflection corresponding to (1234) is the line joining these points, the y-axis; and the perpendicular bisector is the (x, z)-plane, $y = 0$. This is the plane of reflection of the rotatory reflection, so the restriction to this plane is a rotation through $\pi/2$, a quarter-turn. So we expect the section of the tetrahedron by $y = 0$ to be a plane figure with a quarter-turn symmetry. Indeed, the section is a *square*, whose vertices are where $y = 0$ meets the four edges which are not parallel to it, that is, the edges other than 13 and 24. It meets these other edges at their mid-points, and these are $(\pm 1, 0, 0)$ and $(0, 0, \pm 1)$, which do indeed form a square.

6.25: If $\rho \in H$ then $\rho^2 \in H$. So if $\rho^2 \notin H$, then $\rho \notin H$, and $H \cap \rho H = \emptyset$. This means $|H \cup \rho H| = 6 + 6 = 12$, so that $A_4 = H \cup \rho H$, and since $\rho^2 \notin H$, we must have $\rho^2 \in \rho H$, or $\rho^2 = \rho\sigma$ for some $\sigma \in H$. But now $\rho^{-1}\rho^2 = \rho^{-1}\rho\sigma$, or $\rho = \sigma \in H$, a contradiction. So $\rho^2 \in H$, all $\rho \in A_4$. But every 3-cycle ρ in S_4 is in A_4, and $\rho^3 = 1$, so that $\rho = \rho^{-2} = (\rho^{-1})^2 \in H$. But A_4 contains *eight* 3-cycles, and there just isn't room for them all in H: contradiction.

6.26: Take $f(\mathbf{v}) = -\mathbf{v}$, all $\mathbf{v} \in \mathbb{R}^3$, a central inversion with centre $\mathbf{0}$. This swaps each vertex of X with its opposite vertex, so the main diagonals are all f-invariant.

6.27: Let the vertices of one face of X be A, B, C, D, and let the respective opposite vertices be A', B', C', D'. Suppose $f \in S^+(X)$ has AA', BB', CC', DD' as invariant lines, but $f \neq 1$. Then f cannot fix all eight vertices of X, by Corollary 5.6, so suppose $f(A) \neq A$. Then $f(A) = A'$ and $f(A') = A$, and thus f^2 fixes both A and A', and so has AA' as a line of fixed points, and must be a rotation with axis AA'. But f is also a rotation, so its axis is the same as the axis of f^2, and this means that $f(A) = A'$, a contradiction. So $f = 1$. Is this easier?

6.28: $\mathbf{A} = -\mathbf{B}$, where

$$\mathbf{B} = \begin{pmatrix} 0 & 1 & 0 \\ 0 & 0 & 1 \\ 1 & 0 & 0 \end{pmatrix},$$

a permutation matrix corresponding to a 3-cycle. (Indeed, $\mathbf{Be}_1 = \mathbf{e}_3$, $\mathbf{Be}_3 = \mathbf{e}_2$, and $\mathbf{Be}_2 = \mathbf{e}_1$, where $\mathbf{e}_1$, $\mathbf{e}_2$, $\mathbf{e}_3$ are the standard basis of $\mathbb{R}^3$.) So $\mathbf{B}$ has order 3, and it follows that $\mathbf{A}$ has order 6. Since $|\mathbf{A}| = -1$, we have an opposite isometry, not a reflection or central inversion, since the order is 6. So $\mathbf{A}$ must give a rotatory reflection, with angle of rotation $\pi/3$. Its axis is $x = y = z$, the line through the two vertices $\pm(1, 1, 1)$ which it swaps—OK?—and its plane of reflection is the perpendicular bisector of the diagonal through these two vertices, which is $x + y + z = 0$. (Check that this line and plane are invariant under $\mathbf{A}$.)

   Taking $C_2 = \{1, a\}$, the element $x = ((123), a) \in S_4 \times C_2$ has order 6. (Check: $x^2 = ((123)^2, a^2) = ((132), 1)$, $x^3 = ((123)^3, a^3) = (1, a)$, $x^4 = ((123)^4, a^4) = ((123), 1)$, $x^5 = ((123)^5, a^5) = ((132), a)$, $x^6 = ((123)^6, a^6) = (1, 1) = 1$—the '1' of the group $S_4 \times C_2$, that is.)

6.29: The restriction of $\mathbf{A}$ to this plane is a $\frac{1}{6}$-turn, which must be a symmetry of the section; so we expect a regular hexagon. Indeed, of the 12 edges of the cube, three pass through $(1, 1, 1)$, three pass through $(-1, -1, -1)$, and the other six meet $x + y + z = 0$ at their respective mid-points, which are the vertices of the hexagon in question. They are $\frac{1}{2}((1, 1, -1) + (1, -1, -1)) = (1, 0, 1)$ and its successive images under $\mathbf{A}$, that is, $(0, 1, -1)$, $(-1, 1, 0)$, $(-1, 0, 1)$, $(0, -1, 1)$, $(1, -1, 0)$ (and then back to $(1, 1, 0)$). (Don't forget that to work out the images of these vectors, we write them as *columns* and multiply on the *left* by $\mathbf{A}$.)

6.30: The four points $(\pm 1, 0, 0)$, $(0, \pm 1, 0)$ lie at the vertices of a square of side $\sqrt{2}$, and each of the points $(0, 0, \pm 1)$ is distant $\sqrt{2}$ from each vertex of this square. So we have our two square pyramids, base to base, as required.

6.31: Take for example the edge joining $(1, 1, 1, 1)$ to $(-1, 1, 1, 1)$. This edge is an edge of the three cells (cubes) $(\pm 1, \pm 1, \pm 1, 1)$, $(\pm 1, \pm 1, 1, \pm 1)$, and $(\pm 1, 1, \pm 1, \pm 1)$, but *not* of any of the other five cells, $(\pm 1, \pm 1, \pm 1, -1)$, $(\pm 1, \pm 1, -1, \pm 1)$, $(\pm 1, -1, \pm 1, \pm 1)$, $(1, \pm 1, \pm 1, \pm 1)$, or $(-1, \pm 1, \pm 1, \pm 1)$.

6.32: The hyperplane $x_1 + x_2 + x_3 + x_4 = 0$ contains the six vertices $(1, 1, -1, -1)$, $(1, -1, 1, -1)$, $(1, -1, -1, 1)$, $(-1, 1, 1, -1)$, $(-1, 1, -1, 1)$, $(-1, -1, 1, 1)$ of the hypercube; and no two of these vertices lie on the same edge of the hypercube, so no edge of the hypercube lies in the hyperplane. The edge of the hypercube from $(1, p, q, r)$ to $(-1, p, q, r)$ (where $p, q, r \in \{1, -1\}$) consists of all (λ, p, q, r) with $-1 \leq \lambda \leq 1$; but $p + q + r = \pm 3$ or ± 1, so if $\lambda + p + q + r = 0$, then $\lambda = -(p + q + r) = \pm 1$. So this edge (and likewise any other edge) can only meet the hyperplane at a vertex of the hypercube. So our section has six vertices, and we suspect it may be an octahedron. Indeed, the vertices $(1, -1, 1, -1)$, $(1, -1, -1, 1)$ $(-1, 1, -1, 1)$ and $(-1, 1, 1, -1)$ form a square of side $2\sqrt{2}$, and they are all distant $2\sqrt{2}$ from each of $(1, 1, -1, -1)$ and $(-1, -1, 1, 1)$. So we have two square pyramids, base to base: an octahedron.

6.33: This is the 4×4 version of the argument on page 136. The subgroup $\mathcal{S}(X) \subset O_4$ can be written, as a set, as $C_2^4 \times S_4$, the former being the 4×4 diagonal matrices with ± 1's on the diagonal, and the latter being the 4×4 permutation matrices. We get $\mathcal{S}(X) \cong S_4 \wr C_2$, and $|\mathcal{S}(X)| = |C_2^4| \times |S_4| = 16 \times 24 = 384$.

6.34: Joining vertex 1 to vertices $3, 4, \ldots, n - 1$ splits the n-gon up into $n - 2$ triangles, so the internal angles add up to $(n - 2)\pi$, and each of them is therefore equal to $(n - 2)\pi/n$. (Alternatively, considering the change of direction as you 'walk' around the perimeter of an n-gon shows that the *external* angles add up to 2π; so each of them is $2\pi/n$, and each internal angle is thus $\pi - (2\pi)/n$.) If m of these n-gons are fitted together at a vertex, then $(\pi - (2\pi/n))m < 2\pi$, or $(n - 2)m < 2n$, or $(n - 2)(m - 2) < 4$. We know $n \geq 3$ and $m \geq 3$, so $n - 2$ and $m - 2$ are positive. The possibilities are $n - 2 = 1$, $m - 2 = 1$; $n - 2 = 1$, $m - 2 = 2$; $n - 2 = 1$,

$m - 2 = 3$; $n - 2 = 2$, $m - 2 = 1$; $n - 2 = 3$, $m - 2 = 1$, and that is all. These solutions yield $\{3, 3\}$, $\{3, 4\}$, $\{3, 5\}$, $\{4, 3\}$, $\{5, 3\}$ respectively.

6.35: Here we just replace ' $<$ ' in the last exercise by ' $=$ ', so that $(n - 2)(m - 2) = 4$. The solutions are $n - 2 = 1$, $m - 2 = 4$; $n - 2 = 2$, $m - 2 = 2$; $n - 2 = 4$, $m - 2 = 1$, that is, $\{3, 6\}$, $\{4, 4\}$, $\{6, 3\}$, the tilings by equilateral triangles, squares, and regular hexagons, respectively; and no others.

6.36: An octahedron.

6.37: There are 20 triangular faces, so $3 \times 20 = 60$ counts all the edges of all the faces, and so counts each face *twice*. So there are $60/2 = 30$ edges. (*Or* the pentagonal anti-prism has 5 edges for each pentagonal face and 10 edges joining the two pentagons, so that's $5 + 5 + 10 = 20$ edges; then the two pentagonal pyramids add another 5 edges each, giving $20 + 5 + 5 = 30$ edges in all.)

6.38: The faces are pentagons, and 3 meet at each vertex. Counting all the faces meeting at each of the 20 vertices gives $20 \times 3 = 60$, but this has counted each face 5 times (once from each vertex), so the number of faces is $60/5 = 12$. There are 3 edges at each vertex also, and $20 \times 3 = 60$ again, but this counts each edge twice, once from each end, so there are $60/2 = 30$ edges, the same as for an icosahedron. (*Or* each of the 12 faces has 5 edges, and $12 \times 5 = 60$; but this also has counted each edge twice, once from each *side*, so the number of edges is $60/2 = 30$ again.)

6.39: A face can be rotated to any other face, which is 20 choices; and then there are 3 rotational symmetries (the trivial map and two $\frac{1}{3}$-turns) leaving this face invariant, so $20 \times 3 = 60$ rotational symmetries in all. Edges: an edge can be rotated to any other edge, which is 30 choices; and then there are 2 rotational symmetries (the trivial map and a half-turn) leaving this edge invariant, so $30 \times 2 = 60$ rotational symmetries in all, as before.

6.40: The axis of the rotatory reflection is the line ℓ joining the centres of the two pentagonal faces, and the plane of the reflection is the perpendicular bisector τ of this line (segment), which is parallel to the two pentagonal faces but halfway between them. Taking ℓ vertical, if we rotate the upper pentagon about ℓ through $\pi/5$, this places it immediately above the lower pentagon, so that reflection in τ sends it to the lower pentagon; and the same process sends the lower pentagon to the upper one, so is a symmetry of the anti-prism. Calling the resulting rotatory reflection f, we know that f is opposite, so if $f^n = 1$ then n is even; but f^2 is a rotation through $2\pi/5$, a $\frac{1}{5}$-turn, so the least n with $f^n = 1$ is $n = 10$. The possible cycle structures in S_5 are (1) (the trivial permutation), (12), (123), (1234), (12345), (12)(34) and (12)(345); and the corresponding permutations have order 1, 2, 3, 4, 5, 2, 6 respectively. So no element of S_5 has order 10.

6.41: Place the pyramid with the centre O of the pentagon $BCDEF$ at the origin, and put $\mathbf{a} = \overrightarrow{OA}$, $\mathbf{b} = \overrightarrow{OB}$, $\mathbf{c} = \overrightarrow{OC}$, $\mathbf{d} = \overrightarrow{OD}$, $\mathbf{e} = \overrightarrow{OE}$, $\mathbf{f} = \overrightarrow{OF}$. Then $\mathbf{a} \cdot \mathbf{e} = 0$ and $\mathbf{a} \cdot \mathbf{d} = 0$ (OK?), so $\mathbf{a} \cdot (\mathbf{e} - \mathbf{d}) = 0$. Also $\mathbf{b} \cdot \mathbf{e} = \mathbf{b} \cdot \mathbf{d}$ (because $|\mathbf{b}| = |\mathbf{e}| = |\mathbf{d}|$, and the angles are equal), so $\mathbf{b} \cdot (\mathbf{e} - \mathbf{d}) = 0$. Subtracting, $(\mathbf{b} - \mathbf{a}) \cdot (\mathbf{e} - \mathbf{d}) = 0$, or $AB \perp DE$. (What our vector method shows is that if a line ℓ is perpendicular to two lines in a plane τ, not parallel to each other, then ℓ is perpendicular to *every* line in τ: in vectors, if $\mathbf{x} \cdot \mathbf{y} = 0$ and $\mathbf{x} \cdot \mathbf{z} = 0$, then $\mathbf{x} \cdot (\lambda \mathbf{y} + \mu \mathbf{z}) = 0$, for all λ, μ. So this solution can be rewritten: $OA \perp OE$ and $OA \perp OD$ shows $OA \perp DE$; but also $OB \perp DE$—easy property of pentagon—so $AB \perp DE$.)

6.42: Take A as the apex of a pyramid 'chunk' of X, with base $BCDEF$. The opposite pyramid 'chunk' has apex A' and base $B'C'D'E'F'$, where E' and F' are the vertices of X opposite E, F respectively. The third 'chunk' is an anti-prism with pentagonal faces $BCDEF$ and $B'C'D'E'F'$, and since D' is opposite D, it is adjacent to B (and F), so that BD' is an edge (and so is FD').

Apply Exercise 6.41 to the pyramid 'chunk' with apex B and base $ACE'D'F$, and we have $AC \perp BD'$. Then applying Exercise 6.41 to the pyramid with vertex A and base $BCDEF$ we have $AC \perp EF$; but $EF \parallel BD$, so $AC \perp BD$, and thus AC is perpendicular to the plane of the rectangle $BD'B'D$, and this in turn means that the planes of $ACA'C'$ and $BD'B'D$ are perpendicular. Similar arguments show that each of these planes is perpendicular to the plane of the rectangle $EFE'F'$. (Notice that each of the 12 vertices of X occurs as a vertex of exactly one of these three rectangles.)

6.43: Put $|PR|/|PQ| = \lambda$. Let $\overrightarrow{PQ} = \mathbf{u}$ and $\overrightarrow{PT} = \mathbf{v}$, so that $\overrightarrow{TR} = \lambda\mathbf{u}$ and $\overrightarrow{PR} = \mathbf{v} + \lambda\mathbf{u}$; similarly $\overrightarrow{QS} = \lambda\mathbf{v}$ and $\overrightarrow{PS} = \mathbf{u}+\lambda\mathbf{v}$. We then have $\mathbf{u}-\mathbf{v} = \overrightarrow{TQ} = \lambda\overrightarrow{SR} = \lambda((\mathbf{v}+\lambda\mathbf{u})-(\mathbf{u}+\lambda\mathbf{v}))$, whence $(\lambda^2 - \lambda - 1)(\mathbf{u} - \mathbf{v}) = \mathbf{0}$. So $\lambda^2 - \lambda - 1 = 0$, and $\lambda = \frac{1}{2}(1 + \sqrt{5}) = \tau$.

6.44: We have $\tau^2 = \tau + 1$. The point $A = (\tau, 1, 0)$ is distant 2 from each of the five points $(\tau, -1, 0)$, $(0, \tau, \pm 1)$, $(1, 0, \pm\tau)$; and it is distant 2τ from each of the five points $(-\tau, 1, 0)$, $(0, -\tau, \pm 1)$, $(-1, 0, \pm\tau)$; and it is distant $2\sqrt{1 + \tau^2}$ from the remaining point, $(-\tau, -1, 0)$. Notice that $2 < 2\tau < 2\sqrt{1 + \tau^2}$, so there are five points closest to A; and a similar argument applies to each of the 12 points. If we join all these closest pairs, we obtain $\frac{1}{2}(12 \times 5) = 30$ edges, all of length 2. Then, once again, the five points closest to A are $B = (\tau, -1, 0)$, $C = (1, 0, \tau)$, $D = (0, \tau, 1)$, $E = (0, \tau, -1)$, and $F = (1, 0, -\tau)$, and $|BC| = |CD| = |DE| = |EF| = |FA| = 2$, so that the triangles ABC, ACD, ADE, AEF, AFB are congruent equilateral triangles; there are thus five triangles meeting at A; and similarly for each of the other eleven vertices.

6.45: The only elements of S_5 of order 3 are the 3-cycles. Label each edge of X with the number of the set of rectangles to which it belongs. The edges of any triangular face belong to three different sets, so the image of our homomorphism certainly contains 3-cycles. It is now just a matter of going around the icosahedron and checking that every subset of three numbers from $\{1, 2, 3, 4, 5\}$ *does* occur as the edges of some face. Do it!

6.46: If H were such a subgroup, then $\rho^2 \in H$ for all $\rho \in A_5$. But A_5 contains twenty 3-cycles, each the square of its inverse; so these are all in H. A_5 also contains twenty-four 5-cycles, each such ρ satisfying $\rho = (\rho^{-2})^2$, and so in H also: but this is 44 elements in H, which is impossible. (Alternatively, A_5 is generated by the 3-cycles, so once these are in H, so is the whole of A_5.)

6.47: Count the elements of $\mathcal{S}^+(X)$ three ways: (i) Choose a face: there are F faces to which it can be rotated, and n rotations that leave it invariant, so $|\mathcal{S}^+(X)| = nF$. (ii) Choose a vertex: there are V vertices to which it can be rotated, and m rotations leaving it invariant, so $|\mathcal{S}^+(X)| = mV$. (iii) Choose an edge: there are E edges to which it can be rotated, and 2 rotations leaving it invariant, so $|\mathcal{S}^+(X)| = 2E$. Finally, $2 = F + V - E$ gives $4nm = 2nmF + 2nmV - 2nmE = (2m + 2n - nm)|\mathcal{S}^+(X)|$, whence the result.

6.48: We must show $N_0 - N_1 + N_2 - N_3 + N_4 = 1$. Here $N_4 = 1$—there is only the one hypercube!—and, since we are dealing with the 8-cell, $N_3 = 8$. Each of these 8 cubes has 6 faces, each such face belonging to 2 cubes, so $N_2 = \frac{1}{2}(8 \times 6) = 24$. Each of the 8 cubes has 12 edges, each such edge belonging to 3 cubes, by Exercise 6.31, so $N_1 = \frac{1}{3}(8 \times 12) = 32$. Finally, the number of vertices is $2^4 = 16 = N_0$. Then $16 - 32 + 24 - 8 + 1 = 1$, as required.

6.49: False. For example, O_2 contains infinitely many reflections, all of order 2. But SO_2 has only one element of order 2 (the half-turn given by $-\mathbf{I}_2$), so $SO_2 \times C_2$ has only finitely many (in fact, three) elements of order 2. So the groups cannot be isomorphic.

6.50: Let $X = \{\lambda\mathbf{x} : \lambda \in \mathbb{R}\}$, the subspace of $\mathbb{R}^3$ spanned by $\mathbf{x}$. Then by Exercise 6.11, (i) $X^\perp$ is a plane, and (ii) since X is f-invariant for all $f \in G_\mathbf{x}$, so is $X^\perp$. Thus f restricts to a direct isometry of $X^\perp$, and we get an injective homomorphism $G_\mathbf{x} \to \mathcal{I}(\mathbb{R}^2)$ (OK?). $G_\mathbf{x}$ is thus isomorphic to its image, so by Leonardo da Vinci's theorem (or by Proposition 5.43) it is cyclic.

6.51: Let X be the axis of g, and choose $\mathbf{x} \in X$, so that $g(\mathbf{x}) = \mathbf{x}$. Thus $(fgf^{-1})f(\mathbf{x}) = fg(\mathbf{x}) = f(\mathbf{x})$, whence $f(X)$ is f-invariant. Now choose $\mathbf{y} \in X^{\perp}$ with $|\mathbf{y}| = 1$. So $\mathbf{x} \cdot \mathbf{y} = 0$, and if g is a rotation through α, then $\mathbf{y} \cdot g(\mathbf{y}) = \cos\alpha$. Since f is an isometry and $\mathbf{x} \cdot \mathbf{y} = 0$, we have $f(\mathbf{x}) \cdot f(\mathbf{y}) = 0$, so that $f(\mathbf{y}) \in f(X)^{\perp}$, and $|f(\mathbf{y})| = |\mathbf{y}| = 1$. Then $f(\mathbf{y}) \cdot (fgf^{-1})f(\mathbf{y}) = f(\mathbf{y}) \cdot fg(\mathbf{y}) = \mathbf{y} \cdot g(\mathbf{y})$ (by Proposition 6.3(iii)) and this is equal to $\cos\alpha$. So fgf^{-1} is a rotation through α about $f(X)$.

6.52: The orbit of $\mathbf{z}$ under $G_{\mathbf{x}}$ consists of the vertices of a regular r-gon surrounding $\mathbf{x}$. (We deliberately do not say 'with centre $\mathbf{x}$', as $\mathbf{x}$ is not in the plane of this r-gon. The centre of the r-gon is where the r-axis corresponding to $\mathbf{x}$ meets the plane of the r-gon.) By Exercise 6.51, for any $f \in G$ the orbit of $f(\mathbf{z})$ under $G_{f(\mathbf{x})}$ is a regular r-gon surrounding $f(\mathbf{x})$. The closeness of $\mathbf{z}$ to $\mathbf{x}$ means that $f(\mathbf{z})$ is equally close to $f(\mathbf{x})$, so that the r-gons surrounding $\mathbf{x}$ and $f(\mathbf{x})$ have no vertex in common unless $\mathbf{x} = f(\mathbf{x})$. So $\mathbf{z}^{G}$ is the disjoint union of these r-gons, one for each element of $\mathbf{x}^{G}$. Since $\mathbf{z}$ is not a pole, we have $|\mathbf{z}^{G}| = |G|$, and now $|G| = |\mathbf{z}^{G}| = r|\mathbf{x}^{G}| = |G_{\mathbf{x}}||\mathbf{x}^{G}|$.

6.53: Take, for example, $(1\,2)(3\,4\,5) \in S_6$. Its order is $2 \times 3 = 6$.

6.54: Take $(1\,2\,3\,4)(5\,6\,7)(8\,9\,10)(11\,12) \in S_{12}$. Its order is the least common multiple of the cycle lengths 4, 3, 3, 2; that is, 12.

6.55: If ρ is *not* an n-cycle, then its cycle decomposition contains an r-cycle for some r with $1 < r < n$. But now ρ^r has (at least) r fixed points, a contradiction.

6.56: Here $|G| = 4$ and $s = 3$, and the three orbits each consist of two 2-poles. So $G = \{1, f, g, h\}$, where each of f, g, h is a half-turn. Since $fg \in G$, we have $fg = 1$ or f or g or h. The first three give $f = g$, $g = 1$ and $f = 1$ respectively (OK?), and so are impossible. Thus $fg = h$; and similarly $gf = h$. The restrictions of f, g to the plane containing their axes thus give two commuting reflections, and by Exercise 5.31(iv) the axes are orthogonal. Similarly for the other pairs of axes.

6.57: A non-trivial element of G either is of order 3, in which case it fixes at most two of the 3-poles; or it is not of order 3, in which case it fixes *none* of the 3-poles. So the map $G \to S_4$ is injective, and we can identify G with its image. Then G and A_4 are subgroups of S_4, both of order 12. Put $H = G \cap A_4$, so that H is a subgroup of A_4, and $G = A_4$ iff $H = A_4$. So suppose $H \neq A_4$, and choose $\rho, \sigma \in A_4 \setminus H$. Since $\rho, \sigma \in A_4$, we have $\rho A_4 = A_4 = \sigma A_4$. Since $\rho, \sigma \notin G$ (OK?) we have $\rho G \neq G$ and $\sigma G \neq G$; but G has only one non-trivial coset in S_4 (because $|S_4| = 2|G|$), so $\rho G = \sigma G$. Then $\rho H = \rho(G \cap A_4) = (\rho G) \cap (\rho A_4) = (\sigma G) \cap (\sigma A_4) = \sigma(G \cap A_4) = \sigma H$. Thus H has exactly one non-trivial coset in A_4, so that $|A_4| = 2|H|$, or $|H| = 6$. But A_4 has *no* subgroup of order 6 (Exercise 6.25), so we have a contradiction, and $G = A_4$.

6.58: Since $|G| = 24 = |S_4|$, we just need to show our map is injective. Now the non-trivial elements of G are $\frac{1}{2}$-turns, $\frac{1}{3}$-turns and $\frac{1}{4}$-turns. A 2-axis is not also a 3-axis, so a $\frac{1}{2}$-turn cannot leave all the 3-axes invariant unless they lie in a plane orthogonal to its axis; but G contains more than one $\frac{1}{2}$-turn, so this is impossible. The only invariant line of a $\frac{1}{3}$-turn or $\frac{1}{4}$-turn is its own axis, so these cannot map to $1 \in S_4$ either.

6.59: By a similar argument to that used in Exercise 6.58, the map $G \to S_6$ is injective, so the image is a subgroup of S_6 of order 60. The result follows by the argument on page 144.

6.60: Just note that the 3-poles are at the centres of the faces of an octahedron, and so form a cube. Alternatively, each $\frac{1}{4}$-turn in G moves all eight of the 3-poles, and cannot swap two of them (OK?), so it gives the product of two 4-cycles. This means we get two sets $\{P, Q, R, S\}$ and $\{T, U, V, W\}$ of 3-poles, the vertices (in some order) of two squares, in parallel planes (orthogonal to the 4-axis of the $\frac{1}{4}$-turn). The squares corresponding to a *different* 4-axis cannot contain three of $\{P, Q, R, S\}$

or three of $\{T, U, V, W\}$, so must contain two vertices out of each set, say $\{P, Q, T, U\}$ (in some order) and $\{R, S, V, W\}$ (ditto). There is now one remaining 4-axis, and it gives us two more squares, each having two vertices in common with each of the squares found so far. The only way this can be done is $\{P, R, T, V\}$ and $\{Q, S, U, V\}$. There are now enough parallels, right angles and equal edges to deduce that we have a cube, and we leave the remaining details to the reader.

6.61: If they were, then the six 5-axes would be mutually perpendicular (any pair being the diagonals of a square), and this is impossible in $\mathbb{R}^3$.

6.62: Method 1: list the elements. There is an m-pole at the centre of each of the two m-gonal faces, and a 2-pole (i) at the centre of each of the m square faces, and (ii) at the mid-point of each of the m edges joining one m-gon to the other. The result now follows from Proposition 6.25 (where we must be in case 2) and Proposition 6.26.

Method 2: the plane parallel to the m-gonal faces and mid-way between them meets the prism in a section which is another regular m-gon. This plane and this m-gon are invariant under our rotation group, and so we get a homomorphism from our rotation group to the symmetries—not just the rotations—of a regular m-gon in $\mathbb{R}^2$, that is, to D_m. Now show this is an isomorphism. (Look at the kernel to see it is injective; and count to see that it is surjective.)

6.63: There is an m-pole at the centre of each of the two m-gonal faces, and a 2-pole at the mid-point of each of the $2m$ edges joining one m-gon to the other. Finish the argument as in Exercise 6.62. (The second method, as in Exercise 6.62, can be made to work here but is harder, as this time the section by the mid-way plane is a regular $2m$-gon, not a regular m-gon. So we end up with our rotation group D_m appearing as a proper subgroup of the symmetry group D_{2m} of this $2m$-gon in $\mathbb{R}^2$. We leave the reader to investigate this.)

6.64: Take an n-gonal pyramid. This has $n + 1$ vertices, n of which are the vertices of its base, a regular n-gon. It also has another n faces, which are (congruent isosceles) triangles and share the remaining vertex, at the apex of the pyramid. The line joining the apex to the centre of the base is an n-axis for the rotation group, and there are no other axes for this group, which must therefore be C_n. (The triangular faces could be made equilateral when $n \leq 5$, but we choose not to do this because when $n = 3$ this would give a regular tetrahedron, whose rotation group is A_4, not C_3. For $n \geq 6$ we could not make the triangular faces equilateral even if we wanted to. Why?)

6.65: Suppose n is odd. Method 1: $C_{2n} = \{1, a, a^2, \ldots, a^{2n-1}\}$, where a has order $2n$. It follows that a^n has order 2 (easy), and a^2 has order n (also easy); so $\{1, a^n\}$ and $\{1, a^2, a^4, \ldots, a^{2n-2}\}$ are subgroups of C_{2n} isomorphic to C_2 and C_n respectively. Put $n + 1 = 2m$, so that n and m are coprime. We claim that, if $b = a^{2m}$, then b has order n. For $b^n = a^{2mn} = (a^{2n})^m = 1$; and if $b^r = 1$, then $a^{2mr} = 1$, so $2n$ divides $2mr$, and n divides mr, whence n divides r. Then $a = a^n a^{n+1} = a^n b$, so $a^r = (a^n)^r b^r$, the first term being in C_2 and the second in C_n, and this gives our isomorphism $C_{2n} \to C_2 \times C_n$, by $a^r \mapsto ((a^n)^r, b^r)$.

Method 2: use complex numbers. We have $C_{2n} \cong P_{2n}$, the group of complex $2n$th roots of unity. Now use the fact that, since n is odd, every complex $2n$th root of unity is plus or minus an nth root of unity.

For the converse, C_{2n} contains exactly one element of order 2, so if it is isomorphic to $C_2 \times C_n$ then this group also contains exactly one element of order 2. (OK?) But C_2 has an element of order 2, and if n were even, then C_n would also contain an element of order 2, and that would mean $C_2 \times C_n$ had at least two (and in fact three—OK?) elements of order 2, a contradiction.

6.66: This is just an extension of Exercise 6.65. Let $D_{2m} = \{1, a, a^2, \ldots, a^{2m-1}, b, ba, \ldots, ba^{2m-1}\}$, where $\mathrm{order}(a) = 2m$, $\mathrm{order}(b) = 2$ and $ab = ba^{-1}$. Then

$\{1, a^2, \ldots, a^{2m-2}, b, ba^2, \ldots, ba^{2m-2}\}$ is (a subgroup isomorphic to) D_m, and $\{1, a^m\}$ is (a subgroup isomorphic to) C_2; and further, it is easy to check that each element of D_m commutes with each element of C_2. Rather as in Exercise 6.65, the map $D_{2m} \to C_2 \times D_m$ given by $b^r a^s \mapsto (a^{ns}, b^r a^{(n+1)s})$ is an isomorphism.

6.67: The central inversion ι can be written as the composite $\iota = fg$ of a reflection g in the mid-way plane (see solution to Exercise 6.62) and a half-turn f about the m-axis. Here g swaps the two m-gonal faces, so $g \in S(X)$; and f gives a half-turn to each m-gonal face (in its own plane), and is *not* a symmetry of that face, since m is odd. Thus $f \notin S(X)$, and we cannot have $\iota \in S(X)$, else $f = \iota g^{-1} = \iota g \in S(X)$, a contradiction. We know that $S^+(X) = D_m$ (Exercise 6.62), and we are in case 2(b), not 2(a), and $S(X) = D_{2m}$.

6.68: Writing $\iota = fg$ as in the solution to Exercise 6.67, neither f nor g is a symmetry of X. The reflection g moves each m-gonal face of X into the plane of the other, but not into coincidence with it; the coincidence can then achieved by applying the half-turn f, or indeed any rotation about the m-axis through an odd multiple of π/m, of which π is one such. We have $S^+(X) = D_m$ (Exercise 6.63), and we are in case 2(a), not 2(b), so $S(X) = D_m \times C_2$.

6.69: (i) Here $\iota \in S(X)$, so $S(X) = D_m \times C_2$. (ii) Here $\iota \notin S(X)$, so $S(X) = D_{2m}$. Note that these are the other way round from what happens when m is odd.

6.70: Some of these we have already. Going down the column for G, case 1(c) is the n-gonal pyramid, and cases 2(a) and 2(b) are given by the n-gonal prisms and anti-prisms in various ways according to the parity of n. Cases 3(b), 4, and 5 are the tetrahedron, octahedron, and icosahedron, respectively. This leaves 1(a), 1(b), and 3(a), which are subgroups of cases covered already: $C_n \times C_2 \subset D_n \times C_2$; $C_{2n} \subset D_{2n}$; and $A_4 \times C_2 \subset S_4 \times C_2$. The easiest way to achieve the subgroups as symmetry groups is to add patterns to the faces of the various polyhedra to cut out the symmetries we don't want. Given a regular r-gon in $\mathbb{R}^2$, add an r-legged swastika to the polygon so that the resulting figure has C_r as its symmetry group, not D_r, since any of the reflectional symmetries of the original r-gon make the swastika point its feet in the opposite rotational sense. We call this figure (polygon with pattern) a *directed polygon*; and if one is chosen as directed *positively*, its reflection is directed *negatively*. If we direct the two n-gonal faces of an n-gonal prism or anti-prism in opposite senses (viewed from outside the polyhedron), then they still have the same n-axis of symmetry, but they lose all their 2-axes of symmetry because of the patterns. Whichever of the central inversion ι or the reflection in the mid-plane that was a symmetry, is still a symmetry, and this gives cases 1(a) and 1(b).

Case 3(a) is done a little differently. Mark the mid-point of each of the 12 edges of a cube, and draw one line on each face, from the mid-point of one edge to the mid-point of the opposite edge on that face, six lines in all, with no two sharing an end-point. (You have two choices of where to draw the line on the first face, but after that the rest of the pattern is uniquely determined.) This turns the 4-axes of the original cube into 2-axes of the patterned cube, and gives the required symmetry group.

It is also possible to find sets whose symmetry groups G satisfy $G = G^+$, in the various cases. All we have to do is direct all the faces of the n-gonal pyramid, m-gonal prism (or anti-prism), tetrahedron, octahedron and icosahedron, so that all faces are directed in the same sense (viewed from outside), and we get $G (= G^+) = C_n, D_m, A_4, S_4, A_5$, respectively.

6.71: $D_2 = \{1, f, g, gf\}$ where f, g have order 2 and $fg = gf^{-1}$, that is, $fg = gf$, since $f^{-1} = f$. If $C_2 = \{1, a\}$, with order$(a) = 2$, then the isomorphism $C_2 \times C_2 \to D_2$ is given by $(1, 1) \mapsto 1$; $(a, 1) \mapsto f$; $(1, a) \mapsto g$; and $(a, a) \mapsto gf$. (Check this!)

In $\mathbb{R}^2$ we distinguished between the isomorphic groups C_2 (1 and a half-turn) and D_1 (1 and a reflection). In Proposition 6.27, case 1(a), we have $G = C_2 \times C_2$, where the second C_2 is $\{1, \iota\}$, and the first C_2 consists of 1 and a half-turn. The product of ι and the half-turn is a reflection, so G contains 1, a half-turn, ι and a reflection. In case 1(c) $G = D_2$, with $G^+ = C_2$ consisting of 1 and a half-turn, and the other two (opposite) isometries in G both being reflections (in two orthogonal planes). In case 2(a), $G = D_1 \times C_2$, and here $G^+ = D_1$, so this consists of 1 and a half-turn, and G is indistinguishable from case 1(a). Similar remarks apply to case 2(b), which is indistinguishable from case 1(c). So we have just two geometric versions of $C_2 \times C_2$ when $G \neq G^+$; there is a third version, namely case 2(b) with $m = 2$, when $G^+ = D_2$, consisting of 1 and three half-turns, about three mutually orthogonal axes.

7
Circles and other conics

So far, the objects we have been studying have been built up from points, lines, and circles (and, in higher-dimensions than 2, from the higher-dimensional analogues: planes, spheres, and so on). It is time to introduce some new shapes. Let $p = p(x, y)$ be a real polynomial: $p(x, y) = \sum_{n,m} a_{nm} x^n y^m$, where the *coefficients* a_{nm} are real, where the sum is over a finite number of terms with n, m non-negative integers, and where x and y are indeterminates, or 'unknowns'. The *degree* of p, written deg p, is the largest value of $n + m$ for which $a_{nm} \neq 0$; and, for completeness, if $a_{nm} = 0$ for all n, m (so that $p = 0$, the zero polynomial) we write deg $p = -\infty$. A polynomial of degree 1 is *linear* and a polynomial of degree 2 is *quadratic*.

Definition 7.1. *A **curve** is the zero-set of a polynomial p of degree at least 1; or (equivalently) the solution-set of the corresponding polynomial equation $p(x, y) = 0$. That is, given p with deg $p \geq 1$, the curve defined by p is the subset of all $(x, y) \in \mathbb{R}^2$ with $p(x, y) = 0$.*

Example. If p is linear, so that $p(x, y) = a_{10} x^1 y^0 + a_{01} x^0 y^1 + a_{00} x^0 y^0 = a_{10} x + a_{01} y + a_{00}$ (with a_{10}, a_{01} not both zero), then the corresponding curve is the (*straight*) line $a_{10} x + a_{01} y + a_{00} = 0$. So curves do not have to be curvy.

Definition 7.2. *A **conic** is a curve given by a quadratic polynomial.*

A conic, then, is given by a polynomial $p(x, y) = a_{20} x^2 + a_{11} xy + a_{02} y^2 + a_{10} x + a_{01} y + a_{00}$, where a_{20}, a_{11}, a_{02} are not all zero. It will be less cumbersome if we rename the coefficients a_{nm} thus: put $a = a_{20}$, $b = a_{02}$, $c = a_{00}$, $f = \frac{1}{2} a_{01}$, $g = \frac{1}{2} a_{10}$ and $h = \frac{1}{2} a_{11}$, so that

$$p(x, y) = ax^2 + 2hxy + by^2 + 2gx + 2fy + c,$$

with a, b, h not all zero. There *is* method in the madness here: the seemingly perverse choice of letters is chosen with an eye on the following matrix equation, which the reader should verify:

$$p(x, y) = \begin{pmatrix} x & y & 1 \end{pmatrix} \begin{pmatrix} a & h & g \\ h & b & f \\ g & f & c \end{pmatrix} \begin{pmatrix} x \\ y \\ 1 \end{pmatrix}.$$

Note that the matrix $\mathbf{M} = \begin{pmatrix} a & h & g \\ h & b & f \\ g & f & c \end{pmatrix}$ is *symmetric*: $\mathbf{M}^T = \mathbf{M}$. (The $\frac{1}{2}$'s in the definitions of f, g, h were put there to make this happen.) If two lines ℓ_1, ℓ_2 in $\mathbb{R}^3$ meet at a point A, then the surface of revolution formed by rotating ℓ_2 about ℓ_1 is a *cone*, with *apex* A. Conics are so-called because they arise as plane sections of cones: we shall prove this in Section 7.2.

- Ex.7.1:   *Let $m \in \mathbb{R}$. Show that the cone obtained by rotating the line $\mathbf{r} = \lambda(1, 0, m)$ about the z-axis consists of all $(x, y, z) \in \mathbb{R}^3$ with $z^2 = m^2(x^2 + y^2)$.*

The apex of this cone is O; notice that the cone has points on both 'sides' of O, i.e., with $z > 0$ and with $z < 0$. In elementary geometry, a (right circular) cone is a surface formed by joining the points on the rim of a circular disc (the base of the cone) to a point (the apex) lying on the axis of the disc (the line through its centre, perpendicular to its plane). This is the surface of revolution formed by rotating a line *segment* about an axis, and is a bounded object. Our cone is given by rotating the whole line, instead of just a segment of it, and so it is unbounded (no 'base'), and consists of two unbounded parts meeting at the apex (Fig. 7.1).

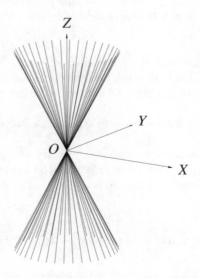

Fig. 7.1   The cone $z^2 = m^2(x^2 + y^2)$

7.1   PROPER AND IMPROPER CONICS

We shall now give lots of examples of conics; the first few are rather strange, even silly.

Example 1. $xy = 0$. Since xy is a polynomial of degree 2, this is a conic within the meanin' of the act; but it just consists of the two lines $x = 0$ and $y = 0$, the axes. More generally, the equation $(\ell x + my + n)(\ell' x + m'y + n') = 0$ gives a conic which consists of the two lines $\ell x + my + n = 0$ and $\ell' x + m'y + n' = 0$: it is a *line-pair*. Note that the plane $y = 0$ meets the cone $z^2 = x^2 + y^2$ in a line-pair, the lines being $z = x$, $y = 0$ (or $\mathbf{r} = \lambda(1, 0, 1)$) and $z = -x$, $y = 0$ (or $\mathbf{r} = \lambda(1, 0, -1)$); so line-pairs do arise as conic sections. Nonetheless, line-pairs are examples of *improper* conics:

Definition 7.3.   *If the real quadratic polynomial $p(x, y)$ does not factorize (over $\mathbb{R}$ or over $\mathbb{C}$) as the product of two linear polynomials, then the conic $p(x, y) = 0$ is* **proper***; otherwise, it is* **improper***.*

A line-pair can consist of two *parallel* lines, $(\ell x + my + n)(\ell x + my + n') = 0$, in which case they are not a section of a cone, but of a *cylinder*, which is the surface of revolution of a line about a parallel line; or, it is the limiting case of a cone as we send its apex off to infinity. A line-pair can even be a *repeated* line, $(\ell x + my + n)^2 = 0$, which *does* arise as a section of a cone:

● Ex.7.2:   *Find a cone in $\mathbb{R}^3$ that meets the plane $z = 0$ where $x^2 = 0$. (That is, the cone is given by a polynomial equation $q(x, y, z) = 0$, and $q(x, y, 0) = x^2$.)*

Example 2. $x^2 + y^2 = 0$. There is only one point satisfying this equation, namely $(0, 0)$, so we have a conic that seems to consist of a single point. This does arise as a section of a cone, for example as the section of $z^2 = x^2 + y^2$ by $z = 0$. Notice that $x^2 + y^2 = (x + iy)(x - iy)$, so we could think of this one-point conic as being a *complex* line-pair, whatever that might mean. The analogy with solving quadratic equations is clear: a real quadratic equation either has two real roots, or a repeated real root, or two complex roots; and a plane through the apex of a cone meets it in either a (real) line-pair, or a repeated (real) line, or a complex line-pair (which, as a real conic, is a single point).

More generally, the conic $(\ell x + my + n)^2 + (\ell' x + m'y + n')^2 = 0$ has only one point, where the two lines $\ell x + my + n = 0$ and $\ell' x + m'y + n' = 0$ meet; but if the two lines are parallel, the conic is empty, for example $(x + y)^2 + (x + y + 1)^2 = 0$ has no points. In all these cases, the polynomial of the conic can be written as the product of two (complex) linear expressions, $(\ell x + my + n) \pm i(\ell' x + m'y + n')$, and the conic is improper.

Example 3. $x^2 + y^2 + 1 = 0$. Since $x^2 + y^2 \geq 0$, for all x, $y \in \mathbb{R}$, this is another empty conic, but for a different reason. Rewriting it as $x^2 + y^2 = (i)^2$, it looks like the circle $x^2 + y^2 = r^2$, of radius $r = i$, and indeed it is called an *imaginary circle*.

● Ex.7.3:   *The imaginary circle $x^2 + y^2 + 1 = 0$ is a proper conic!*

That is the last of the rather silly conics; the ones that follow are both proper and non-empty.

Example 4. $x^2 + y^2 = 1$. This is a circle (centre $(0, 0)$, radius 1), and arises (obviously) as a plane section of a (circular) cone, for example where the plane $z = 1$ meets the cone $z^2 = x^2 + y^2$. More generally, the circle $(x - \alpha)^2 + (y - \beta)^2 = r^2$ (with centre (α, β) and radius r, where $\alpha, \beta, r \in \mathbb{R}$) is a conic.

• **Ex.7.4:**   *Given a circle, with the centre not marked, describe a ruler-and-compasses construction to find the centre of the circle.*

Example 5.   We now give the *focus–directrix* definition of some conics. Let $a, e \in \mathbb{R}$ with $a > 0$, $e > 0$, and $e \neq 1$. Let F be the point $(ae, 0)$, the *focus*, and let d be the line $x = a/e$, the *directrix*, and let the point $P = (x, y)$ move so that its distance from the point F and its (perpendicular) distance from the line d are in the ratio $e : 1$. We shall see that the locus of P is a conic. Let Q be the foot of the perpendicular from P onto d, so that $Q = ((a/e), y)$. Thus $|PF|^2 = (x - ae)^2 + y^2$ and $|PQ|^2 = (x - (a/e))^2 + 0^2$. Since $|PF| = e|PQ|$, or $|PF|^2 = e^2|PQ|^2$, we have $(x - ae)^2 + y^2 = e^2(x - (a/e))^2 = (ex - a)^2$, or $x^2 - 2aex + a^2e^2 + y^2 = e^2x^2 - 2aex + a^2$, or $(1 - e^2)x^2 + y^2 = a^2(1 - e^2)$, or

$$\frac{x^2}{a^2} + \frac{y^2}{a^2(1 - e^2)} = 1, \tag{7.1}$$

which is a conic, Γ. Notice that Γ is symmetrical under reflection in the x-axis ($(x, y) \mapsto (x, -y)$)—this is expected from its definition, as both F and d are invariant under this reflection—and it is also symmetrical under reflection in the y-axis ($(x, y) \mapsto (-x, y)$), which is unexpected. It follows that there is a second focus, $F' = (-ae, 0)$, and a second directrix d': $x = -(a/e)$ that could have been used to define Γ. The half-turn about O, the *centre* of Γ, is also a symmetry of Γ, and the symmetry group of Γ is the dihedral group D_2. The two (orthogonal) lines of symmetry, $x = 0$ and $y = 0$, are the *principal axes* of Γ. The positive real number e is the *eccentricity* of Γ, which is a very different shape according as $e < 1$ or $e > 1$:

Case 1: Suppose the eccentricity lies between 0 and 1: $0 < e < 1$. Here $1 > 1 - e^2 > 0$, so we can find $b > 0$ with $b^2 = a^2(1 - e^2)$, and then $a > b$. The equation of Γ is

$$\frac{x^2}{a^2} + \frac{y^2}{b^2} = 1,$$

an *ellipse*. Notice that, if $P = (x, y)$ is on Γ, then

$$b^2 = \left(\frac{b}{a}\right)^2 x^2 + y^2 \leq x^2 + y^2 \leq x^2 + \left(\frac{a}{b}\right)^2 y^2 = a^2,$$

so that $b \leq |OP| \leq a$. The curve is bounded and oval-shaped (Fig. 7.2), being nearest to O where it meets the y-axis, at $(0, \pm b)$, and furthest from O where it meets the x-axis, at $(\pm a, 0)$. The line segment joining $(\pm a, 0)$ is the *major axis* and the line segment joining

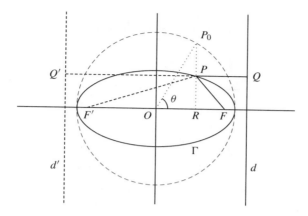

Fig. 7.2 The ellipse Γ: $\frac{x^2}{a^2} + \frac{y^2}{b^2} = 1$

$(0, \pm b)$ is the *minor axis* of the ellipse; and the quantities a, b are the lengths of the *semi-major* and *semi-minor* axes, respectively.

- Ex.7.5: *Let $A = (a, 0)$ and $A' = (-a, 0)$ be the ends of the major axis, and let the major axis meet the directrix d at H. Show that $\{A, A'; F, H\} = -1$.*

The circle $x^2 + y^2 = a^2$ has the same centre as Γ, lies wholly outside Γ, and touches Γ at the ends of its major axis. It is called the *auxiliary circle* of Γ. The point $P_0 = (a \cos \theta, a \sin \theta)$, as θ varies, moves around this circle—indeed, the old-fashioned name for the sine and cosine is the *circular functions*—and the perpendicular $P_0 R$ from P_0 onto the x-axis meets Γ at $P = (a \cos \theta, b \sin \theta)$, which gives a parametric representation $x = a \cos \theta$, $y = b \sin \theta$ of Γ. The parameter θ, coming from the auxiliary circle, is called the *auxiliary angle* of P; note that it is *not* $\angle P O R$, but is in fact $\angle P_0 O R$ (Fig. 7.2 again).

If the eccentricity e is close to 1, then b is small compared to a, and the ellipse is rather thin. On the other hand, if e is close to 0, then b is close to a and the ellipse is nearly circular; indeed, if we let $e \to 0$, then $b \to a$, and in the limit the ellipse becomes $\frac{x^2}{a^2} + \frac{y^2}{a^2} = 1$, or $x^2 + y^2 = a^2$, a circle. So we may regard the circle as a conic with eccentricity zero. (However, the foci F, F' have now moved into coincidence at O, and the directrices d, d' have gone off to infinity, so one cannot give the same focus–directrix definition for a circle.)

Case 2: Suppose the eccentricity is greater then 1: $e > 1$. Here $e^2 - 1 > 0$, so we can find $b > 0$ with $b^2 = a^2(e^2 - 1)$. This time there is nothing about the comparative sizes of a and b: we can have $b < a$ or $b = a$ or $b > a$, according as $e < \sqrt{2}$ or $e = \sqrt{2}$ or $e > \sqrt{2}$. The equation of Γ is

$$\frac{x^2}{a^2} - \frac{y^2}{b^2} = 1,$$

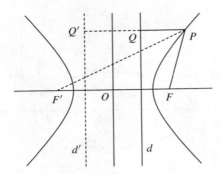

Fig. 7.3   The hyperbola $\frac{x^2}{a^2} - \frac{y^2}{b^2} = 1$

a *hyperbola* (Fig. 7.3). For $P = (x, y)$ on Γ, we have

$$a^2 = x^2 - \left(\frac{a}{b}\right)^2 y^2 \leq x^2 \leq x^2 + y^2,$$

so that P is nearest to O where it meets the x-axis, at $(\pm a, 0)$. It does not meet the y-axis at all, and is in fact in two parts, called *branches*. Indeed, writing Γ as

$$\frac{x^2}{a^2} = 1 + \frac{y^2}{b^2}, \tag{7.2}$$

we see that (i) $x^2/a^2 \geq 1$, so that $|x| \geq a$; and (ii) for any $y \in \mathbb{R}$ we can solve for x, so that Γ is unbounded. If we make $|y|$ large, then $|x|$ is also large, and we can neglect the 1 in (7.2) and write $x^2/a^2 \approx y^2/b^2$. The equation $x^2/a^2 = y^2/b^2$ is a line-pair, the lines being $x/a = y/b$ and $x/a = -y/b$, the *asymptotes* of the hyperbola, two lines which the hyperbola approaches for large values of $|x|$ and $|y|$.

The quantity a is still the length of a semi-axis (though perhaps the use of the word *major* is now inappropriate). It is harder to 'see' the quantity b in a diagram of Γ. The ellipse was parametrized (via the auxiliary circle) using the circular functions, so it will be no surprise if the hyperbola is parametrized using the hyperbolic trigonometric functions cosh and sinh. Indeed, since $\cosh^2\theta - \sinh^2\theta = 1$, for all θ, the point $P = (x, y)$, where $x = a \cosh\theta$, $y = b \sinh\theta$, lies on Γ. But $\cosh\theta > 0$, for all θ, so this parametrizes only the right-hand branch, and points on the left-hand branch are given by $x = -a \cosh\theta$, $y = b \sinh\theta$. Alternatively, we can use the fact that $\sec^2\varphi - \tan^2\varphi = 1$ to give the parametrization $x = a \sec\varphi$, $y = b \tan\varphi$, which does *both* branches, jumping from one to the other by going 'through infinity' when $\varphi = \pm\pi/2$.

A special case is the *rectangular* hyperbola, when the asymptotes are orthogonal. This happens when $a = b$, or the eccentricity is $e = \sqrt{2}$, and the equation of Γ is then $x^2/a^2 - y^2/a^2 = 1$, or $x^2 - y^2 = a^2$.

Example 6.   The parabola. We take $a > 0$, the focus F at $(a, 0)$, and the directrix d as $x = -a$. The parabola Γ is the locus of $P = (x, y)$ moving so that its distance

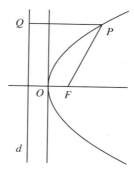

Fig. 7.4 The parabola $y^2 = 4ax$

from F is *equal* to its perpendicular distance from d. (So we are filling in the case of the previously 'missing' value of the eccentricity, $e = 1$.) The equation of Γ is thus $(x - a)^2 + y^2 = (x + a)^2$, or $y^2 = 4ax$ (Fig. 7.4). No trigonometric functions are needed this time: the parametrization is $x = at^2$, $y = 2at$. There is only one axis of symmetry, the x-axis, and no centre, no second focus and no second directrix; and the symmetry group is D_1, cyclic of order 2.

It is possible to see the parabola as a limiting case of an ellipse/hyperbola as $e \to 1$. Write equation (7.1) as

$$x^2(1 - e^2) + y^2 = a^2(1 - e^2)$$

and then translate by the vector $(a, 0)$ to put the end of the major axis at O. This gives $(x - a)^2(1 - e^2) + y^2 = a^2(1 - e^2)$, or $(x^2 - 2ax)(1 - e^2) + y^2 = 0$. Put $\alpha = a(1 - e)$ and we get $x^2(1 - e^2) - 2ax(1 + e) + y^2 = 0$. Now let $e \to 1$, fixing α (so that $a = \frac{\alpha}{1-e} \to \infty$), which gives $y^2 = 4\alpha x$, a parabola.

- Ex.7.6: *Explain why the parabola has only one focus and one directrix, and no centre, by tracking what happens to each of the two foci, the two directrices and the centre of the above ellipse/hyperbola as $e \to 1$.*

7.2 PLANE SECTIONS OF A CONE

In this section we show that a plane meets a cone in a conic. One can produce a proof by algebra: a cone is a particular sort of *quadric* surface, being given by a polynomial equation $f(x, y, z) = 0$ where the degree of f is 2. If we ask where this cone meets the plane $ax + by + cz = d$ then, as long as $c \neq 0$, we can eliminate z between the two equations to get $f(x, y, (d - ax - by)/c) = 0$, and it is easy to see that this is an equation of degree 2. So surely it is a conic?

This argument is just a little too glib: apart from the fact that we brushed aside the possibility that $c = 0$, there is the problem that the lines $x = 0$ and $y = 0$ in the plane $ax + by + cz = d$ are not orthogonal in general, so we do not have the *cartesian* equation of the 'conic'. One could apply a suitable transformation (an *affine* transformation) to put this right, or (better) start by applying an isometry to $\mathbb{R}^3$ that transforms the plane $ax + by + cz = d$ to the plane $z = 0$. Since this involves a linear substitution for x, y, and z, (Lemma 6.2), the transformed cone still has equation given by a second-degree polynomial in x, y and z; but now the cartesian equation of the intersection is just given by putting $z = 0$ in this equation, which clearly yields a conic.

We shall now give a geometrical argument for the same result. We shall do the case of the ellipse; the reader is invited to construct the appropriate arguments (and draw the diagrams) for the hyperbola and the parabola. First, the *focus–focus* property of an ellipse:

Proposition 7.4.   *Let P move on an ellipse with foci F, F'. Then $|PF| + |PF'|$ is constant.*

Proof.   Let the two directrices be d, d', and let the feet of the perpendiculars from P onto d, d' be Q, Q' respectively (Fig. 7.2). If the ellipse has eccentricity e, then

$$|PF| + |PF'| = e|PQ| + e|PQ'| = e(|PQ| + |PQ'|) = e|QQ'|,$$

which is e times the distance between the lines d, d', and so is constant.   □

- Ex.7.7:   *Prove the converse of Proposition 7.4.*

This gives a practical way to draw an ellipse: fix two pins in a board, at the foci, and place a loop of string around the pins, with a pencil in the loop to keep it taut. Then the pencil traces out an ellipse. The author once constructed an elliptical flower-bed in his front garden, using two cricket-stumps, a loop of washing-line and a spade, to the amusement (and puzzlement) of his neighbours.[1]

- Ex.7.8:   *What is the corresponding focus–focus property of a hyperbola?*

- Ex.7.9:   *Hanging pictures on a wall. Suppose a picture hangs symmetrically from a nail in a wall, by means of a picture-cord and two hooks. Suppose that the centre of mass of the picture is the mid-point of the line segment joining the hooks, that the cord is weightless, and that cord, nail and wall are frictionless. (All pretty unlikely, really, but then this is applied mathematics.) Show that, as everyone who has dusted a picture knows, the slightest touch will make the picture hang crooked.*

The reader is invited to investigate the conditions for stability when (as is more usually the case) the centre of mass of the picture is *below* the line joining the hooks.

Now consider the tangents from a point V to a circle. The two tangents are of equal length, and if we rotate the diagram about the line joining V to the centre of the circle, we obtain a sphere with a cone of tangents to the sphere. The cone has apex V, and all the tangents to the sphere from V have the same length. Further, the points of contact of

---

[1] I had thought of adding a joke here about how my eccentricity was cultivated, but perhaps I'll leave that as an exercise for the reader.

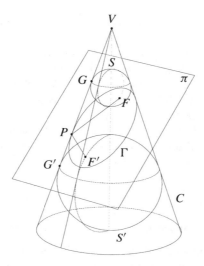

Fig. 7.5   A plane π meeting a cone C in an ellipse Γ

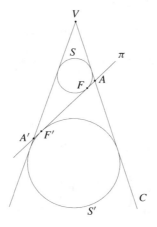

Fig. 7.6   Section of cone C by the plane through V perpendicular to π

these tangents lie on a circle, the circle of contact of sphere and cone, and this circle lies in a plane perpendicular to the axis of the cone.

Take a cone C, apex V, with a plane π meeting C in a closed curve Γ (Fig. 7.5). There are two spheres having a circle of contact with C and also touching π. For consider the section of C by the plane through V perpendicular to π (Fig. 7.6). This meets the figure in a triangle VAA' say, where A, A' lie in $\pi \cap C$. The incircle of this triangle, and its excircle opposite V, rotated about the axis of C, give the required spheres, S and S' respectively (Fig. 7.5). Let these spheres touch π at F and F' respectively, and let P lie on Γ. We shall show that $|PF| + |PF'|$ is constant, whence Γ is an ellipse with foci F, F'.

Note that every line through V lying in C is a tangent to both S and S'; every line in π through F is a tangent to S; and every line in π through F' is a tangent to S'.

The line VP lies in C; let it meet the circle of contact of S and C at G, and the circle of contact of S' and C at G'. Then $|PF| = |PG|$ (equal tangents to S from P), and $|PF'| = |PG'|$ (equal tangents to S' from P), whence $|PF| + |PF'| = |PG| + |PG'| = |GG'| = |VG'| - |VG|$. But $|VG|$ is the length of a tangent to S from V, so does not depend on the position of P on Γ, and likewise $|VG'|$ is the length of a tangent from V to S', and is also constant; so $|PF| + |PF'|$ is constant, as claimed, and Γ is an ellipse.

- Ex.7.10: *Let the circles of contact of S, S' with C lie in the planes τ, τ' respectively. Show that the lines $\pi \cap \tau$, $\pi \cap \tau'$ are the directrices of Γ.*

7.3   CLASSIFICATION OF CONICS UNDER ISOMETRY

In this section we are going to find an isometry that puts a given conic into a standard form; or (equivalently) we are going to move our axes so that (in the new coordinates) the equation of the conic makes it easy to recognize what sort of conic it is. Let Γ be the conic given by $p(x, y) = 0$, where

$$p(x, y) = ax^2 + 2hxy + by^2 + 2gx + 2fy + c,$$

and a, h, b are not all zero. Let $d = |\mathbf{M}|$, the determinant of the matrix $\mathbf{M}$ of p, that is,

$$d = \begin{vmatrix} a & h & g \\ h & b & f \\ g & f & c \end{vmatrix}.$$

When we change variables, we shall have $p(x, y) = p_1(x_1, y_1)$, say, where

$$\begin{pmatrix} x \\ y \\ 1 \end{pmatrix} = \mathbf{P} \begin{pmatrix} x_1 \\ y_1 \\ 1 \end{pmatrix}$$

for some matrix $\mathbf{P}$ corresponding to a direct plane isometry, so that $\det \mathbf{P} = 1$. This means that if $\mathbf{M}$ and $\mathbf{M}_1$ are the matrices of p and p_1, then $\mathbf{M}_1 = \mathbf{P}^T \mathbf{M} \mathbf{P}$, so that on taking determinants, $|\mathbf{M}_1| = |\mathbf{M}|$, or $|\mathbf{M}_1| = d$.

We first attempt to find the *centre* of Γ:

Definition 7.5.   *The point K is a **centre** of Γ if the half-turn about K is a symmetry of Γ.*

Lemma 7.6.   *O is a centre of Γ iff $g = f = 0$.*

Proof.   The image of Γ under the half-turn about O is the conic $p(-x, -y) = 0$, or $ax^2 + 2hxy + by^2 - 2gx - 2fy + c = 0$. This is the same as Γ iff $g = f = 0$. $\square$

Lemma 7.7. *The point (α, β) is a centre of Γ iff $a\alpha + h\beta + g = 0$ and $h\alpha + b\beta + f = 0$.*

Proof. Change variables (translate) to put the origin at (α, β). The equation of Γ becomes $p(x + \alpha, y + \beta) = 0$, or

$$a(x + \alpha)^2 + 2h(x + \alpha)(y + \beta) + b(y + \beta)^2 + 2g(x + \alpha) + 2f(y + \beta) + c = 0, \text{ or}$$
$$ax^2 + 2hxy + by^2 + 2(a\alpha + h\beta + g)x + 2(h\alpha + b\beta + f)y + p(\alpha, \beta) = 0.$$

By Lemma 7.7, we just need to equate the coefficients of x and of y to zero. □

Definition 7.8. Γ *is a* **central** *conic if it has a centre.*

Proposition 7.9. *If $h^2 \neq ab$ then Γ is a central conic, with a unique centre.*

Proof. This is just the condition that the equations $a\alpha + h\beta + g = 0$ and $h\alpha + b\beta + f = 0$ have a unique solution for α, β. □

Provided $h^2 \neq ab$, then, our conic Γ now takes the form $q(x, y) = 0$, where

$$q(x, y) = p(x + \alpha, y + \beta) = ax^2 + 2hxy + by^2 + c_1,$$

and $c_1 = p(\alpha, \beta)$.

• Ex.7.11: *Prove that $d = (ab - h^2)c_1$.*

Let us try the effect of rotating Γ about O, its centre, through an angle θ. We obtain $q(x \cos\theta - y \sin\theta, x \sin\theta + y \cos\theta) = 0$, or

$$a(x \cos\theta - y \sin\theta)^2 + 2h(x \cos\theta - y \sin\theta)(x \sin\theta + y \cos\theta)$$
$$+ b(x \sin\theta + y \cos\theta)^2 + c_1 = 0.$$

The coefficient of xy here is

$$-2a \cos\theta \sin\theta + 2h(\cos^2\theta - \sin^2\theta) + 2b \sin\theta \cos\theta = (b - a) \sin 2\theta + 2h \cos 2\theta.$$

Choose θ such that $\tan 2\theta = \frac{2h}{a-b}$ and the conic becomes $a_1 x^2 + b_1 y^2 + c_1 = 0$, where

$$\begin{pmatrix} \cos\theta & \sin\theta \\ -\sin\theta & \cos\theta \end{pmatrix} \begin{pmatrix} a & h \\ h & b \end{pmatrix} \begin{pmatrix} \cos\theta & -\sin\theta \\ \sin\theta & \cos\theta \end{pmatrix} = \begin{pmatrix} a_1 & 0 \\ 0 & b_1 \end{pmatrix}. \qquad (7.3)$$

Taking determinants, $ab - h^2 = a_1 b_1$, so that $a_1 \neq 0$ and $b_1 \neq 0$; and also $d = (ab - h^2)c_1 = a_1 b_1 c_1$.

• Ex.7.12: *Show that a and a_1 have the same sign if $ab > h^2$.*

Let us now consider the possible shapes of our conic $a_1 x^2 + b_1 y^2 + c_1 = 0$, in terms of the quantities a, $ab - h^2$, and d. Since a and a_1 have the same sign (when $ab - h^2 > 0$), and $ab - h^2 = a_1 b_1$, and $d = a_1 b_1 c_1$, knowledge of a, $ab - h^2$, and d helps determine the signs of a_1, b_1, and c_1, and hence the shape of Γ.

Case 1: $a > 0$, $ab - h^2 > 0$, $d > 0$. Then $a_1 > 0$, $b_1 > 0$, and $c_1 > 0$, and Γ is an imaginary ellipse (or circle).

Case 2: $a > 0$, $ab - h^2 > 0$, $d < 0$. Then $a_1 > 0$, $b_1 > 0$, and $c_1 < 0$, and Γ is a (real) ellipse (or circle).

Case 3: $a > 0$, $ab - h^2 > 0$, $d = 0$. Then $a_1 > 0$, $b_1 > 0$, and $c_1 = 0$, and Γ is an imaginary line-pair (a single real point).

Case 4: $ab - h^2 < 0$, $d \neq 0$. In this (and the next) case we don't know the sign of a_1; but we do know that a_1 and b_1 have opposite signs, and $c_1 \neq 0$, so we have a hyperbola.

Case 5: $ab - h^2 < 0$, $d = 0$. Then again a_1 and b_1 have opposite signs, and this time $c_1 = 0$, so we have a (real) line-pair, not parallel or coincident.

We leave the reader to deal with the version of cases 1–3 when $a < 0$. We now examine what happens when $ab - h^2 = 0$. In this case, $ax^2 + 2hxy + by^2 = \pm(a_0x + b_0y)^2$ for some a_0, b_0, and thus $p(x, y) = \pm(a_0x + b_0y)^2 + 2gx + 2fy + c$.

• Ex.7.13:   *Show that under a suitable rotation, Γ takes the form $ky^2 = \ell x + 2my + n$ for some constants k, ℓ, m, n with $k \neq 0$.*

Now write this last equation as $k(y - m/k)^2 = 2\ell x + n_1$, where $n_1 = n + m^2/k$. If $\ell \neq 0$, a translation puts this in the form $ky^2 = 2\ell x$, a parabola. If $\ell = 0$ but $n_1 \neq 0$, it is $ky^2 = n_1$, two parallel straight lines, real if $kn_1 > 0$ and imaginary if $kn_1 < 0$. If $\ell = n_1 = 0$ we have $ky^2 = 0$, a repeated straight line. Note that, going back to the equation $ky^2 = 2\ell x + n_1$, we have

$$
d = \begin{vmatrix} 0 & 0 & -\ell \\ 0 & k & 0 \\ -\ell & 0 & n_1 \end{vmatrix} = -k\ell^2
$$

so that we have a parabola when $d \neq 0$ and two parallel (possibly imaginary, or coincident) lines when $d = 0$. Let us summarize our results:

Theorem 7.10.   *Let the conic Γ: $ax^2 + 2hxy + by^2 + 2gx + 2fy + c = 0$ have determinant d. Then Γ is a proper conic if $d \neq 0$: an ellipse (or circle, and possibly imaginary) if $ab > h^2$; a parabola if $ab = h^2$; and a hyperbola if $ab < h^2$. Next, Γ is an improper conic if $d = 0$: an imaginary (non-parallel) line-pair if $ab > h^2$ (that is, a single real point); a parallel or coincident line-pair (possibly imaginary) if $ab = h^2$; and a non-parallel real line-pair if $ab < h^2$.*   □

Note that if $d = 0$ and $ab = h^2$, the conic has infinitely many centres; cf. Proposition 7.9. The only conic without a centre is the parabola.

We have shown, then, that every proper real conic is congruent to (that is, can be mapped by an isometry to) one of the conics

$$
\frac{x^2}{a^2} \pm \frac{y^2}{b^2} = 1 \quad \text{or} \quad y^2 = 4ax.
$$

Here the a, b are not the same as in the original equation of the conic $(ax^2 + 2hxy + \ldots)$, but have the meanings given in Section 7.1, examples 5 and 6, to do with lengths of semi-axes. Indeed, equation (7.1) shows that every proper conic is congruent to one of the conics

$$\frac{x^2}{a^2} + \frac{y^2}{a^2(1 - e^2)} = 1 \quad \text{or} \quad y^2 = 4ax, \tag{7.4}$$

where $a > 0$ and $e \geq 0$ with $e \neq 1$. Since the parabola corresponds to the case $e = 1$ (see page 171), we have:

Theorem 7.11. Classification of conics (I). *Every proper real conic is congruent to one of the forms (7.4), and two such conics are congruent if and only if they give rise as above to the same two real numbers a, e. Here $a > 0$ and $e \geq 0$, and we have the first or second equation (7.4) according as $e \neq 1$ or $e = 1$.* □

We shall give other theorems classifying conics (under other groups of transformations) in the next chapter.

We finish this section by showing how, given an conic, we can use ruler and compasses to find its centre (if it has one), its axes and its foci, and hence how we can construct the quantities a and e for our conic.

Definition 7.12. *A **chord** of a conic Γ is a line segment whose end-points lie on Γ.*

Let Γ be an ellipse in its standard form, given parametrically by $x = a \cos \theta$, $y = b \sin \theta$. Let $P = (a \cos \theta, b \sin \theta)$ and $Q = (a \cos \varphi, b \sin \varphi)$ be two points of Γ. The gradient of PQ is

$$\frac{b(\sin \theta - \sin \varphi)}{a(\cos \theta - \cos \varphi)} = \frac{2b \cos((\theta + \varphi)/2) \sin((\theta - \varphi)/2)}{-2a \sin((\theta + \varphi)/2) \sin((\theta - \varphi)/2)} = -\frac{b}{a} \cot \frac{\theta + \varphi}{2} = m,$$

say. Then the mid-point R of PQ is

$$\left(\frac{a \cos \theta + a \cos \varphi}{2}, \frac{b \sin \theta + b \sin \varphi}{2} \right)$$

$$= \left(a \cos \frac{\theta + \varphi}{2} \cos \frac{\theta - \varphi}{2}, b \sin \frac{\theta + \varphi}{2} \cos \frac{\theta - \varphi}{2} \right),$$

so the gradient of OR is

$$\frac{b \sin((\theta + \varphi)/2) \cos((\theta - \varphi)/2)}{a \cos((\theta + \varphi)/2) \cos((\theta - \varphi)/2)} = \frac{b}{a} \tan \frac{\theta + \varphi}{2} = -\frac{b^2}{a^2 m},$$

which depends only on the gradient of PQ, and not on the particular choice of P and Q. So if we construct two parallel chords of our ellipse, bisect them, and join their mid-points, we have a diameter of the ellipse, that is, a chord through its centre. Bisecting this diameter yields the centre, O. (We shall find a much neater argument to get this same result in the next chapter, Section 8.3.)

Now a circle centre O has equation $x^2 + y^2 = c^2$, say, and this meets $(x^2/a^2) + (y^2/b^2) = 1$ in four points $(\pm c_1, \pm c_2)$, say, provided $a > c > b$. They can now be joined in pairs to give two diameters of Γ, equally inclined to the axes, so bisect the angles to give the axes of Γ. So we have constructed, from Γ, the points $(0, 0)$, $(\pm a, 0)$ and $(0, \pm b)$; so we now know the lengths a and b. To find the focus $(ae, 0)$, note that the distance from $(0, b)$ to $(ae, 0)$ is $\sqrt{a^2 e^2 + b^2} = a$, so we can find the foci by seeing where a circle, centre $(0, b)$ and radius a, meets the major axis.

For the hyperbola $(x^2/a^2) - (y^2/b^2) = 1$, a similar construction with parallel chords will find the centre, and a similar construction with a circle will find the axes. Details are left to the reader. At this point we know the length a (because we have found the points $(0, 0)$ and $(\pm a, 0)$), but the length b is not visible in our figure. Since we know the points $(0, 0)$ and $(a, 0)$, we can construct (a, a) and hence $(a\sqrt{2}, 0)$. The line $x = a\sqrt{2}$ meets the hyperbola $(x^2/a^2) - (y^2/b^2) = 1$ at $(a\sqrt{2}, \pm b)$, so now we can 'see' b; and then we can construct the lines $y = b$ and $x = a$, which meet on the asymptote $y/b = x/a$; thus we construct the asymptotes. To construct the foci, note that $a^2 + b^2 = a^2 e^2$, so use Pythagoras.

For the parabola $y^2 = 4ax$, take any two points $P = (ap^2, 2ap)$ and $Q = (aq^2, 2aq)$ of the parabola. The slope of PQ is

$$\frac{2ap - 2aq}{ap^2 - aq^2} = \frac{2}{p+q} = m,$$

say, and the mid-point of PQ has y-coordinate

$$\frac{2ap + 2aq}{2} = a(p+q) = \frac{2a}{m}.$$

So put in two parallel chords, and join their mid-points to get a line ℓ parallel to the axis. The mid-point of a chord perpendicular to ℓ lies on the axis, and thus we can construct the axis, $x = 0$, and the vertex $(0, 0)$. It is easy to construct a line through the vertex with gradient 2, that is, the line $y = 2x$, and this meets $y^2 = 4ax$ at $(a, 2a)$, so the focus $(a, 0)$ is the foot of the perpendicular from this point onto the axis.

7.4   CIRCLES: ANGLE THEOREMS

We return now to that most familiar conic of all, the circle, to prove a few standard theorems without which no introductory book on geometry would be complete. These theorems are (mostly) about *cyclic tetragrams*, more usually known as *cyclic quadrilaterals*. The relevant terminology was given in Definitions 4.16, 4.17, and 4.18, which we repeat informally here. (see Fig. 4.13.) Recall that a *tetragram* $ABCD$ consists of four points A, B, C, D (no three collinear) in the named order, and it is *cyclic* if they lie on a circle. If two of the opposite *sides*, the line segments AB and CD, or the line segments BC and DA, meet internally, we have a *crossed* tetragram; and if the *diagonals*, the line-segments AC and BD, meet internally, we have a *convex* tetragram. There is a third type of tetragram, a *re-entrant* tetragram, where one of the vertices is inside the triangle formed by the other three; but this cannot occur for a cyclic tetragram. One could take

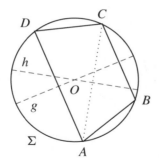

Fig. 7.7 Lemma 7.13

the view that cyclic permutations and reversal of the vertex list give the 'same' tetragram eight times: $ABCD = BCDA = CDAB = DABC = DCBA = CBAD = BADC = ADCB$, since all of these have the same sides and the same diagonals. From this point of view, the full set of permutations of vertices give rise to $4!/8 = 3$ tetragrams, $ABCD$, $ACBD$, and $ABDC$. (This is just a matter of whether the vertex opposite A is C, B or D.) Either two of these tetragrams are crossed and one is convex, or else all three are re-entrant.

Lemma 7.13. *Let $ABCD$ be a cyclic tetragram, on a circle with centre O. Then there is a rotation f about O with $f(A) = B$ and $f(C) = D$ if and only if $AD \parallel BC$.*

Proof. Let g, h be the reflections in the diameters of Σ orthogonal to AD, AC respectively. Thus g swaps A and D, and h swaps A and C (Fig. 7.7).

Suppose $AD \parallel BC$. Then g *also* swaps B and C. Put $f = gh$. Then $f(A) = gh(A) = g(C) = B$ and $f(C) = gh(C) = g(A) = D$; and f is a rotation about O, since it is direct, and both g and h (and hence also f) fix O.

Conversely, suppose we are given f as in the statement of the lemma. Then $f(C) = D$ and also $gh(C) = g(A) = D$, so that $f^{-1}gh$ fixes both C and O. But a non-trivial direct isometry either has no fixed points (if it is a translation), or just one fixed point (if it is a rotation). Therefore $f^{-1}gh = 1$, or $f = gh$. It follows that $g(C) = gh(A) = f(A) = B$, so BC is orthogonal to the axis of g, and hence is parallel to AD. □

Proposition 7.14. *Let $ABCD$ be a crossed cyclic tetragram. Then $\angle ABC = \angle ADC$.*

Proof. Let O be the centre of Σ, and let f be a rotation about O with angle to be chosen; let $f(A) = A'$, $f(D) = D'$, and $f(C) = C'$. Thus $\triangle ADC \equiv \triangle A'D'C'$ and so $\angle ADC = \angle A'D'C'$, regardless of the angle of rotation.

Choose the angle of rotation so that $A'D' \parallel AB$ (Fig. 7.8). By Lemma 7.13 there is a rotation about O that sends A to A' and D' to B. But f is a rotation about O with $f(A) = A'$; and so $f(D') = B$. But also $f(C) = C'$, so from Lemma 7.13 again, $BC \parallel D'C'$. Since we already have $AB \parallel A'D'$, we deduce $\angle ABC = \angle A'D'C'$, which we already know is equal to $\angle ADC$. □

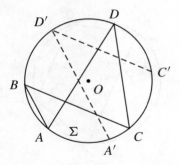

Fig. 7.8   Proposition 7.14

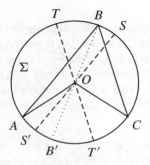

Fig. 7.9   Proposition 7.15

Proposition 7.14 is often said in words as: angles in the same segment are equal. Here the *segment* of Σ in question is that part of Σ (or rather, of Σ and its interior) that lie on the same side of the chord AC as B and D.

Proposition 7.15.   *Let A, B, C lie on a circle Σ, centre O. Then $\angle AOC = 2\angle ABC$. (But see the remarks following the end of the proof.)*

Proof.   Let S, T lie on Σ, on the same side of AC as B, such that $OS \parallel AB$ and $OT \parallel CB$. Let B', S', T' lie on Σ, diametrically opposite B, S, T respectively. Note that $AB \parallel S'O$ and $BC \parallel OT'$, so that $\angle ABC = \angle S'OT'$ (Fig. 7.9).

Now $AB \parallel S'S$, so by Lemma 7.13 there is a rotation f about O with $f(A) = S'$ and $f(S) = B$; and since S', B' are diametrically opposite S, B we also have $f(S') = B'$. Thus $f^2(A) = f(S') = B'$.

Next, $CB \parallel T'T$, so by Lemma 7.13 there is a rotation g about O with $g(T') = C$ and $g(B) = T$; and since B', T' are diametrically opposite B, T we also have $g(B') = T'$. Thus $g^2(B') = g(T') = C$.

Now f, g are both rotations about O, so gf and $(gf)^2$ are rotations about O, with the latter having twice the angle of rotation of the former. Further, $fg = gf$, so $(gf)^2 = g^2 f^2$. But $gf(S') = g(B') = T'$, so the angle of rotation of gf is $\angle S'OT' = \angle ABC$;

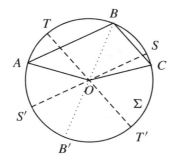

Fig. 7.10 Proposition 7.15, reflex case

and $(gf)^2(A) = g^2 f^2(A) = g^2(B') = C$, so the angle of rotation of $(gf)^2$ is $\angle AOC$.
□

Proposition 7.15 is often said in words as: the angle at the centre is twice the angle at the circumference.

A certain amount of sleight of hand has taken place in the above proof. If $\angle ABC < \pi/2$, there is no problem. If $\angle ABC > \pi/2$, then the result claims that $\angle AOC > \pi$, so we are looking at the *reflex* angle in this case; or, working modulo 2π, it is the *signed* $\angle AOC$ and $2\angle ABC$ that are equal (Fig. 7.10).

In the case $\angle ABC = \pi/2$, we must have $\angle AOC = \pi$, so that AC is a diameter of Σ; and conversely. Thus we obtain as a corollary a result we have met before (see Fig. 1.2, Exercise 2.19, and the solution to Exercise 3.25), that the angle in a semicircle is a right angle:

Corollary 7.16. *If A, B, C lie on a circle Σ such that AC is a diameter of Σ, then $\angle ABC = \pi/2$.*
□

- Ex.7.14: *Deduce Proposition 7.14 from Proposition 7.15.*
- Ex.7.15: *Criticise the following proof of Proposition 7.15: Let B' be on Σ, diametrically opposite B. Then $\angle AOB' = \angle OAB + \angle ABO = 2\angle ABO$ (because $\triangle OAB$ is isosceles); and $\angle B'OC = \angle BCO + \angle OBC = 2\angle OBC$ (because $\triangle OBC$ is isosceles). So $\angle AOC = \angle AOB' + \angle B'OC = 2\angle ABO + 2\angle OBC = 2(\angle ABO + \angle OBC) = 2\angle ABC.$*
□
- Ex.7.16: *Given a circle Σ and a point A outside Σ, give a ruler-and-compasses construction for the tangents from A to Σ.*

Proposition 7.17. *Let ABCD be a convex cyclic tetragram. Then $\angle ABC + \angle CDA = \pi$. Also, if R is on CD produced, then $\angle ABC = \angle ADR$ (Fig. 7.11).*

Proof. $\angle ABC = \angle ABD + \angle DBC = \angle ACD + \angle DAC$, by two applications of Proposition 7.14. But the angles of $\triangle ACD$ add up to π, and the result follows. (As an alternative, let O be the centre of the circle. By Proposition 7.15, $2\angle ABC + 2\angle CDA =$

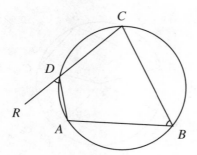

Fig. 7.11 Proposition 7.17

$\angle AOC + \angle COA = 2\pi$. Here exactly one of $\angle AOC$ and $\angle COA$ is a reflex angle, or else both are flat.) The second statement follows since $\angle CDA + \angle ADR = \pi$. □

Proposition 7.17 can be stated in words as: in a convex cyclic tetragram, opposite (internal) angles are supplementary ($\angle ABC + \angle CDA = \pi$), and each internal angle is equal to the opposite external angle ($\angle ABC = \angle ADR$).

Let us compare Propositions 7.14 and 7.17. Each is about a cyclic tetragram $ABCD$ and the angles ABC and ADC. In Proposition 7.14, B and D lie on the same side of AC and $ABCD$ is a *crossed* tetragram; whereas in Proposition 7.17, B and D lie on opposite sides of AC and $ABCD$ is a *convex* tetragram. In the first case, $\angle ABC = \angle ADC$; and in the second case, $\angle ABC + \angle CDA = \pi$, or, using signed angles, $\angle ABC \equiv \angle ADC$ modulo π. It is clear that the two results are intimately related. We explore them a little further, using complex numbers.

Since similarities preserve angles, we may as well make life easier by assuming that A, B, C, D lie on the circle $x^2 + y^2 = 1$, or $|z| = 1$. Let our four points be represented by the complex numbers a, b, c, d respectively. So a, b, c, d are distinct, and $|a| = |b| = |c| = |d| = 1$, from which $\bar{a} = a^{-1}$, and similarly for the others. Put

$$\lambda = \left(\frac{c-b}{a-b}\right)\left(\frac{a-d}{c-d}\right).$$

We have

$$\bar{\lambda} = \left(\frac{\bar{c}-\bar{b}}{\bar{a}-\bar{b}}\right)\left(\frac{\bar{a}-\bar{d}}{\bar{c}-\bar{d}}\right) = \left(\frac{c^{-1}-b^{-1}}{a^{-1}-b^{-1}}\right)\left(\frac{a^{-1}-d^{-1}}{c^{-1}-d^{-1}}\right).$$

Multiplying numerator and denominator of the last expression by $abcd$, we get

$$\bar{\lambda} = \left(\frac{b-c}{b-a}\right)\left(\frac{d-a}{d-c}\right) = \lambda,$$

so that $\lambda \in \mathbb{R}$. But this means, using polar form, that if

$$\frac{c-b}{a-b} = re^{i\theta} \quad \text{and} \quad \frac{c-d}{a-d} = se^{i\varphi},$$

then $e^{i(\theta-\varphi)} \in \mathbb{R}$, or $\theta \equiv \varphi$ modulo π. But θ, φ are the signed angles $\angle ABC$, $\angle ADC$, respectively, so we have a combined proof of Propositions 7.14 and 7.17; to separate the two results we would have to distinguish according to the cyclic order of the four points on the circle, and we leave the reader to explore this.

- Ex.7.17: *Use complex numbers to give another proof of Proposition 7.15.*

The complex number techniques used above are part of a larger picture which will be explored further in Section 9.1.

- Ex.7.18: *Let P, Q, R, S, T, U lie on a circle, in the given order. Prove that if PQ ∥ TS, and QR ∥ UT, then also RS ∥ PU. (Hint: join PS and look at the two cyclic tetragrams PQRS and STUP.)*

This exercise is a special case of the theorem known as *Pascal's mystic hexagram*, which we shall meet again later. A *hexagram*, of course, is just six points in a given order.

We shall now prove the converse of Proposition 7.14 (and of Proposition 7.17). One method might be as follows: let $ABCD$ be a crossed tetragram with $\angle ABC = \angle ADC$, and let Σ be the circumcircle of $\triangle ABC$. Then we must show that D lies on Σ. So let CD meet Σ at $(C$ and$)$ D'. By Proposition 7.14, $\angle ABC = \angle AD'C$, provided B and D' are on the same side of AC. Must this be so? (Or, indeed, might CD be a *tangent* to Σ, so that $D' = C$?) If nonetheless all is well, we now have $\angle ADC = \angle AD'C$, which either means $D = D'$ (which is what we want, and now $ABCD$ is cyclic), or else D' is the reflection of D in the line through A orthogonal to CD. Could *this* happen? Such difficulties cannot just be swept under the carpet, and we leave the reader to worry about them. We shall instead take a cowardly approach, and use complex numbers again.

Proposition 7.18 (Converse of Propositions 7.14 and 7.17). *Let ABCD be a tetragram. If either ABCD is a crossed tetragram with $\angle ABC = \angle ADC$, or ABCD is a convex tetragram with $\angle ABC = \angle ADR$ (where R is on CD produced), then ABCD is cyclic.*

Proof. Let Σ be the circumcircle of $\triangle ABC$. By applying a suitable similarity, we may as well assume that Σ is the circle $|z| = 1$. So if A, B, C, D are represented by the complex numbers a, b, c, d respectively, we have $|a| = |b| = |c| = 1$, and we want to prove that $|d| = 1$ also. Using polar form, let

$$\frac{c-b}{a-b} = re^{i\theta} \quad \text{and} \quad \frac{c-d}{a-d} = se^{i\varphi}.$$

The given conditions imply that $\theta = \varphi$, so if

$$\lambda = \left(\frac{c-b}{a-b}\right)\left(\frac{a-d}{c-d}\right),$$

then $\lambda = r/s \in \mathbb{R}$. But

$$\bar{\lambda} = \left(\frac{\bar{c}-\bar{b}}{\bar{a}-\bar{b}}\right)\left(\frac{\bar{a}-\bar{d}}{\bar{c}-\bar{d}}\right) = \left(\frac{c^{-1}-b^{-1}}{a^{-1}-b^{-1}}\right)\left(\frac{a^{-1}-d_1^{-1}}{c^{-1}-d_1^{-1}}\right),$$

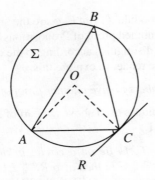

Fig. 7.12   Proposition 7.19

where $d_1 = (\bar{d})^{-1}$. Multiplying numerator and denominator by $abcd_1$, we have

$$\bar{\lambda} = \left(\frac{b-c}{b-a}\right)\left(\frac{d_1-a}{d_1-c}\right) = \left(\frac{c-b}{a-b}\right)\left(\frac{a-d_1}{c-d_1}\right).$$

But $\lambda = \bar{\lambda}$, so we deduce

$$\frac{c-d}{a-d} = \frac{c-d_1}{a-d_1}, \quad \text{or} \quad \frac{c-d}{c-d_1} = \frac{a-d}{a-d_1} = \frac{(c-d)-(a-d)}{(c-d_1)-(a-d_1)} = 1,$$

by the well-known theorem on equal ratios;[2] so that $d = d_1$. Thus $d = (\bar{d})^{-1}$, or $d\bar{d} = 1$ and $|d| = 1$, as required. $\qquad\square$

Proposition 7.19 (The alternate segment theorem).   *Let A, B, C lie on a circle Σ, and let R lie on the tangent to Σ at C, with R and B on opposite sides of AC. Then $\angle ABC = \angle ACR$.*

Proof.   One method would be to take the limiting case of Proposition 7.14 *or* Proposition 7.17 as D approaches C, and this nicely illustrates again the connection between those two propositions. We leave the details of this to the reader, and instead take a more direct approach.

Let $\angle ACR = \alpha$, and let O be the centre of Σ. The proof splits into three cases, according as α is acute, obtuse, or a right angle. If α is acute, then R and O are on opposite sides of AC, and since $OC \perp CR$ we have $\angle OCA = (\pi/2) - \alpha$ (Fig. 7.12). But $\triangle OAC$ is isosceles ($|OA| = |OC|$), so that $\angle AOC = 2\alpha$, and therefore $\angle ABC = \alpha$, by Proposition 7.15, and we have finished.

If α is obtuse, choose S on RC produced, put $\beta = \angle SCA$, and choose D on Σ in the other segment from B (Fig. 7.13). Thus α and β are supplementary; and so are $\angle ABC$ and $\angle CDA$, by Proposition 7.17. Then β is acute, so $\beta = \angle CDA$, as above, and thus $\alpha = \angle ABC$.

[2] This says that if $k = x_1/y_1 = x_2/y_2 = x_3/y_3 = \ldots$, then $k = \sum_i \lambda_i x_i / \sum_i \lambda_i y_i$ also, for any $\lambda_1, \lambda_2, \ldots$ (provided $\sum \lambda_i y_i \neq 0$, of course). Prove it. (Hint: $x_i = ky_i$, for all i, so substitute.)

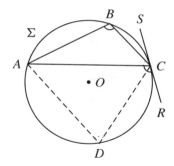

Fig. 7.13 Proposition 7.19, obtuse case

Finally, if $\alpha = \pi/2$, then AC is a diameter of Σ, so $\angle ABC = \pi/2$ also, by Corollary 7.16. □

In words, the angle between tangent and chord is equal to the angle in the alternate segment, that is, the segment on the other side of the chord.

- **Ex.7.19:** *Prove the converse of the alternate segment theorem: if B, S lie on opposite sides of AC and $\angle ABC = \angle ACS$, then CS is the tangent to the circumcircle Σ of $\triangle ABC$ at C.*

7.5 CIRCLES: RECTANGULAR PROPERTIES

The theorems in this section are also about cyclic tetragrams, but are about lengths rather than angles.

Proposition 7.20 (Rectangular properties of a circle (1)). *Let $ABCD$ be a tetragram, and suppose its diagonal lines AC and BD meet at X. Then $ABCD$ is cyclic if and only if $(XA)(XC) = (XB)(XD)$.*

Proof. Here $(XA)(XC)$ means the product of the *signed* lengths, so it is negative if X lies between A and C, and positive otherwise.

Suppose $ABCD$ is cyclic. If it is a convex tetragram, then X lies between A and C, and also between B and D, and thus both $(XA)(XC)$ and $(XB)(XD)$ are negative (Fig. 7.14). Further, the angles subtended by AB at C and D are equal: $\angle ACB = \angle ADB$, or $\angle ADX = \angle BCX$. But also $\angle AXD = \angle BXC$, and we deduce that $\triangle ADX \sim \triangle BCX$ (*similar* triangles, case AAA, see Section 4.7), so that $|XA|/|XB| = |XD|/|XC|$, and the result follows. If on the other hand $ABCD$ is a crossed tetragram, then X does not lie between A and C or between B and D, so that both $(XA)(XC)$ and $(XB)(XD)$ are positive (Fig. 7.15). Suppose C lies between A and X, and B between D and X. (If not, swap labels of A with C and/or B with D.) Then $\angle BAC = \angle BDC$, or $\angle BAX = \angle CDX$. Since also $\angle BXA = \angle CXD$, we deduce $\triangle ABX \sim \triangle DCX$ (case AAA), and $|XA|/|XD| = |XB|/|XC|$, and the result follows again.

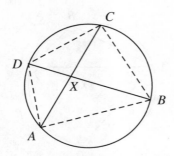

Fig. 7.14   Proposition 7.20 for a convex tetragram

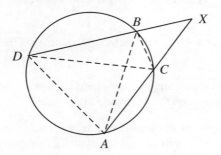

Fig. 7.15   Proposition 7.20 for a crossed tetragram

Now for the converse. If $(XA)(XC) = (XB)(XD) < 0$, then $ABCD$ is a convex tetragram, and $|XA|/|XB| = |XD|/|XC|$. Since also $\angle AXD = \angle BXC$, we deduce $\triangle ADX \sim \triangle BCX$ (case SAS), and thus $\angle ADX = \angle BCX$, or $\angle ACB = \angle ADB$. It follows that $ABCD$ is cyclic, by Proposition 7.18. If on the other hand $(XA)(XC) = (XB)(XD) > 0$, then $ABCD$ is a crossed tetragram, and as before it makes no difference if we swap labels (if necessary) to ensure that C lies between A and X, and B between D and X. Now $|XA|/|XD| = |XB|/|XC|$, and $\angle BXA = \angle CXD$, so $\triangle ABX \sim \triangle DCX$ (case SAS), so that $\angle BAX = \angle CDX$, or $\angle BAC = \angle BDC$. It follows that $ABCD$ is cyclic, by Proposition 7.18 again, and we have finished.   □

Proposition 7.21 (Rectangular properties of a circle (2)).   *Let Σ be the circumcircle of $\triangle ABC$, and let X be on CA produced. Then BX is the tangent to Σ at B if and only if $(XA)(XC) = (XB)^2$.*

Proof.   If X lies on the tangent (Fig. 7.16), then $\angle ABX = \angle ACB$, by the alternate segment theorem, that is, $\angle ABX = \angle BCX$. But also $\angle AXB = \angle BXC$, so $\triangle ABX \sim \triangle BCX$ (case AAA), from which $|XA|/|XB| = |XB|/|XC|$, and the result follows.

For the converse, if $(XA)(XC) = (XB)^2$, then $|XA|/|XB| = |XB|/|XC|$; and also $\angle AXB = \angle BXC$, so that $\triangle ABX \sim \triangle BCX$ (case SAS), and thus $\angle ABX = \angle BCX$, or $\angle ABX = \angle ACB$. By Exercise 7.19, X is on the tangent to Σ at B.   □

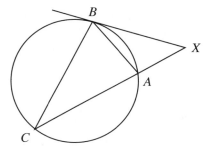

Fig. 7.16 Proposition 7.21

Let X be a point and Σ a circle. If a line ℓ through X meets Σ in A and B, then we have just proved that the quantity $(XA)(XB)$ depends only on X and Σ, and not on ℓ. It is called the *power* of X with respect to Σ; it is positive if X is outside Σ, negative if X is inside Σ, and zero if X is on Σ.

● Ex.7.20: *Let Σ have equation $p(x, y) = 0$, where $p(x, y) = x^2 + y^2 + 2gx + 2fy + c$, and let $X = (x_0, y_0)$. Prove that the power of X with respect to Σ is just $p(x_0, y_0)$.*

Let Σ_i be the circle $p_i(x, y) = 0$, where $p_i(x, y) = x^2 + y^2 + 2g_i x + 2f_i y + c_i$, for $i = 1, 2$. Assume that either $g_1 \neq g_2$ or $f_1 \neq f_2$ (or both). Then $p_1(x, y) - p_2(x, y) = 2(g_1 - g_2)x + 2(f_1 - f_2)y + (c_1 - c_2)$, so that the curve $p_1(x, y) - p_2(x, y) = 0$ is actually a line. It is called the *radical axis* of Σ_1 and Σ_2, and consists of all points whose powers with respect to Σ_1, Σ_2 are equal.

● Ex.7.21: *Σ_1 and Σ_2 do not have a radical axis if $g_1 = g_2$ and $f_1 = f_2$. What else is special about the circles in this exceptional case?*

● Ex.7.22: *Assume we are not in the exceptional case. Show that the radical axis of Σ_1, Σ_2 is orthogonal to the line joining their centres.*

● Ex.7.23: *Let the circles Σ_1 and Σ_2 meet at A and B. Show that the radical axis of Σ_1 and Σ_2 is the line AB.*

● Ex.7.24: *Suppose the circles Σ_1 and Σ_2 touch at A. Where is their radical axis?*

● Ex.7.25: *Show that the radical axes of three circles, taken in pairs, are either concurrent or all parallel.*

● Ex.7.26: *Suppose the circles Σ_1 and Σ_2 do not meet, and that we are not in the exceptional case. Give a ruler-and-compasses construction for their radical axis.*

● Ex.7.27: *Given a circle Σ_1 and two points A, B outside Σ_1, give a ruler-and-compasses construction for a circle Σ_2 through A and B that also touches Σ_1.*

• Ex.7.28: *Let Σ_1, Σ_2 be circles with centres O_1, O_2 respectively, and suppose the circles touch externally at A (so that A is between O_1 and O_2). Let O_3 be on Σ_2, and let the tangents to Σ_2 at A and at O_3 meet at X. Let P be on Σ_1 such that O_3P is a tangent to Σ_1, and let Y be the mid-point of O_3P. Prove that $XY \perp O_1O_3$.*

We shall now use the rectangular properties of a circle to deal with a more general case of Pascal's mystic hexagram (cf. Exercise 7.18).

Proposition 7.22. *Let X_1, X_2, X_3, X_4, X_5, X_6 lie on a circle in the given order. Let $L = X_1X_2 \cdot X_4X_5$ (the meet), $M = X_2X_3 \cdot X_5X_6$, and $N = X_3X_4 \cdot X_6X_1$. Then L, M, N are collinear.*

Proof. Suppose first that no two of the lines X_1X_2, X_3X_4, X_5X_6 are parallel. Let $A = X_3X_4 \cdot X_5X_6$, $B = X_5X_6 \cdot X_1X_2$, and $C = X_1X_2 \cdot X_3X_4$ (Fig. 7.17). Applying Menelaus' theorem three times to $\triangle ABC$, we obtain

$$\frac{(CL)(BX_5)(AX_4)}{(LB)(X_5A)(X_4C)} \times \frac{(BM)(AX_3)(CX_2)}{(MA)(X_3C)(X_2B)} \times \frac{(AN)(CX_1)(BX_6)}{(NC)(X_1B)(X_6A)} = (-1)^3 = -1.$$

But by the rectangular properties, we have $(AX_3)(AX_4) = (AX_5)(AX_6) = (X_5A)(X_6A)$, and similarly $(BX_5)(BX_6) = (X_1B)(X_2B)$ and $(CX_1)(CX_2) = (X_3C)(X_4C)$.

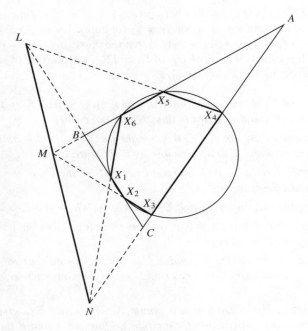

Fig. 7.17 Proposition 7.22

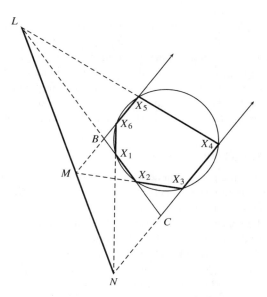

Fig. 7.18   Proposition 7.22, with $X_3X_4 \parallel X_5X_6$

Substituting back and cancelling now gives

$$\frac{(CL)(BM)(AN)}{(LB)(MA)(NC)} = -1,$$

so that L, M, N are collinear, by Menelaus' theorem.

Now let us deal with the case where $X_3X_4 \parallel X_5X_6$ (Fig. 7.18), so that A is not defined (but B and C are as before). We now have lots of similar triangles: $\triangle CLX_4 \sim \triangle BLX_5$, $\triangle BMX_2 \sim \triangle CX_3X_2$, and $\triangle BX_6X_1 \sim \triangle CNX_1$. Thus

$$\frac{(CL)(BX_5)}{(LB)(X_4C)} \times \frac{(BM)(CX_2)}{(X_3C)(X_2B)} \times \frac{(CX_1)(BX_6)}{(NC)(X_1B)} = (1)^3 = 1.$$

Applying the rectangular properties again, and cancelling, gives

$$\frac{(CL)(BM)}{(LB)(NC)} = 1,$$

whence $\triangle CLN \sim \triangle BLM$, so that $\angle CLN = \angle BLM$, and once again L, M, N are collinear.   □

The two parts the above proof were deliberately written for comparison: notice that the second part is a limiting case of the first part as $A \to \infty$, when $(AX_4)/(X_5A) \to -1$, $(AX_3)/(MA) \to -1$, and $(AN)/(X_6A) \to -1$. Exercise 7.18 can be interpreted as another limiting case, saying that if $L \to \infty$ and $M \to \infty$, then $N \to \infty$ also. We leave the reader to explore what happens if we just have $L \to \infty$ (that is, if X_1X_2 and X_3X_4 are parallel, but neither is parallel to X_5X_6), and also to vary the order of the six points on the circle. The number of different diagrams required to cover all cases is quite overwhelming, and it is clear that some new technique is needed to take care of all possible configurations; we shall outline such a technique in Section 10.1. Pascal discovered his wonderful theorem at the age of 16; his original proof is lost, but there are grounds for believing that it may have been like the one given above.

7.6   STEREOGRAPHIC PROJECTION

We have met stereographic projection before, in Exercise 6.13. In this section, we show that stereographic projection sends each circle on a sphere to a circle or line in a plane, and that, in a sense to be made precise, it preserves angles.

In $\mathbb{R}^3$, let Σ be the sphere $x^2 + y^2 + z^2 = 1$, with 'north pole' $N = (0, 0, 1)$, and let π be the 'equatorial' plane $z = 0$. The *stereographic projection* map $f : \Sigma \to \pi$ is defined as follows: for $A \in \Sigma$, let NA meet π at A', and put $f(A) = A'$ (Fig. 7.19). (Strictly, this is a map $\Sigma \setminus \{N\} \to \pi$; alternatively, note that $A' \to \infty$ as $A \to N$, so adjoin an extra 'point' called ∞ to $\mathbb{R}^3$ and write $f(N) = \infty$, so that f is a map from Σ to $\pi \cup \{\infty\}$.) Referring back to Exercise 6.13, we see that if $A = (x, y, z)$, $A \neq N$, then $A' = (x/(1 - z), y/(1 - z), 0)$.

Now a plane and a sphere meet in a circle, provided of course that the distance from the centre of the sphere to the plane is less than the radius of the sphere. So suppose that the plane $ax + by + cz + d = 0$ meets Σ in the circle Γ. Then $\Gamma' = f(\Gamma)$ is a subset of $\pi \cup \{\infty\}$.

- Ex.7.29:   *Prove that, if $c + d \neq 0$, then $\Gamma' \subset \pi$, and that Γ' is a circle. What happens if $c + d = 0$? (Method: let $(x, y, z) \in \Gamma$, so that $x^2 + y^2 + z^2 = 1$ and $ax + by + cz + d = 0$. Put $X = x/(1 - z)$ and $Y = y/(1 - z)$, so substitute $X(1 - z)$ for x and $Y(1 - z)$ for y in the two equations, and then eliminate z to get an equation in X and Y.)*

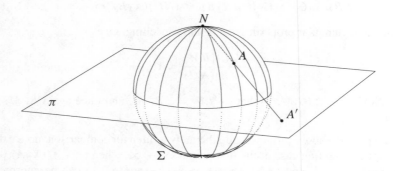

Fig. 7.19   Stereographic projection

It is possible to use the theory of quadrics to give a rather posh proof of the last exercise, which we sketch here. Recall that a quadric is a surface in $\mathbb{R}^3$ given by a second-degree polynomial equation, so that (for example) spheres, cones, and pairs of planes are examples of quadrics. Let our sphere have equation $\Sigma = 0$, and let the tangent plane at N have equation $P = 0$. (So Σ, if you insist, stands for the polynomial $x^2 + y^2 + z^2 - 1$, and P stands for the polynomial $z - 1$.) The equatorial plane π has equation $P + \lambda = 0$, for some scalar λ. The circle Γ is given by intersecting Σ with a plane $Q = 0$.

There is a cone with apex N containing Γ; it is the union of all lines joining N to points of Γ. (It is *not* a right circular cone, in general; but it *is* a quadric.) This cone meets $P + \lambda = 0$ in the curve Γ', which we wish to show is a circle.

Now our cone meets the quadric $\Sigma = 0$ in its intersection with the planes $P = 0$ and $Q = 0$, from which it can be shown that the cone has equation $\Sigma + \mu P Q = 0$, for some scalar μ. Then the equation $(\Sigma + \mu P Q) - \mu(P + \lambda)Q = 0$ represents another quadric, meeting the cone in its intersections with the planes $Q = 0$ and $P + \lambda = 0$, that is, in Γ and Γ'. But the equation of this last quadric simplifies to $\Sigma - \mu \lambda Q = 0$ which, since Q is linear, is another sphere; so Γ', being the intersection of this sphere with the equatorial plane, is a circle. As a bonus, the circles Γ and Γ' lie on a sphere.

- Ex.7.30: *What happened to the special case when $c + d = 0$?*

We now investigate what stereographic projection does to angles. If two (smooth) curves (such as circles) meet at a point, then the *angle* between them at that point is defined to be the angle between their respective tangents at that point.

Let Σ, π, N, f be as before, and let π' be the tangent plane to Σ at N, so that the planes π and π' are parallel. This means that any other plane (not also parallel) will meet π and π' in two parallel lines.

Now for $i = 1, 2$, let Γ_i be a circle on Σ through the point A; then if $f(A) = A'$ and $f(\Gamma_i) = \Gamma_i'$, then Γ_i' is a circle or line in π, through A'. Let ℓ_i be the tangent line to Γ_i at A. If Γ_i does not pass through N, then Γ_i' is a circle. There is a plane π_i containing both ℓ_i and N, and this plane meets Σ in a circle Λ_i, through N and A, and it meets π in $\Lambda_i' = f(\Lambda_i)$, which is a line. In fact, ℓ_i is the tangent to Λ_i at A (so that Γ_i and Λ_i touch at A), and Λ_i' is the tangent (line) to the circle Γ_i' at A'. Further, $\ell_i' = \pi' \cap \pi_i$ is the tangent line to Λ_i at N; and since the planes π and π' are parallel, so are the lines Λ_i' and ℓ_i'. If on the other hand Γ_i *does* pass through N, then Γ_i lies in a plane which meets π in the line Γ_i', and meets π' in the tangent ℓ_i' to Γ_i' at N. In this case we write $\Lambda_i' = \Gamma_i'$, so that in both cases we have $\Lambda_i' \parallel \ell_i'$.

Now the angle α between Γ_1 and Γ_2 at A is, by definition, the angle between their tangents ℓ_1 and ℓ_2 at A. Then, reflection in the plane which is the perpendicular bisector of AN leaves both Γ_1 and Γ_2 invariant, and swaps A with N, so it swaps the respective tangents at A with the tangents at N. Therefore α is also the angle between ℓ_1' and ℓ_2'. But $\Lambda_i' \parallel \ell_i'$, $i = 1, 2$, so that α is the angle between Λ_1' and Λ_2', and by definition this is the angle between Γ_1' and Γ_2' at A'. We have proved:

Proposition 7.23. *Stereographic projection sends circles on a sphere to circles or lines in a plane, and it preserves angles.* □

If $g : \Sigma \to \Sigma$ is a rotation about O, then g, being an isometry, sends circles to circles and also preserves angles. Consequently, the composite map $fgf^{-1} : \pi \cup \{\infty\} \to \pi \cup \{\infty\}$ sends each circle and each line to a circle or a line (but not necessarily in that order), and it preserves angles. This is one example of a type of map called a *Möbius transformation*, which we shall look at in more detail in Section 9.1.

ANSWERS TO EXERCISES

7.1: Rotate $\lambda(1, 0, m)$ about the z-axis to get

$$\begin{pmatrix} \cos\theta & -\sin\theta & 0 \\ \sin\theta & \cos\theta & 0 \\ 0 & 0 & 1 \end{pmatrix} \begin{pmatrix} \lambda \\ 0 \\ \lambda m \end{pmatrix} = \begin{pmatrix} \lambda\cos\theta \\ \lambda\sin\theta \\ \lambda m \end{pmatrix},$$

and now $(\lambda m)^2 = m^2((\lambda\cos\theta)^2 + (\lambda\sin\theta)^2)$. Conversely, given (x, y, z) with $z^2 = m^2(x^2 + y^2)$, we can find λ and θ with $x = \lambda\cos\theta$ and $y = \lambda\sin\theta$ (OK?), and now $\lambda(1, 0, m)$ rotates (as before) to (x, y, z). (If this gives $(x, y, -z)$ instead, then replace λ by $-\lambda$ and θ by $\theta + \pi$.)

7.2: Take the cone $z^2 = x^2 + y^2$ and rotate through $\pi/4$ about the x-axis, giving $((y + z)/\sqrt{2})^2 = x^2 + ((y - z)/\sqrt{2})^2$, or $2yz = x^2$. On putting $z = 0$ we get $x^2 = 0$, a repeated line.

7.3: Suppose $x^2 + y^2 + 1 = (\ell x + my + n)(\ell' x + m'y + n')$. From the x^2-terms, we have $\ell\ell' = 1$, so we may divide the first linear expression by ℓ and the second by ℓ' to get $x^2 + y^2 + 1 = (x + sy + t)(x + s'y + t')$, say. From the y^2-terms, $ss' = 1$, but from the xy-terms, $s + s' = 0$, so $1 = ss' = -s^2$ and $s = i$, $s' = -i$ (or vice versa). From the x-terms, $t + t' = 0$, and from the y-terms, $it - it' = 0$, so $t = t' = 0$; but from the constant terms, $tt' = 1$, a contradiction.

7.4: Take any two points of the circle, A and B. Reflection in the diameter CD of the circle orthogonal to AB is a symmetry of the circle, and also of the line AB, so it must swap A and B. So CD is the perpendicular bisector of AB, and this we can construct. Finally, construct the mid-point of CD, which is the required centre. (*Or* repeat the construction with another two points of the circle to give a second diameter, and the centre is where the two diameters meet.)

7.5: We have $(OF)(OH) = (ae)(ae^{-1}) = a^2 = (OA)^2$, so the result follows by Exercise 4.33.

7.6: $(ae, 0)$ translates to $(a + ae, 0) = (\alpha(1 + e)/(1 - e), 0)$, which goes to infinity as $e \to 1$; whereas $(-ae, 0)$ translates to $(a - ae, 0) = (\alpha, 0)$. The directrix $x = a/e$ translates to $x - a = a/e$, or $x = a(1 + 1/e) = (\alpha/(1 - e))(1 + 1/e)$, which goes to infinity as $e \to 1$; whereas $x = -a/e$ translates to $x - a = -a/e$, or $x = a(1 - 1/e) = -\alpha/e$, which has limit $x = -\alpha$ as $e \to 1$. Finally, the centre O translates to $(a, 0)$, which goes to infinity as $e \to 1$.

7.7: With foci $(\pm ae, 0)$ and $|PF| + |PF'| = 2a$, we obtain $\sqrt{(x - ae)^2 + y^2} + \sqrt{(x + ae)^2 + y^2} = 2a$. Squaring, $(x - ae)^2 + (x + ae)^2 + 2y^2 + 2\sqrt{(x^2 + y^2 + a^2e^2 - 2aex)(x^2 + y^2 + a^2e^2 + 2aex)} = 4a^2$, or, dividing by 2 and squaring again, $(x^2 + y^2 + a^2e^2 - 2a^2)^2 = (x^2 + y^2 + a^2e^2)^2 - 4a^2e^2x^2$, whence $4a^2e^2x^2 = (x^2 + y^2 + a^2e^2)^2 - (x^2 + y^2 + a^2e^2 - 2a^2)^2 = 2a^2(2x^2 + 2y^2 + 2a^2e^2 - 2a^2)$, or $(1 - e^2)x^2 + y^2 = a^2(1 - e^2)$.

7.8: Here the two directrices are on the *same* side of P; if P is nearer to d' than d, then $|PF| - |PF'| = e|PQ| - e|PQ'| = e(|PQ| - |PQ'|) = e|QQ'|$, which is constant on one branch of the hyperbola; and on the other branch also, with a change of sign.

7.9: If you keep the picture still and move the wall about—applied mathematicians do this sort of thing all the time—the locus of the nail in the wall is an ellipse with foci at the hooks and centre at the centre of mass of the picture. In the symmetrical position, the nail is at one end of the minor axis of the ellipse, and so the centre of mass is at its highest point. Any disturbance lowers the centre of mass, so the equilibrium is unstable.

7.10: Note that the lines $\pi \cap \tau$ and AA' are perpendicular. Let them meet at H. Then $\{A, A'; F, H\} = -1$, by Exercise 4.34, and the result follows by Exercise 7.5. Similarly for the other directrix.

7.11: Noting that

$$\begin{pmatrix} x + \alpha \\ y + \beta \\ 1 \end{pmatrix} = \begin{pmatrix} 1 & 0 & \alpha \\ 0 & 1 & \beta \\ 0 & 0 & 1 \end{pmatrix} \begin{pmatrix} x \\ y \\ 1 \end{pmatrix},$$

we have

$$\begin{pmatrix} 1 & 0 & 0 \\ 0 & 1 & 0 \\ \alpha & \beta & 1 \end{pmatrix} \begin{pmatrix} a & h & g \\ h & b & f \\ g & f & c \end{pmatrix} \begin{pmatrix} 1 & 0 & \alpha \\ 0 & 1 & \beta \\ 0 & 0 & 1 \end{pmatrix} = \begin{pmatrix} a & h & 0 \\ h & b & 0 \\ 0 & 0 & c_1 \end{pmatrix}.$$

Take determinants.

7.12: Consider the expression $ax^2 + 2hxy + by^2$. Putting $x = 1$ and $y = 0$ gives a; and putting $x = \cos\theta$ and $y = \sin\theta$ gives a_1. If a and a_1 have opposite signs, then there will be values of x, y, not both zero, for which $ax^2 + 2hxy + by^2 = 0$; but this entails $ab \leq h^2$.

7.13: Choose θ with $\tan\theta = a_0/b_0$, so that $a_0 x + b_0 y = \sqrt{a_0^2 + b_0^2}(x\sin\theta + y\cos\theta)$.

7.14: $\angle ABC = \frac{1}{2}\angle AOC = \angle ADB$—done.

7.15: The proof is fine if B, B' lie on opposite sides of AC; but the signs of the angles need adjusting if B, B' lie in the same segment (determined by AC); and the whole thing simplifies dramatically if $B' = A$ or $B' = C$. (Check the details!) This illustrates the danger of writing down proofs from a diagram: can one be sure that all possible cases have been covered?

7.16: Let O be the centre of Σ which, if not given, can be constructed. (OK?) Let the circle on AO as diameter meet Σ at P and Q. Then $\angle APO = \angle AQO = \pi/2$, so AP and AQ are the required tangents.

7.17: Let $|a| = |b| = |c| = 1$, and put $\mu = ((c - b)/(a - b))^2(a/c)$. Then $\overline{\mu} = ((c^{-1} - b^{-1})/(a^{-1} - b^{-1}))^2(a^{-1}/c^{-1})$. Multiplying numerator and denominator by $a^2 b^2 c^2$, we have $\overline{\mu} = \mu$, so $\mu \in \mathbb{R}$. If $\angle ABC = \theta$ and $\angle AOC = \varphi$, then $(c - b)/(a - b) = re^{i\theta}$, say, and $c/a = e^{i\varphi}$. (OK?) Since $\mu \in \mathbb{R}$, we have $e^{i(2\theta - \varphi)} \in \mathbb{R}$, so $2\theta \equiv \varphi$ modulo π. Is that enough?

7.18: Because of the given parallels, we have $\angle PQR = \angle STU$. (OK?) Then $\angle RSP = \pi - \angle PQR = \pi - \angle STU = \angle UPS$, whence the result.

7.19: Put R on the tangent to Σ at C, as in Proposition 7.19. Then R and S are on the same side of AC and, using Proposition 7.19, $\angle ACR = \angle ABC = \angle ACS$. It follows that S lies on CR, as required.

7.20: Let Σ have centre $O = (\alpha, \beta)$ and radius r, so that $p(x, y) = (x - \alpha)^2 + (y - \beta)^2 - r^2$. If X is outside Σ, and XT is a tangent to Σ at T, then the power of X is $(XT)^2 = (OX)^2 - (OT)^2 = (x_0 - \alpha)^2 + (y_0 - \beta)^2 - r^2 = p(x_0, y_0)$. If X is inside Σ and $X \neq O$, let ST be the chord of Σ through X and orthogonal to OX. Then the power of X is $(XS)(XT) = -(XT)^2 = (OX)^2 - (OT)^2 = p(x_0, y_0)$ as before. If $X = O$ or X is on Σ, the proof is trivial.

7.21: They are concentric, i.e. they have the same centre.

7.22: Reflection in the line of centres sends each circle to itself, and if it maps X to Y, then X and Y have the same power with respect to Σ_1, and the same power with respect to Σ_2. So the radical axis is invariant; since it is not the line of centres (OK?), it must be orthogonal to it.

7.23: A has power 0 with respect to both circles, so it lies on the radical axis; and similarly for B; so the radical axis is the common chord, AB.

7.24: Let B lie on the common tangent at A. Then the power of B with respect to either circle is $(BA)^2$, so that B lies on the radical axis. Thus the radical axis is the common tangent.

7.25: Let the circles be Σ_i, $i = 1, 2, 3$, and assume that each pair *has* a radical axis. If the three radical axes are not all parallel, then two of them must meet, so suppose without loss of generality that the radical axis of Σ_1 and Σ_2, and the radical axis of Σ_2 and Σ_3, meet at X. Then the power of X with respect to Σ_1 and Σ_2 is the same, and the power of X with respect to Σ_2 and Σ_3 is the same. Thus the power of X with respect to Σ_1 and Σ_3 is the same, so that X also lies on the radical axis of Σ_1 and Σ_3.

7.26: Use Exercise 7.25. Draw any circle Σ_3, meeting Σ_1 at A and B, and Σ_2 at C and D. If AB and CD meet at X, then X lies on the required radical axis. (If AB turns out to be parallel to CD, try again. If these lines *always* turn out to be parallel, then Σ_1 and Σ_2 must be concentric, which was banned.) Repeat with yet another circle to get a second point Y on the required radical axis, and join XY.

7.27: Use Exercise 7.25 again. Draw any circle Σ_3 through A and B, so that Σ_3 and Σ_1 meet at C and D. It may happen that $AB \parallel CD$, in which case the perpendicular bisector of AB is a diameter PQ of Σ_1. Then Σ_2 is the circumcircle of $\triangle ABP$ or, as an alternative, the circumcircle of $\triangle ABQ$. So now suppose instead that AB and CD meet at X. Construct P on Σ_1 so that XP is the tangent from X to Σ_1. (See Exercise 7.16.) Then Σ_2 is the circumcircle of $\triangle ABP$. (There are two choices for P, and so two choices for Σ_2, once again.)

7.28: This is a trick. Let Σ_3 be the *point* O_3, thought of as a circle of zero radius: $(x - \alpha)^2 + (y - \beta)^2 = 0$, where $O_3 = (\alpha, \beta)$. The radical axis of Σ_1 and Σ_2 is the common tangent at A, and the radical axis of Σ_2 and Σ_3 is the tangent to Σ_2 (and Σ_3!) at O_3. These meet at X, so by Exercise 7.25, X lies on the radical axis of Σ_1 and Σ_3. Now $(YP)^2 = (YO_3)^2$, so that Y also lies on the radical axis of Σ_1 and Σ_3, which must therefore be XY. The result follows by Exercise 7.22.

7.29: We have $1 - z^2 = x^2 + y^2 = (X^2 + Y^2)(1 - z)^2$ so that, since $z \neq 1$ (if $A \neq N$) we have $1 + z = (X^2 + Y^2)(1 - z)$. Then $ax + by + cz + d = 0$ gives $(aX + bY)(1 - z) + cz + d = 0$, so that $(aX + bY + k)(1 - z) + (c + k)z + (d - k) = 0$ any k, or $(aX + bY + k)(1 - z) + (c + k)(1 + z) + (d - c - 2k) = 0$. Put $k = \frac{1}{2}(d - c)$ and multiply through by 2 to get $(2aX + 2bY + d - c)(1 - z) + (c + d)(1 + z) = 0$; but $1 + z = (X^2 + Y^2)(1 - z)$, so $(2aX + 2bY + d - c)(1 - z) + (c + d)(X^2 + Y^2)(1 - z) = 0$. Then $z \neq 1$, so the equation of Γ' is $(c + d)(X^2 + Y^2) + 2aX + 2bY + d - c = 0$, which is a circle provided $c + d \neq 0$, and a line if $c + d = 0$. The latter occurs precisely when the plane $ax + by + cz + d = 0$ passes through the point $(0, 0, 1)$, that is, N; and here Γ' is the intersection of this plane with the equatorial plane—a line, obviously—together with $f(N) = \infty$.

7.30: Shifting the origin to N, that is, replacing z by $z + 1$, we have $\Sigma + \mu PQ = x^2 + y^2 + z^2 + 2z + \mu z(ax + by + cz + c + d)$, and we get a cone if we can get rid of the linear terms, for then the remaining terms are all quadratic, and if (x, y, z) lies on the cone, so does (vx, vy, vz), for all v. So we must find μ such that $2 + \mu(c + d) = 0$, and this can be done precisely when $c + d \neq 0$. When $c + d = 0$, it is impossible to construct the cone.

8

Beyond isometry

8.1   KLEIN'S DEFINITION OF GEOMETRY

Most of this book has been about the transformations called isometries, the distance-preserving maps on $\mathbb{R}^2$ (or $\mathbb{R}^n$). These transformations form a group, $\mathcal{I}(\mathbb{R}^2)$, and we have been studying the properties preserved by the transformations in this group: distance, angle, collinearity, etc., and for a conic (in the last chapter), eccentricity and axis length.

In 1872, Felix Klein gave his inaugural address at the University of Erlangen, in which he *defined* geometry in exactly these terms. To do geometry, he said, we select a set of points—in our case, $\mathbb{R}^2$—and a group of transformations—in our case, $\mathcal{I}(\mathbb{R}^2)$—and study the properties of the set that are invariant under the elements of the group. Vary the group, or the set of points, and you get a different sort of geometry. The geometry we have been doing, with isometries of $\mathbb{R}^2$, is sometimes called *distance geometry*, because it is distances that are preserved by our group. In this chapter we shall briefly explore two more 'geometries'.

8.2   EUCLIDEAN GEOMETRY: SIMILARITIES

We have met similarities before, in Section 4.7. So recall that a transformation $f : \mathbb{R}^2 \to \mathbb{R}^2$ is a *(plane) similarity* if there is a positive real number k (the *scale factor* of f) such that, if $f(P) = P'$ and $f(Q) = Q'$, then $|P'Q'| = k|PQ|$. So f multiplies all distances by the same amount, and applying f to a figure will in general make it larger (if $k > 1$) or smaller (if $k < 1$). The set of all similarities of $\mathbb{R}^2$ is a group, $\mathcal{E}(\mathbb{R}^2)$, the Euclidean group, and we are now (finally!) doing *Euclidean geometry*.

Example 1.   Every isometry is a similarity, with scale factor $k = 1$; and indeed $\mathcal{I}(\mathbb{R}^2)$ is a subgroup of $\mathcal{E}(\mathbb{R}^2)$.

Example 2.   The map $f : \mathbb{C} \to \mathbb{C}$ given by $f(z) = kz$, where k is a positive real number, is called a *central dilation*, with centre 0 (Fig. 8.1). It is obviously a similarity, with scale factor k, and 0 is a fixed point of f. Every line through 0 is invariant under f, and points on such a line get moved along the line, away from 0 if $k > 1$, and towards 0 if $k < 1$.

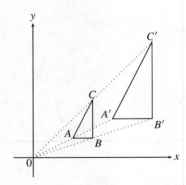

Fig. 8.1   The central dilation $z \mapsto 2z$

Example 3.   The map $f : \mathbb{C} \to \mathbb{C}$ given by $f(z) = az + b$, where $a, b \in \mathbb{C}$ with $a \neq 0$. Here we have $|f(z_1) - f(z_2)| = |(az_1 + b) - (az_2 + b)| = |az_1 - az_2| = |a||z_1 - z_2|$, so that f is a similarity; with scale factor $k = |a|$. This f is called a *direct* similarity; cf. Definition 5.22. By a similar argument, if we put $f(z) = a\bar{z} + b$ we obtain a similarity (with scale factor $k = |a|$, again), and this is called an *opposite* similarity.

In fact the last example gives us all possible similarities:

Proposition 8.1.   *Every similarity $f : \mathbb{C} \to \mathbb{C}$ is given either by $f(z) = az + b$ or else by $f(z) = a\bar{z} + b$, for some $a, b \in \mathbb{C}$ with $a \neq 0$.*

Proof.   Let f be a similarity with scale factor k. Define g by $g(z) = f(k^{-1}z)$, for all $z \in \mathbb{C}$, so that $f(z) = g(kz)$. Now $|g(z_1) - g(z_2)| = |f(k^{-1}z_1) - f(k^{-1}z_2)| = k|k^{-1}z_1 - k^{-1}z_2| = |z_1 - z_2|$, so that g is an isometry. By Proposition 5.21, there exist θ, b such that $g(z) = e^{i\theta}z + b$ (or else $e^{i\theta}\bar{z} + b$), and now $f(z) = g(kz) = ke^{i\theta}z + b$ (or else $ke^{i\theta}\bar{z} + b$). Put $a = ke^{i\theta}$. $\square$

8.2.1   Classification of similarities

We have classified our similarities into two types, direct and opposite, just as we did previously for isometries, in Section 5.3.1. In that section, we then went on to classify the non-trivial direct isometries into two types (translations and rotations), and the opposite isometries into two types (reflections and glide-reflections). We shall now attempt something similar(!) for similarities.

Let $f(z) = az + b$, a direct similarity. If $a = 1$, we have a translation; so let us now assume that $a \neq 1$. We can solve $f(w) = w$ to find a fixed point of f: we have $aw + b = w$, so that $w = \frac{b}{1-a}$. Thus f has a unique fixed point. For any z,

$$f(z) - w = (az + b) - w = (az + b) - (aw + b) = a(z - w) = ke^{i\theta}(z - w),$$

where $a = ke^{i\theta}$. If we now define f_1 and f_2 by $f_1(z) - w = e^{i\theta}(z - w)$ and $f_2(z) - w = k(z - w)$, then f_1 is a rotation about w through the angle θ, and f_2 is a central dilation

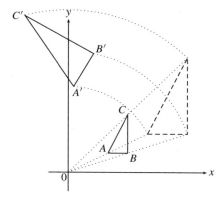

Fig. 8.2 The spiral similarity $z \mapsto 2e^{i\pi/3}z$

with centre w. Further,

$$f_1 f_2(z) - w = e^{i\theta}(f_2(z) - w) = e^{i\theta}k(z - w) = a(z - w),$$

so that $f = f_1 f_2$. This composite of a rotation about w and a central dilation with centre w is called a *spiral similarity* or (in some books) a *dilative rotation* (Fig. 8.2). As special cases, if $k = 1$, then $f = f_1$, a rotation, and if $\theta = 0$ then $f = f_2$, a central dilation. We have proved:

Proposition 8.2. *Every non-trivial direct plane similarity is either a translation or else a spiral similarity.* □

Now suppose $f(z) = a\bar{z} + b$, an opposite similarity. If $|a| = 1$, we have an opposite isometry, that is, a reflection or a glide-reflection. So let us assume $|a| \neq 1$, that is, $a\bar{a} \neq 1$, and look for a fixed point, w. The equation $w = f(w)$ reads $w = a\bar{w} + b$, and is nasty to solve because it contains both w and $\bar{w}$. But if $f(w) = w$, then also $f(f(w)) = f(w) = w$, that is,

$$w = f(f(w)) = f(a\bar{w} + b) = a\overline{(a\bar{w} + b)} + b$$
$$= a\left(\bar{a}w + \bar{b}\right) + b = a\bar{a}w + (a\bar{b} + b),$$

from which $w = \frac{a\bar{b}+b}{1-a\bar{a}}$. This shows there is *at most one* fixed point; but we must still check that it *is* fixed by f (and not just by f^2). We have

$$f(w) = a\overline{\left(\frac{a\bar{b} + b}{1 - a\bar{a}}\right)} + b = a\left(\frac{\bar{a}b + \bar{b}}{1 - a\bar{a}}\right) + b$$
$$= \frac{a\bar{a}b + a\bar{b} + b - a\bar{a}b}{1 - a\bar{a}} = \frac{a\bar{b} + b}{1 - a\bar{a}} = w,$$

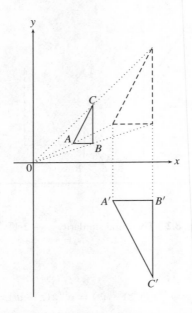

Fig. 8.3   The dilative reflection $z \mapsto 2\overline{z}$

so f does indeed have a unique fixed point, w. Then for any z,

$$f(z) - w = (a\overline{z} + b) - (a\overline{w} + b) = a\overline{(z - w)} = ke^{i\theta}\overline{(z - w)},$$

where $a = ke^{i\theta}$. If we now define f_1 and f_2 by $f_1(z) - w = e^{i\theta}\overline{(z - w)}$ and $f_2(z) - w = k(z - w)$, then f_1 is reflection in some line through w, and f_2 is a central dilation with centre w. Further,

$$f_1 f_2(z) - w = e^{i\theta}\overline{(f_2(z) - w)} = e^{i\theta}\overline{k(z - w)} = a\overline{(z - w)},$$

so that $f = f_1 f_2$. This composite of reflection in a line and a central dilation with centre on the line is called a *dilative reflection* (Fig. 8.3). As a special case, if $k = 1$, this is just a reflection. We have proved:

Proposition 8.3.   *Every opposite plane similarity is either a glide-reflection or else a dilative reflection.*

Corollary 8.4.   *A plane similarity which is* not *an isometry has a unique fixed point, and is either a spiral similarity (if it direct) or else a dilative reflection (if it is opposite).*

8.2.2 Classification of conics (again)

In the last chapter, we classified conics under isometry (Theorem 7.11), and found that every proper real conic is congruent to one of the forms

$$\frac{x^2}{a^2} + \frac{y^2}{a^2(1 - e^2)} = 1 \quad \text{or} \quad y^2 = 4ax. \tag{8.1}$$

This means that there are two real numbers a, e associated with each such conic, and two conics are congruent if and only if they give rise to the same two real numbers. We now apply a central dilation, centre 0, to our conics, with scale factor a^{-1}: specifically, we replace x by ax and y by ay in the above equations, yielding

$$x^2 + \frac{y^2}{1 - e^2} = 1 \quad \text{or} \quad y^2 = 4x, \tag{8.2}$$

where $e \geq 0$, the first equation occurring if $e \neq 1$ and the second if $e = 1$. We have

Theorem 8.5. (**Classification of conics (II)**) *Every proper real conic is similar to one of the conics (8.2). Two such conics are similar iff they have the same eccentricity, e.*

Proof. The first sentence of the theorem is proved already. Suppose we are given two proper real conics. Apply suitable isometries to bring each of them into the form (8.1), and the two new conics obtained will be similar iff the original conics were similar, and will have the same eccentricities as the original conics. Now apply suitable central dilations to get yet another pair of conics, in the form (8.2), again preserving similarity and eccentricity. If the original conics had the same eccentricity, then we have finished. For the converse, if the conics are similar, then we have two ellipses, two hyperbolas, or two parabolas. (OK?) Two ellipses (or hyperbolas) with the same major axis can only be similar if they are congruent (because a similarity that fixes $(0, 0)$ and fixes or swaps $(\pm 1, 0)$ must be an isometry), so they have the same eccentricity; and of course all parabolas have the same eccentricity anyway. □

Our list (8.2) of typical conics is now shorter than before: it is still an infinite list, but we now have only one conic of each possible eccentricity, instead of infinitely many, as in (8.1). This is what we expect to happen when we make our group of transformations larger: the classification is coarser, because similar conics are not necessarily congruent. Of course, congruent conics are always similar, because $\mathcal{E}(\mathbb{R}^2) \supset \mathcal{I}(\mathbb{R}^2)$.

- Ex.8.1: *Theorem 8.5 says that any two parabolas are similar. Prove this directly, from the focus–directrix definition of a parabola.*

8.2.3 Centres of similitude and the nine-point circle

Theorem 8.5 also includes the rather obvious statement that all circles are similar. Given two circles Σ, Σ' with centres O, O' and radii r, r' respectively ($r \neq r'$), we can construct an explicit central dilation f with $f(\Sigma) = \Sigma'$. First apply a central dilation g with centre O and scale factor $k = r'/r$. The image $g(\Sigma)$ is now congruent to Σ', so if

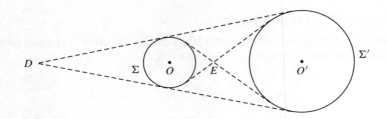

Fig. 8.4   External and internal centres of similitude of Σ and Σ'

we let h be the translation by $\overrightarrow{OO'}$, then $hg(\Sigma) = \Sigma'$. So put $f = hg$, and note that f is a central dilation. (OK?) The fixed point of f is called the *external centre of similitude* of the circles, and if neither circle is wholly inside the other, it is the point where their *external* common tangents meet. Call this point D (Fig. 8.4).

If in the above calculation, we replace g by a spiral similarity with angle π (and the same centre and scale factor as before), then $f = hg$ still sends Σ to Σ', and is also a spiral similarity with angle π. (OK?) The fixed point of f is called the *internal centre of similitude* of the circles, and if neither circle has points inside the other (so that there is 'clear water' between them) then it is the point where their *internal* common tangents meet. Call this point E (Fig. 8.4 again).

Now consider an arbitrary similarity f such that $f(\Sigma) = \Sigma'$. The scale factor of f must still be $k = r'/r$; suppose its fixed point is P. Where might P be? Since $f(O) = O'$ and $f(P) = P$, we have $|O'P| = k|OP|$. Representing O, O', P by the complex numbers a, b, z respectively, we have $|z-b| = k|z-a|$; and if $a = a_1+ia_2$, $b = b_1+ib_2$, and $z = x + iy$, we obtain $(x - b_1)^2 + (y - b_2)^2 = k^2 \left((x - a_1)^2 + (y - a_2)^2\right)$, which is of the form $(k^2 - 1)(x^2 + y^2) + \text{linear terms} = 0$, and so is a circle, provided $k \neq 1$. (What if $k = 1$?) Note that D and E lie on this circle, and indeed DE is a diameter of the circle. (Also O and O' lie on the line DE.)

There is a more geometric and less algebraic way of obtaining the above result, using the following exercise:

● Ex.8.2:   *Given $\triangle ABC$, let the bisector of $\angle BAC$ meet BC at X. Then $|BX|/|XC| = |BA|/|AC|$.*

This exercise works equally well for the internal or external bisector of the angle. Returning to our circles, the internal and external bisectors of $\angle OPO'$ meet OO' in two points which divide OO' internally and externally in the ratio $1 : k$, which is independent of the choice of (f and hence of) P. The two points in question are of course E and D, and since EP and DP bisect $\angle OPO'$, we have $EPD = \frac{\pi}{2}$. Thus P lies on the circle with DE as diameter. (This circle, constructed as the locus of P moving so that $|O'P| = k|OP|$, is called the *circle of Apollonius*.)

● Ex.8.3:   *Given two circles, not concentric and of different size, describe how to construct their two centres of similitude with ruler and compasses.*

We shall now prove an amusing proposition about *three* circles, and their various centres of similitude. First a lemma:

Lemma 8.6. *Let f_1, f_2, f_3 be three non-trivial central dilations, with $f_1 f_2 = f_3$. Then the three respective fixed points are collinear.*

Proof. Let $f_j(z) = a_j z + b_j$, $j = 1, 2, 3$. Here $a_j \in \mathbb{R}$ and $a_j \neq 1$, for all j. We have $f_3(z) = f_1 f_2(z)$, for all z, that is, $a_3 z + b_3 = a_1(a_2 z + b_2) + b_1$, whence $a_3 = a_1 a_2$ and $b_3 = a_1 b_2 + b_1$. From the second equation,

$$\frac{b_3}{1 - a_3} = \frac{a_1 b_2}{1 - a_3} + \frac{b_1}{1 - a_3} = \left(\frac{a_1(1 - a_2)}{1 - a_3} \right) \left(\frac{b_2}{1 - a_2} \right) + \left(\frac{1 - a_1}{1 - a_3} \right) \left(\frac{b_1}{1 - a_1} \right).$$

The result follows on noting that the coefficients of the centres $b_j/(1 - a_j)$ are real, and add up to 1:

$$\frac{a_1(1 - a_2)}{1 - a_3} + \frac{1 - a_1}{1 - a_3} = \frac{1 - a_1 a_2}{1 - a_3} = 1,$$

since $a_1 a_2 = a_3$. □

Proposition 8.7. *Let Σ_1, Σ_2, Σ_3 be three circles, no two the same size, and let the external centres of similitude of Σ_2, Σ_3, of Σ_3, Σ_1, and of Σ_1, Σ_2 be O_1, O_2, O_3 respectively. Then O_1, O_2, O_3 are collinear* (Fig. 8.5).

Proof. There are central dilations f_1, f_2, f_3 with $f_1(\Sigma_3) = \Sigma_2$, $f_2(\Sigma_1) = \Sigma_3$, and $f_3(\Sigma_1) = \Sigma_2$. It follows that $f_1 f_2(\Sigma_1) = \Sigma_2$ also, so that, since $f_1 f_2$ and f_3 are both central dilations, $f_1 f_2 = f_3$. The result is immediate from the lemma. □

• Ex.8.4: *Give a second proof, by Menelaus' theorem.*

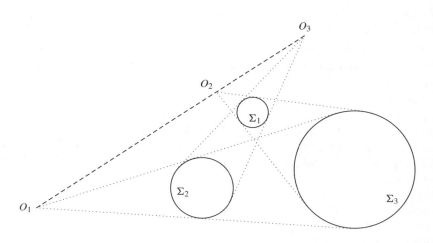

Fig. 8.5 Proposition 8.7

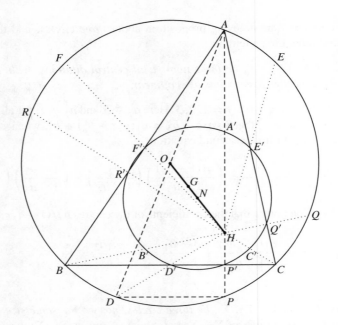

Fig. 8.6   The nine-point circle and Euler line

- Ex.8.5:   *Give a third proof, using the fact that the lines invariant under f_i are precisely the lines through O_i, each i.*

- Ex.8.6:   *What can be said about the* internal *centres of similitude?*

We now use similarities to prove the *nine-point circle theorem*, which we have met before, in Exercise 2.39. Refer to Fig. 8.6 throughout. Take $\triangle ABC$ with origin at O, the circumcentre, so that $|\mathbf{a}| = |\mathbf{b}| = |\mathbf{c}|$, where $\overrightarrow{OA} = \mathbf{a}$, $\overrightarrow{OB} = \mathbf{b}$, and $\overrightarrow{OC} = \mathbf{c}$. Put $\mathbf{h} = \mathbf{a} + \mathbf{b} + \mathbf{c}$ and define H by $\overrightarrow{OH} = \mathbf{h}$. Then $\mathbf{h} - \mathbf{a} = \mathbf{b} + \mathbf{c}$, which is orthogonal to $\mathbf{b} - \mathbf{c}$, by Exercise 3.25. So $AH \perp BC$, and similarly $BH \perp CA$ and $CH \perp AB$, so H is the orthocentre of $\triangle ABC$.

Let D be the point on the circumcircle diametrically opposite A, so that $\overrightarrow{OD} = -\mathbf{a}$. The mid-point of HD is thus $\frac{1}{2}(\mathbf{h} + (-\mathbf{a})) = \frac{1}{2}(\mathbf{b} + \mathbf{c})$, the mid-point of BC. Call this point D'. Likewise, the points on the circumcircle diametrically opposite B and C are E and F respectively; the mid-points of HE and of CA coincide at E', and the mid-points of HF and of AB coincide at F'.

Let AH meet the circumcircle again at P, and let AH meet BC at P'. Since AD is a diameter of the circumcircle, $\angle DPA = \pi/2$, from which $D'P' \parallel DP$, both lines being orthogonal to AP. Since D' is the mid-point of HD, P' must be the mid-point of HP. Similarly BH, CH meet the circumcircle again in Q, R, and meet CA, AB in Q', R', the mid-points of HQ, HR, respectively.

Let A', B', C' be the mid-points of HA, HB, HC respectively. Apply a central dilation with centre H and scale factor $\frac{1}{2}$ to the circumcircle. Since this circle passes through A, B, C, D, E, F, P, Q, R, the image circle must pass through the mid-points of HA, HB, $HC, HD, HE, HF, HP, HQ, HR$, that is, through $A', B', C', D', E', F', P', Q', R'$. The new circle is the *nine-point circle* of $\triangle ABC$, and its centre N, the *nine-point centre* of $\triangle ABC$, is the mid-point of HO, so that if $\overrightarrow{ON} = \mathbf{n}$, then $\mathbf{n} = \frac{1}{2}\mathbf{h} = \frac{1}{2}(\mathbf{a} + \mathbf{b} + \mathbf{c})$. Recall that the centroid G of $\triangle ABC$ is given by $\mathbf{g} = \frac{1}{3}(\mathbf{a} + \mathbf{b} + \mathbf{c})$, where $\mathbf{g} = \overrightarrow{OG}$. It follows that O, G, N, H are collinear, with $OG : GN : NH = 2 : 1 : 3$. We have proved:

Theorem 8.8.   *Given $\triangle ABC$ with circumcentre O and orthocentre H, there is a circle through the mid-points of AB, BC, CA, HA, HB, HC and the feet of the altitudes of the triangle. This is the* nine-point circle *of the triangle. If its centre is N, and the centroid of $\triangle ABC$ is G, then O, G, N, H lie on a line, the* Euler line *of the triangle, and $OG : GN : NH = 2 : 1 : 3$. The radius of the nine-point circle is half the circumradius.*   □

Notice that A, G, D' are collinear, with $AG : GD' = 2 : 1$ (see Section 4.1), and similarly for the other medians BE', CF'. So a dilative half-turn, centre G, sends A, B, C to D', E', F' respectively, and hence sends the circumcircle of $\triangle ABC$ to the circumcircle of $\triangle D'E'F'$, which is none other than the nine-point circle of $\triangle ABC$. So H and G are the external and internal centres of similitude of the two circles.

- Ex.8.7:   *Where is the nine-point circle of $\triangle HBC$?*

- Ex.8.8:   *Prove that, if the circumcircle of $\triangle ABC$ is reflected, in turn, in the three sides of the triangle, then the three reflected circles have a point in common.*

8.2.4   Dilatations

Definition 8.9.   *The transformation $f : \mathbb{R}^2 \to \mathbb{R}^2$ is a **dilatation** if, whenever $f(P) = P'$ and $f(Q) = Q'$, we have $PQ \parallel P'Q'$. The set of (plane) dilatations is denoted $\mathcal{D}(\mathbb{R}^2)$.*

- Ex.8.9: $\mathcal{D}(\mathbb{R}^2)$ *is a group.*

Recall from Definition 5.15 that a *collineation* is a transformation that sends lines to lines, and the group of all (plane) collineations is denoted $\mathcal{C}(\mathbb{R}^2)$.

- Ex.8.10: $\mathcal{D}(\mathbb{R}^2) \subset \mathcal{C}(\mathbb{R}^2)$. *That is, every dilatation is a collineation.*

It is rather too easy to confuse the word *dilatation* with the word *dilation*; we shall try to limit the confusion by always saying *central* dilation for the latter.

Examples. Every translation is a dilatation, and so is every half-turn. Also, central dilations and dilative half-turns are dilatations.

- Ex.8.11:   *Verify the above statements by showing that, if $f : \mathbb{C} \to \mathbb{C}$ is given by $f(z) = az + b$ with a real (and non-zero), then f is a dilatation.*

It is not obvious from the definition that a dilatation must be a similarity, but this is in fact the case, as we shall now show; and in consequence there are no dilatations other than the examples given above.

Lemma 8.10. *Let f be a dilatation, and let P be a fixed point of f. Then every line through P is an invariant line of f.*

Proof. Let ℓ be a line through P, and choose $Q \in \ell$. Let $f(Q) = Q'$, so that $PQ \parallel PQ'$. But this means P, Q, Q' are collinear, or in other words, $Q' \in \ell$. □

Corollary 8.11. *Let f be a dilatation with two fixed points. Then $f = 1$.*

Proof. Let P, Q be fixed points of f. Pick any point R not on PQ. Then the lines RP and RQ are invariant, by Lemma 8.10, whence R is fixed. So every point not on PQ is fixed by f. If now we take S on PQ, we have that both PQ and SR are invariant, and hence S is fixed. The result follows. □

Proposition 8.12. *Every (plane) dilatation is a direct similarity.*

Proof. Let $f : \mathbb{C} \to \mathbb{C}$ be a dilatation, and let $f(0) = b$. Put $g(z) = f(z) - b$, all $z \in \mathbb{C}$, and note that g, being the composite of two dilatations (f and a translation) is itself a dilatation, and furthermore $g(0) = 0$. Put $g(1) = a$. By Lemma 8.10, a is real, and of course $a \neq 0$. (Why?) Put $h(z) = a^{-1}g(z)$, and note that h, being the composite of two dilatations (g, and a central dilation or dilative half-turn, according as $a > 0$ or $a < 0$) is itself a dilatation. But $h(0) = 0$ and $h(1) = 1$, so $h = 1$ by Corollary 8.11. Thus $h(z) = z$, for all z, so that $g(z) = az$ and $f(z) = az + b$, for all $z \in \mathbb{C}$. □

8.3　AFFINE GEOMETRY

Every plane similarity is of the form $z \mapsto az + b$ or $z \mapsto a\bar{z} + b$ for some $a, b \in \mathbb{C}$. Putting $a = ke^{i\theta}$, we can write our similarity in matrix form as

$$\begin{pmatrix} x \\ y \end{pmatrix} \mapsto k \begin{pmatrix} \cos\theta & \mp\sin\theta \\ \cos\theta & \pm\sin\theta \end{pmatrix} \begin{pmatrix} x \\ y \end{pmatrix} + \begin{pmatrix} b_1 \\ b_2 \end{pmatrix},$$

or indeed simply as

$$\begin{pmatrix} x \\ y \end{pmatrix} \mapsto \begin{pmatrix} a_1 & \mp a_2 \\ a_2 & \pm a_1 \end{pmatrix} \begin{pmatrix} x \\ y \end{pmatrix} + \begin{pmatrix} b_1 \\ b_2 \end{pmatrix}$$

for some $a_1, a_2, b_1, b_2 \in \mathbb{R}$, with a_1, a_2 not both zero. There is now an obvious way to generalize this:

Definition 8.13. *The map*

$$\begin{pmatrix} x \\ y \end{pmatrix} \mapsto \begin{pmatrix} a_{11} & a_{12} \\ a_{21} & a_{22} \end{pmatrix} \begin{pmatrix} x \\ y \end{pmatrix} + \begin{pmatrix} b_1 \\ b_2 \end{pmatrix},$$

or $\mathbf{v} \mapsto \mathbf{Av} + \mathbf{b}$, *where* $\det \mathbf{A} \neq 0$, *is called an* **affine transformation**. *It is the composite of a non-singular linear transformation* $\mathbf{v} \mapsto \mathbf{Av}$ *and a translation* $\mathbf{v} \mapsto \mathbf{v} + \mathbf{b}$. *The set of all (plane) affine transformations is denoted* $\mathcal{A}(\mathbb{R}^2)$.

- Ex.8.12: $\mathcal{A}(\mathbb{R}^2)$ *is a group.*

Note that affine transformations can also be written using 3×3 matrices:

$$\begin{pmatrix} x \\ y \\ 1 \end{pmatrix} \mapsto \begin{pmatrix} a_{11} & a_{12} & b_1 \\ a_{21} & a_{22} & b_2 \\ 0 & 0 & 1 \end{pmatrix} \begin{pmatrix} x \\ y \\ 1 \end{pmatrix},$$

so that $\mathcal{A}(\mathbb{R}^2)$ can be thought of as the subgroup of all matrices in $GL_3(\mathbb{R})$ with last row $(0\ 0\ 1)$.

According to Klein, we should now study whatever is preserved by affine transformations. Isometries preserve lines, angles, parallels, distances, areas, conics. Similarities preserve some of these things: lines, angles and parallels, but *ratio* of distances and areas, not the actual values, and a similarity sends a conic to another conic with the same *eccentricity*, but possibly of different size.

So what does an affine transformation preserve? First, it does preserve lines:

Proposition 8.14. *Every affine transformation is a collineation, that is,* $\mathcal{A}(\mathbb{R}^2) \subseteq \mathcal{C}(\mathbb{R}^2)$.

Proof. Given an affine transformation $\mathbf{v} \mapsto \mathbf{Av} + \mathbf{b}$ and two points $\mathbf{u}$ and $\mathbf{v}$, then every point collinear with these is of the form $\mathbf{w} = \lambda\mathbf{u} + \mu\mathbf{v}$, where $\lambda + \mu = 1$. But then

$$\lambda(\mathbf{Au} + \mathbf{b}) + \mu(\mathbf{Av} + \mathbf{b}) = \mathbf{A}(\lambda\mathbf{u} + \mu\mathbf{v}) + (\lambda + \mu)\mathbf{b} = \mathbf{Aw} + \mathbf{b}, \qquad (8.3)$$

so that the three image points are collinear. □

- Ex.8.13: *Affine transformations preserve* non-*collinearity: if points are not collinear, neither are their images.*

It is easy to see that affine transformations do not preserve lengths, or even ratios of lengths, in general: for example, $f : (x, y) \mapsto (2x, y)$ is affine (OK?); putting $O = (0, 0)$, $P = (1, 0)$, $Q = (0, 1)$, and $f(O) = O'$, $f(P) = P'$, and $f(Q) = Q'$, we have $OP/OQ = 1/1 = 1$ but $O'P'/O'Q' = 2/1 = 2$. However, the proof of the last proposition shows that, for collinear points, ratios of lengths *are* preserved:

Corollary 8.15. *Let* f *be an affine transformation, let* P, Q, R *be collinear points, and let* $f(P) = P'$, $f(Q) = Q'$, *and* $f(R) = R'$. *Then* $PQ/QR = P'Q'/Q'R'$.

Proof. Representing P, R, Q by $\mathbf{u}$, $\mathbf{v}$, $\mathbf{w}$ as in the proof of Proposition 8.14, we have $PQ/QR = \mu/\lambda = P'Q'/Q'R'$, by (8.3). □

Note that in particular an affine transformation sends mid-points to mid-points.

- Ex.8.14: *Let $\triangle ABC$ have centroid G, and let the images of A, B, C, G under an affine transformation f be A', B', C', G' respectively. Then G' is the centroid of $\triangle A'B'C'$.*

- Ex.8.15: *Recall that a set is* convex *if for every two of its points it contains the line segment they define. Show that, if X is a convex set and f is an affine transformation, then $f(X)$ is convex.*

Affine transformations do not preserve angles: consider again the map $(x, y) \mapsto (2x, y)$ and note that whereas the angle between the lines $y = 0$ and $y = x$ is $\pi/4$, the angle between the image lines $y = 0$ and $y = x/2$ (OK?) most certainly is not $\pi/4$. However, parallels are preserved:

Proposition 8.16.   *An affine transformation maps parallel lines to parallel lines.*

Proof.   Let ℓ, m be distinct lines with $\ell \parallel m$, and suppose the images $f(\ell)$, $f(m)$ meet at P. Then $f^{-1}(P)$ lies on both ℓ and m, a contradiction. $\qquad\square$

- Ex.8.16: *Suppose $PQ \parallel RS$ and let the images of these four points under the affine transformation f be P', Q', R', S'. Show that $PQ/RS = P'Q'/R'S'$.*

Finally, in this section, we show that affine transformations preserve ratios of areas; more explicitly, the map $\mathbf{v} \mapsto \mathbf{Av} + \mathbf{b}$ multiplies all areas by $|\mathbf{A}|$:

Proposition 8.17.   *Let $f : \mathbf{v} \mapsto \mathbf{Av} + \mathbf{b}$ be an affine transformation, and let $A_1 A_2 A_3$ be a triangle, with $f(A_i) = A'_i$, $i = 1, 2, 3$. Then $\operatorname{area}(A'_1 A'_2 A'_3) = |\mathbf{A}|\operatorname{area}(A_1 A_2 A_3)$.*

Proof.   Let $A_i = (x_i, y_i)$ and $A'_i = (x'_i, y'_i)$, $i = 1, 2, 3$. Then

$$\begin{pmatrix} x'_i \\ y'_i \\ 1 \end{pmatrix} = \begin{pmatrix} a_{11} & a_{12} & b_1 \\ a_{21} & a_{22} & b_2 \\ 0 & 0 & 1 \end{pmatrix} \begin{pmatrix} x_i \\ y_i \\ 1 \end{pmatrix}$$

for $i = 1, 2, 3$, so that

$$\begin{pmatrix} x'_1 & x'_2 & x'_3 \\ y'_1 & y'_2 & y'_3 \\ 1 & 1 & 1 \end{pmatrix} = \begin{pmatrix} a_{11} & a_{12} & b_1 \\ a_{21} & a_{22} & b_2 \\ 0 & 0 & 1 \end{pmatrix} \begin{pmatrix} x_1 & x_2 & x_3 \\ y_1 & y_2 & y_3 \\ 1 & 1 & 1 \end{pmatrix}.$$

Take determinants, and divide through by 2. The result follows, by Exercises 5.40 and 5.41. $\qquad\square$

The general result about areas then follows for any area that can be dissected into triangles, or which is a limit of such areas.

8.3.1   Affine regular polygons

Recall that two subsets of $\mathbb{R}^2$ are *congruent* if one can be mapped to the other by an isometry, and *similar* if one can be mapped to the other by a similarity.

Definition 8.18. *The subsets X, X' of $\mathbb{R}^2$ are* **affinely equivalent** *if there is an affine transformation f with $f(X) = X'$.*

Proposition 8.19. *All triangles are affinely equivalent.*

Proof. Given $\triangle PQR$, let $\mathbf{p}$, $\mathbf{q}$, $\mathbf{r}$ be vectors representing its vertices, and let $\mathbf{e}_1 = (1, 0)$ and $\mathbf{e}_2 = (0, 1)$. The vectors $\mathbf{q} - \mathbf{p}$ and $\mathbf{r} - \mathbf{p}$ are linearly independent, so if $\mathbf{A}$ is the 2×2 matrix having these two vectors as its columns, then $\mathbf{A}$ is non-singular, and further $\mathbf{A}\mathbf{e}_1 = \mathbf{q} - \mathbf{p}$ and $\mathbf{A}\mathbf{e}_2 = \mathbf{r} - \mathbf{p}$, all vectors being written as columns, as usual. So if we define f by $\mathbf{v} \mapsto \mathbf{A}\mathbf{v} + \mathbf{p}$, then $f(\mathbf{0}) = \mathbf{p}$, $f(\mathbf{e}_1) = \mathbf{q}$, and $f(\mathbf{e}_2) = \mathbf{r}$. Thus the arbitrarily chosen $\triangle PQR$ is affinely equivalent to the triangle with vertices $(0, 0)$, $(1, 0)$, $(0, 1)$, and the result follows. $\square$

The last step of this proof used implicitly the fact that triangles affinely equivalent to the same triangle are affinely equivalent to each other, which follows from the fact that $\mathcal{A}(\mathbb{R}^2)$ is a group. (In fact affine equivalence is an equivalence relation; cf. Exercise 5.16.)

- Ex.8.17: *Show that, if the affine transformation f has three non-collinear fixed points, then $f = 1$.*

- Ex.8.18: *Given $\triangle PQR$ and $\triangle P'Q'R'$, show that there is a* unique *affine transformation f with $f(P) = P'$, $f(Q) = Q'$, and $f(R) = R'$.*

- Ex.8.19: *The group $GL_2(\mathbb{Z})$ consists of all matrices $\mathbf{A} = (a_{ij}) \in GL_2(\mathbb{R})$ such that $a_{ij} \in \mathbb{Z}$, for all i, j, and $\det \mathbf{A} = \pm 1$. By considering the affine transformations that permute the three points $(1, 0)$, $(0, 1)$, and $(-1, -1)$, find six matrices in $GL_2(\mathbb{Z})$ that form a subgroup isomorphic to the permutation group S_3. (Cf. Exercise 6.21.)*

Definition 8.20. *A polygon is* **affine regular** *if it is affinely equivalent to a regular polygon.*

So Proposition 8.19 says that *all* triangles are affine regular.

- Ex.8.20: *Show that a 4-gon (or tetragram; see Definition 4.16) is affine regular if and only if it is a parallelogram.*

What about affine regular pentagons? In a *regular* pentagon, each of the five sides is parallel to the 'opposite' diagonal, that is, the diagonal not through either of its endpoints. By Proposition 8.16, the same will be true if the pentagon is merely *affine* regular. We shall now show that the converse is true, and in fact we can even drop one of the five parallel requirements:

Proposition 8.21. *The convex pentagon $PQRST$ is affine regular if and only if*

$$PQ \parallel TR, \quad PR \parallel TS, \quad SQ \parallel TP, \quad and \quad SR \parallel TQ. \tag{8.4}$$

Proof. If $PQRST$ is affine regular, then (8.4) holds, as noted above. So now suppose (8.4) holds, and put $U = PR \cdot TQ$, $V = PR \cdot QS$, and $W = TR \cdot QS$, the meets (Fig. 8.7).

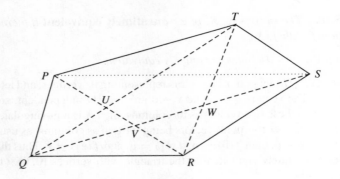

Fig. 8.7   Proposition 8.21

First note that $PVST$, $RSTU$ and $PQWT$ are parallelograms, so $|PV| = |TS| = |UR|$, $|QW| = |PT| = |VS|$, $|PQ| = |TW|$, and $|TU| = |SR|$. Then put

$$\tau = \frac{|PV|}{|VR|} = \frac{|UR|}{|PU|} = \frac{|TR|}{|PQ|} \quad \text{(since triangles } TRU, QPU \text{ are similar)}$$

$$= \frac{|TR|}{|TW|}$$

$$= \frac{|PR|}{|PV|} \quad \text{(since triangles } TRP, WRV \text{ are similar)}$$

$$= \frac{|PV| + |VR|}{|PV|} = 1 + \left(\frac{|PV|}{|VR|}\right)^{-1} = 1 + \tau^{-1}$$

so that $\tau = 1 + \tau^{-1}$, or $\tau^2 - \tau - 1 = 0$. Thus $\tau = \frac{1+\sqrt{5}}{2}$, the *golden number*. (We met this before, in Exercise 6.43.) So $|PV| : |VR| = \tau : 1 = |RU| : |UP|$, and similarly $|SV| : |VQ| = \tau : 1 = |QW| : |WS|$.

Let $P'Q'R'S'T'$ be a regular pentagon. By Proposition 8.19, there is an affine transformation f with $f(P) = P'$, $f(Q) = Q'$, and $f(R) = R'$. Our proof will be complete if we can show that $f(S) = S'$ and $f(T) = T'$ also.

Put $U' = P'R' \cdot T'Q'$, $V' = P'R' \cdot Q'S'$, and $W' = T'R' \cdot Q'S'$, the meets. Exactly as above, we have $|P'V'| : |V'R'| = \tau : 1 = |R'U'| : |U'P'|$, from which it follows that $f(U) = U'$ and $f(V) = V'$, by Corollary 8.15. Then again, $|S'V'| : |V'Q'| = \tau : 1$, so that $f(S) = S'$, as required; and $|Q'W'| : |W'S'| = \tau : 1$, so that $f(W) = W'$. Finally, $T = QU \cdot RW$, the meet, and $T' = Q'U' \cdot R'W'$, so that $f(T) = T'$.   □

- Ex.8.21:   *With the above notation, show that $|PU| : |UV| = \tau : 1$.*

- Ex.8.22:   *Since $PQRST$ is affine regular, we also have that $PS \parallel QR$. Deduce this directly from (8.4), using areas.*

- Ex.8.23:   *Must we include the word 'convex' in the statement of Proposition 8.21?*

- Ex.8.24:   *Give a ruler-and-compasses construction for a regular pentagon.*

Of course, if a regular n-gon has been constructed, then a regular $2n$-gon can be constructed by bisecting angles.

- Ex.8.25: *How can a regular 15-gon be constructed?*
- Ex.8.26: *What is the smallest angle which is a whole number of degrees and which can be constructed with ruler and compasses?*

Gauss, at the age of 19, proved that a regular n-gon can be constructed with ruler and compasses if and only if the odd prime factors of n are distinct *Fermat primes*, that is, primes of the form $2^{2^k} + 1$. The only known Fermat primes are 3, 5, 17, 257, and 65 537, corresponding to $k = 0$, 1, 2, 3, and 4. The first two n for which the corresponding polygon cannot be constructed with ruler and compasses are $n = 7$ (because 7 is not a Fermat prime) and $n = 9$ (because of the repeated odd factor 3). For more details, see [16], [9], and [30].

8.3.2 Classification of conics (yet again)

We have classified conics twice before, under the group $\mathcal{I}(\mathbb{R}^2)$ (in Section 7.3) and under the larger group $\mathcal{E}(\mathbb{R}^2)$ (in Section 8.2.2). We now classify conics yet again, under the still larger group $\mathcal{A}(\mathbb{R}^2)$. In Theorem 7.11, the 'typical' conic was

$$\frac{x^2}{a^2} + \frac{y^2}{a^2(1 - e^2)} = 1 \text{ or } y^2 = 4ax,$$

one conic for each $e \geq 0$ and each $a > 0$. In Theorem 8.5, the list was shorter: the 'typical' conic was $x^2 + y^2/(1 - e^2) = 1$ or $y^2 = 4x$, one conic for each $e \geq 0$. Here our list will be shorter still, and will in fact be a finite list. But first we need to be sure that, if Γ is a conic and f is an affine transformation, then $f(\Gamma)$ is a conic also. The swift answer is that this is obviously true, since Γ is given by a quadratic equation, and f involves a linear change of variables, so that the transformed equation must surely be quadratic, and hence $f(\Gamma)$ is a conic. However, we need just a little more detail than this.

Some thought is needed to get the change of variables right. As an example, the transform of the line $x = 1$ under the translation $f : (x, y) \mapsto (x + 1, y)$ (which moves every point one unit in the direction of x increasing) is clearly the line $x = 2$. This equation is obtained from the equation $x = 1$ *not* (as one might initially suppose) by substituting $x + 1$ for x (and y for y), but by substituting $x - 1$ for x: in other words, by using the *inverse* translation $f^{-1} : (x, y) \mapsto (x - 1, y)$ to give the formula for the change of variables. In a similar manner, if Γ has equation

$$(x \quad y \quad 1) \, \mathbf{M} \begin{pmatrix} x \\ y \\ 1 \end{pmatrix} = 0$$

for some 3×3 symmetric matrix $\mathbf{M}$, and the affine transformation f is such that

$$f^{-1} : \begin{pmatrix} x \\ y \\ 1 \end{pmatrix} \mapsto \mathbf{A} \begin{pmatrix} x \\ y \\ 1 \end{pmatrix}$$

for some matrix $\mathbf{A} \in GL_3(\mathbb{R})$ with last row $(0\,0\,1)$, then the equation of $f(\Gamma)$ is obtained by substituting

$$\mathbf{A} \begin{pmatrix} x \\ y \\ 1 \end{pmatrix} \text{ for } \begin{pmatrix} x \\ y \\ 1 \end{pmatrix} \text{ and } \begin{pmatrix} x & y & 1 \end{pmatrix} \mathbf{A}^T \text{ for } \begin{pmatrix} x & y & 1 \end{pmatrix}$$

in the equation of Γ, giving the equation of $f(\Gamma)$ as

$$\begin{pmatrix} x & y & 1 \end{pmatrix} \mathbf{A}^T \mathbf{M} \mathbf{A} \begin{pmatrix} x \\ y \\ 1 \end{pmatrix} = 0.$$

If Γ is a proper conic, then $|\mathbf{M}| \neq 0$, so that $|\mathbf{A}^T \mathbf{M} \mathbf{A}| \neq 0$, and $f(\Gamma)$ is a proper conic also. Thus:

Proposition 8.22.   *The image of a proper conic under an affine transformation is a proper conic.*   □

Theorem 8.23 (Classification of conics (III)).   *Every proper real conic is affinely equivalent to the circle $x^2 + y^2 = 1$ or to the rectangular hyperbola $x^2 - y^2 = 1$ or to the parabola $y^2 = 4x$.*

Proof.   We know that every proper real conic is similar, and hence affinely equivalent, to $x^2 + y^2/(1 - e^2) = 1$ or $y^2 = 4x$, that is, to

$$x^2 + \frac{y^2}{b^2} = 1 \quad \text{or} \quad x^2 - \frac{y^2}{b^2} = 1 \quad \text{or} \quad y^2 = 4x, \tag{8.5}$$

where $b^2 = 1 - e^2$ if $0 \leq e < 1$ and $b^2 = e^2 - 1$ if $e > 1$. The affine transformation $f : (x, y) \mapsto (x, y/b)$ has the effect of substituting by for y (and x for x), since $f^{-1} : (x, y) \mapsto (x, by)$, and so f transforms the first two equations in (8.5) to $x^2 + y^2 = 1$ and $x^2 - y^2 = 1$, respectively.   □

So our list of 'typical' conics is down to three. It is easy to see that no two of the conics in this list are affinely equivalent. For if Γ is a conic with centre O, and if f is an affine transformation, then since f sends mid-points to mid-points, $f(O)$ must be the centre of $f(\Gamma)$. Thus a central conic such as a circle or a hyperbola cannot be affinely equivalent to a parabola, which has no centre. Again, if Γ is a circle with centre O, then every line through O meets Γ in two points, so if f is an affine transformation, every line through the centre $f(O)$ of $f(\Gamma)$ must meet $f(\Gamma)$ in two points, and thus $f(\Gamma)$ cannot be a hyperbola.

Corollary 8.24. *Two proper real conics are affinely equivalent if and only if both are ellipses or both are hyperbolas or both are parabolas. (Here a circle is regarded as a special sort of ellipse, of course.)*

8.3.3 A selection of affine theorems

We now give a short selection of some theorems that 'belong' to affine geometry.

Proposition 8.25. *Let a hexagon $PQRSTU$ be inscribed in an ellipse, Γ. If $PQ \parallel TS$ and $QR \parallel UT$, then also $RS \parallel PU$ (Fig. 8.8).*

Proof. The method is to transform the ellipse into a circle, prove the theorem for the circle, and then transform back. In detail, there is an affine transformation f such that $f(\Gamma)$ is a circle, and the image $P'Q'R'S'T'U'$ of the hexagon is a hexagon inscribed in the circle. Since affine transformations preserve parallels, we have $P'Q' \parallel T'S'$ and $Q'R' \parallel U'T'$. If we can prove that $R'S' \parallel P'U'$, then the result follows—OK?—but this was done already, in Exercise 7.18. □

Proposition 8.26. *Let P, Q, R, S lie on an ellipse Γ, such that $PQ \parallel SR$. Let O be the centre of Γ. Let the tangents to Γ at P and Q meet at T, and let the tangents to Γ at R and S meet at U. Let the mid-point of PQ be V and let the mid-point of RS be W. Let $X = PR \cdot QS$ (the meet), and $Y = PS \cdot QR$. Then the seven points O, T, U, V, W, X, Y lie on a line, ℓ (Fig. 8.9).*

Proof. Transform Γ into a circle, and the images of the seven points lie on the diameter perpendicular to the images of PQ and RS. □

- Ex.8.27: *Show that the tangents to Γ where ℓ meets Γ are both parallel to PQ. Let m be the diameter of Γ parallel to PQ. Show that the tangents to Γ where m meets Γ are both parallel to ℓ. (ℓ and m are called* conjugate diameters *of Γ.)*

- Ex.8.28: *Let the tangents to Γ at P and R meet at A, and let the tangents to Γ at Q and S meet at B. Then $AB \parallel PQ$, and the mid-point of AB lies on ℓ. A similar result holds if we interchange R and S.*

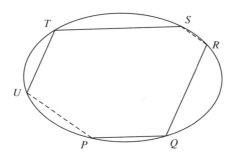

Fig. 8.8 Proposition 8.25

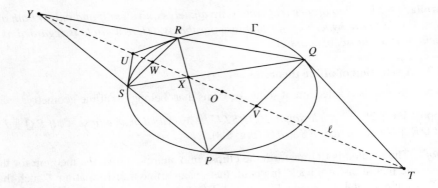

Fig. 8.9   Proposition 8.26

There is a second way of proving Proposition 8.26. The affine transformation $f : \mathbf{v} \mapsto A\mathbf{v}$, where

$$A = \begin{pmatrix} a & 0 \\ 0 & b \end{pmatrix} \begin{pmatrix} \cos\alpha & -\sin\alpha \\ \sin\alpha & \cos\alpha \end{pmatrix} \begin{pmatrix} a^{-1} & 0 \\ 0 & b^{-1} \end{pmatrix}$$

$$= \begin{pmatrix} \cos\alpha & -ab^{-1}\sin\alpha \\ a^{-1}b\sin\alpha & \cos\alpha \end{pmatrix},$$

maps the ellipse $x^2/a^2 + y^2/b^2 = 1$ to itself: in fact $f(a\cos\theta, b\sin\theta) = (a\cos(\theta + \alpha), b\sin(\theta + \alpha))$. ($f$ is the composite of three affine transformations, as shown, that turn the ellipse into a circle, rotate the circle through α, and then turn the circle back into the ellipse.) Using such a map, we can transform the chords PQ, RS of Proposition 8.26 into chords parallel to one of the axes of Γ, and now the images of the seven points clearly lie on the *other* axis.

- Ex.8.29: *State and prove the corresponding result for a hyperbola* ...
- Ex.8.30: ... *and for a parabola.*
- Ex.8.31: *A point P moves on a hyperbola Γ. Prove that, as P moves, the triangle formed by the tangent to Γ at P and the asymptotes of Γ is of constant area.*

Proposition 8.27.   *The area of the ellipse $x^2/a^2 + y^2/b^2 = 1$ is πab.*

Proof.   The affine map $\mathbf{v} \mapsto A\mathbf{v}$, with $A = \begin{pmatrix} a & 0 \\ 0 & b \end{pmatrix}$, has $|A| = ab$, so it multiplies areas by ab. It turns the unit circle $x^2 + y^2 = 1$, which has area π, into our ellipse, which therefore has area πab.   □

Proposition 8.28.   The mid-point ellipse. *Given $\triangle ABC$, there is an ellipse which touches the sides BC, CA, AB at their mid-points (Fig. 8.10). Its area is $(\pi/3\sqrt{3})$ area(ABC).*

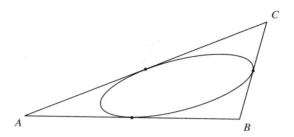

Fig. 8.10 The mid-point ellipse

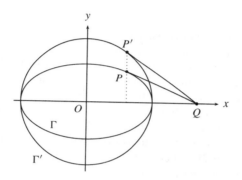

Fig. 8.11 Proposition 8.29

Proof. Choose an affine transformation f such that, if $f(A) = A'$, $f(B) = B'$, and $f(C) = C'$, then $\triangle A'B'C'$ is equilateral. (Proposition 8.19.) Let Γ' be the inscribed circle of $\triangle A'B'C'$ which, since $\triangle A'B'C'$ is equilateral, touches $B'C'$, $C'A'$, $A'B'$ at their mid-points. Then $\Gamma = f^{-1}(\Gamma')$ is the required ellipse, and $(\text{area}(\Gamma)/\text{area}(ABC)) = (\text{area}(\Gamma')/\text{area}(A'B'C')) = (\pi/3\sqrt{3})$. $\qquad\qquad\square$

Proposition 8.29. *Let Γ be the ellipse $(x^2/a^2) + (y^2/b^2) = 1$ and let Γ' be the auxiliary circle $x^2 + y^2 = a^2$. Let $P = (a\cos\theta, b\sin\theta)$ lie on Γ, not on either axis, and let $P' = (a\cos\theta, a\sin\theta)$. Then the tangent ℓ to Γ at P and the tangent ℓ' to Γ' at P' meet on the major axis of Γ (Fig. 8.11).*

Proof. This is an old chestnut from school examinations, when the expected (boring) method was to find the equations of the two tangents and show that they cut $y = 0$ at the same point. Very bright candidates might save half the work by spotting that the calculation for Γ' is the same as that for Γ, with b replaced by a throughout, and this is the essence of our (affine) proof. Let $f : (x, y) \mapsto (x, \frac{a}{b}y)$, so that $f(\Gamma) = \Gamma'$, $f(P) = P'$, and $f(\ell) = \ell'$. Now f fixes every point of $y = 0$, so if ℓ, ℓ' meet $y = 0$ at Q, Q' respectively, then $Q = f(Q) = Q'$, and we are done. $\qquad\qquad\square$

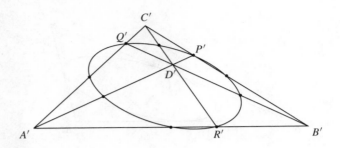

Fig. 8.12   The nine-point ellipse

We finish this section by seeing what happens when an affine transformation is applied to the diagram of the nine-point circle theorem. Let D be the orthocentre of $\triangle ABC$, so that A, B, C, D are an orthocentric tetrad, each point being the orthocentre of the triangle formed by the other three. If $\triangle ABC$ is acute-angled, then D is inside $\triangle ABC$. If $\triangle ABC$ has an obtuse angle at A, then A is inside $\triangle BCD$, and similarly if there is an obtuse angle at B or at C. (We shall assume that $\triangle ABC$ is not a right-angled triangle, so that A, B, C, D are distinct.) So one of A, B, C, D is inside the triangle formed by the other three. Let f be an affine transformation, and let $f(A) = A'$, $f(B) = B'$, $f(C) = C'$, and $f(D) = D'$. Of course, A', B', C', D' are not an orthocentric tetrad, in general, because f does not preserve angles. However, a triangle together with its interior is a convex set, so, by the above remarks and Exercise 8.15, it follows that one of A', B', C', D' is inside the triangle formed by the other three.

The nine-point circle Γ of $\triangle ABC$ passes through the meets $AB \cdot CD$, $AC \cdot BD$, $AD \cdot BC$, and the mid-points of AB, AC, AD, BC, BD, CD, so, $\Gamma' = f(\Gamma)$ is an ellipse through the meets $A'B' \cdot C'D'$, $A'C' \cdot B'D'$, $A'D' \cdot B'C'$, and the mid-points of $A'B'$, $A'C'$, $A'D'$, $B'C'$, $B'D'$, $C'D'$. This is half of the proof of:

Proposition 8.30 (The nine-point ellipse).   *Let A', B', C', D' be four distinct points, no three being collinear. Then there is an ellipse through the meets $A'B' \cdot C'D'$, $A'C' \cdot B'D'$, $A'D' \cdot B'C'$, and the mid-points of $A'B'$, $A'C'$, $A'D'$, $B'C'$, $B'D'$, $C'D'$ if and only if one of A', B', C', D' is inside the triangle formed by the other three* (Fig. 8.12).

Proof.   Suppose, without loss of generality, that D' is inside $\triangle A'B'C'$. We must find an orthocentric tetrad A, B, C, D and an affine transformation f with $f(A) = A'$, $f(B) = B'$, $f(C) = C'$, and $f(D) = D'$. If Γ is the nine-point circle of $\triangle ABC$, then $\Gamma' = f(\Gamma)$ is the required ellipse.

Let $P' = A'D' \cdot B'C'$ and $Q' = A'C' \cdot B'D'$. Since D' is inside $\triangle A'B'C'$, then P' lies between B' and C', and Q' lies between A' and C'. Choose a line ℓ and B, C, P on ℓ with $BP : PC = B'P' : P'C'$. (So P is between B and C.) Let m be the line through P with $m \perp \ell$, and let Σ be the circle on BC as diameter. Let Σ' be the image of Σ under the central dilation with centre C and scale factor $k = A'C'/Q'C'$. Let Σ' meet m at A, and let AC meet Σ at Q. Then $AC/QC = k$, by the central dilation, so that

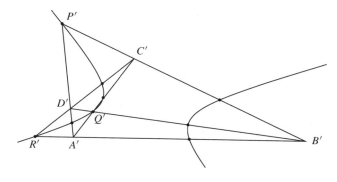

Fig. 8.13   The nine-point hyperbola

$AQ : QC = A'Q' : Q'C'$. By Proposition 8.19 there is an affine transformation f with $f(A) = A'$, $f(B) = B'$, and $f(C) = C'$. Since $BP : PC = B'P' : P'C'$ we must have $f(P) = P'$, and since $AQ : QC = A'Q' : Q'C'$ we must have $f(Q) = Q'$, by Corollary 8.15; and now $D = AP \cdot BQ$ and $D' = A'P' \cdot B'Q'$, so $f(D) = D'$. Finally, $\angle APB = \pi/2$ (since $\ell \perp m$) and $\angle BQC = \pi/2$ (angle in a semicircle), so A, B, C, D are an orthocentric tetrad, and we have finished.   □

It is reasonable to ask what happens to this proposition if *none* of A', B', C', D' lies inside the triangle formed by the other three, and perhaps it is no real surprise that in this case the nine points (the three meets and the six mid-points, as before) lie on a hyperbola (Fig. 8.13). If we fix A', B', C' and move D' around, then this nine-point conic is an ellipse when D' is inside $\triangle A'B'C'$ and becomes a hyperbola when D' crosses (say) $A'C'$. At the moment of crossing, when D' lies on $A'C'$, the conic is not a parabola, but degenerates into two parallel straight lines. (There are other degenerate cases also, for example if $A'B' \parallel C'D'$.) To get a parabola as the nine-point conic, one has to take the limit as D' goes off to infinity; but then the mid-points of $A'D'$, $B'D'$, $C'D'$ go to infinity as well, so we get a *six*-point parabola for this configuration (Fig. 8.14). All these results are special cases of the *eleven-point conic* theorem, which is a result of projective geometry, and details can be found in [23].

8.3.4   Collineations

Recall that a collineation of $\mathbb{R}^2$ is a transformation that sends lines to lines, that is, a set of collinear points is mapped to a set of collinear points. We have already seen (Proposition 8.14) that every affine transformation is a collineation: $\mathcal{A}(\mathbb{R}^2) \subseteq \mathcal{C}(\mathbb{R}^2)$. In this section we show that the converse is true: every collineation is an affine transformation, so that $\mathcal{A}(\mathbb{R}^2) = \mathcal{C}(\mathbb{R}^2)$.

Lemma 8.31.   *Let f be a collineation. Then:*

(i) *The line joining two fixed points of f is an invariant line of f.*
(ii) *The meet of two invariant lines of f is a fixed point of f.*

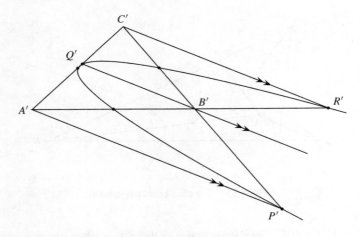

Fig. 8.14   The six-point parabola

(iii) *If ℓ, m are parallel lines, then so are f(ℓ), f(m).*
(iv) *If the line ℓ is parallel to an invariant line of f, it is parallel to f(ℓ) also.*
 (v) *If the line ℓ is parallel to an invariant line of f, and if ℓ contains a fixed point of f, then ℓ is an invariant line of f.*

Proof. (i) and (ii) are obvious. For (iii), if $\ell \neq m$ and $f(\ell)$, $f(m)$ meet at X, then $f^{-1}(X)$ lies on both ℓ and m, so they are not parallel. Then (iv) follows from (iii), with $m = f(m)$. For (v), $\ell \parallel f(\ell)$ by (iv); but ℓ and $f(\ell)$ both contain the fixed point, so $\ell = f(\ell)$. □

Lemma 8.32. *Let $f : \mathbb{R}^2 \to \mathbb{R}^2$ be a collineation which fixes the points $(0, 0)$, $(1, 0)$, and $(0, 1)$. Then f fixes the points $(1, 1)$ and $(\frac{1}{2}, \frac{1}{2})$ also; and the six lines $x = 0$, $y = 0$, $x = 1$, $y = 1$, $x = y$, and $x + y = 1$ are invariant lines of f.*

Proof. The lines $x = 0$ and $y = 0$ are invariant, by Lemma 8.31(i). Then the lines $x = 1$ and $y = 1$ are invariant, by Lemma 8.31(v), and now the point $(1, 1)$ is fixed, by Lemma 8.31(ii). Then the lines $x = y$ and $x + y = 1$ are invariant, by Lemma 8.31(i), and finally the point $(\frac{1}{2}, \frac{1}{2})$ is fixed, by Lemma 8.31(ii). □

Lemma 8.33. *Let f be as in Lemma 8.32. Then there is a function $\alpha : \mathbb{R} \to \mathbb{R}$ such that $f(a, b) = (\alpha(a), \alpha(b))$, for all $a, b \in \mathbb{R}$.*

Proof. The line $y = 0$ is invariant (Lemma 8.32), and $(a, 0)$ lies on $y = 0$, so $f(a, 0) = (a', 0)$ for some $a' \in \mathbb{R}$. Define $\alpha : \mathbb{R} \to \mathbb{R}$ by $f(a, 0) = (\alpha(a), 0)$, for all $a \in \mathbb{R}$, that is, $\alpha(a) = a'$. Similarly, the line $x = 0$ is invariant (Lemma 8.32), so define $\beta : \mathbb{R} \to \mathbb{R}$ so that $f(0, b) = (0, \beta(b))$, for all $b \in \mathbb{R}$.

Since the line $x = 0$ is invariant, the image of the parallel line $x = a$ is the line $x = \alpha(a)$, by Lemma 8.31(iv). Similarly, since the line $y = 0$ is invariant, the image of the parallel line $y = b$ is the line $y = \beta(b)$, by Lemma 8.31(iv) again. So $f(a, b) =$

$(\alpha(a), \beta(b))$, for all $a, b \in \mathbb{R}$. In particular, $f(a, a) = (\alpha(a), \beta(a))$, for all $a \in \mathbb{R}$; but (a, a) lies on the line $y = x$ which is invariant by Lemma 8.32, and thus $\alpha(a) = \beta(a)$, for all $a \in \mathbb{R}$. So $\alpha = \beta$, and $f(a, b) = (\alpha(a), \alpha(b))$, for all $a, b \in \mathbb{R}$. $\square$

Lemma 8.34. *Let f, α be as in Lemma 8.33. Then* (i) $\alpha(0) = 0$; (ii) $\alpha(1) = 1$; (iii) $\alpha(a + b) = \alpha(a) + \alpha(b)$, *for all $a, b \in \mathbb{R}$; and* (iv) $\alpha(ab) = \alpha(a)\alpha(b)$, *for all a, $b \in \mathbb{R}$.*

Proof. We have $(1, 0) = f(1, 0) = (\alpha(1), \alpha(0))$, whence (i) and (ii). Then $(\frac{1}{2}, \frac{1}{2}) = f(\frac{1}{2}, \frac{1}{2}) = (\alpha(\frac{1}{2}), \alpha(\frac{1}{2}))$, so that $\alpha(\frac{1}{2}) = \frac{1}{2}$. Now, for any $a \in \mathbb{R}$, the points $(0, 0)$, $(1, a)$ and $(\frac{1}{2}, \frac{1}{2}a)$ are collinear, whence $f(0, 0)$, $f(1, a)$, and $f(\frac{1}{2}, \frac{1}{2}a)$ are collinear, that is, $(0, 0)$, $(1, \alpha(a))$, and $(\frac{1}{2}, \alpha(\frac{1}{2}a))$ are collinear. It follows that $\alpha(\frac{1}{2}a) = \frac{1}{2}\alpha(a)$, for all $a \in \mathbb{R}$.

Now for any $a, b \in \mathbb{R}$, the points $(0, a)$, $(1, b)$, and $(\frac{1}{2}, \frac{1}{2}(a + b))$ are collinear, whence $f(0, a)$, $f(1, b)$, and $f(\frac{1}{2}, \frac{1}{2}(a + b))$ are collinear, that is, $(0, \alpha(a))$, $(1, \alpha(b))$, and $(\frac{1}{2}, \alpha(\frac{1}{2}(a + b)))$ are collinear. It follows that $\frac{1}{2}(\alpha(a) + \alpha(b)) = \alpha(\frac{1}{2}(a + b))$; but this last expression is equal to $\frac{1}{2}\alpha(a+b)$, as in the preceding paragraph, and (iii) follows.

For (iv), note that for any $a, b \in \mathbb{R}$, the points $(0, 0)$, $(a, 1)$, and (ab, b) are collinear, whence $f(0, 0)$, $f(a, 1)$, and $f(ab, b)$ are collinear, that is, $(0, 0)$, $(\alpha(a), 1)$, and $(\alpha(ab), \alpha(b))$ are collinear. It is immediate that $\alpha(ab) = \alpha(a)\alpha(b)$. $\square$

Lemma 8.35. *Let α be as in the preceding lemmas. Then* (i) *for all $q \in \mathbb{Q}$ (the rational numbers), we have $\alpha(q) = q$;* (ii) *for all $a, b \in \mathbb{R}$ with $a < b$, we have $\alpha(a) < \alpha(b)$; and* (iii) *α is continuous.*

Proof. (i): We have $\alpha(1) = 1$; and if $\alpha(n) = n$, then $\alpha(n+1) = \alpha(n)+\alpha(1) = n+1$; so by induction, α fixes all positive integers. Then $\alpha(-n) = -\alpha(n)$ (OK?), and $\alpha(0) = 0$, so α fixes every $n \in \mathbb{Z}$. Finally, if $q = n/m$ with $n, m \in \mathbb{Z}$ and $m \neq 0$, then $n = qm$, so $\alpha(n) = \alpha(qm) = \alpha(q)\alpha(m)$, or $n = \alpha(q)m$, so that $\alpha(q) = n/m = q$.

(ii): If $a < b$ then $b - a > 0$, so $b - a = c^2$ for some $c \in \mathbb{R}$, $c \neq 0$. Thus $\alpha(b) - \alpha(a) = \alpha(b - a) = \alpha(c^2) = \alpha(c)^2 > 0$. (Note that $\alpha(c) \neq 0$ because $c \neq 0$ and f is bijective.)

(iii): (Readers unfamiliar with the notion of continuity can either read it up in an elementary analysis text such as [4], or omit this section and use the alternative proof of Lemma 8.36 below.) Given $x \in \mathbb{R}$ and $\varepsilon > 0$, choose $n \in \mathbb{Z}$, $n \geq 1/\varepsilon$, and put $\delta = 1/n$; so $\delta \leq \varepsilon$. If $|x - x_0| < \delta$, then $x_0 - \delta < x < x_0 + \delta$, so $\alpha(x_0 - \delta) < \alpha(x) < \alpha(x_0 + \delta)$, whence $\alpha(x_0) - \alpha(\delta) < \alpha(x) < \alpha(x_0) + \alpha(\delta)$, or $|\alpha(x) - \alpha(x_0)| < \alpha(\delta)$. But $\delta \in \mathbb{Q}$, so $\alpha(\delta) = \delta$, and $\delta \leq \varepsilon$, so $|\alpha(x) - \alpha(x_0)| < \varepsilon$. $\square$

Lemma 8.36. *Let f, α be as before. Then $\alpha(x) = x$, for all $x \in \mathbb{R}$, and so $f = 1$.*

Proof. Given $x \in \mathbb{R}$, find $x_n \in \mathbb{Q}$ with $x_n \to x$ as $n \to \infty$. Since α is continuous, we have $\alpha(x_n) \to \alpha(x)$ as $n \to \infty$; but $\alpha(x_n) = x_n$, for all n, so $\alpha(x) = \lim_{n \to \infty} \alpha(x_n) = \lim_{n \to \infty} x_n = x$.

Here is an alternative argument, avoiding the use of continuity. Suppose $\alpha(x) \neq x$ for some $x \in \mathbb{R}$. If $x < \alpha(x)$, then find $q \in \mathbb{Q}$ with $x < q < \alpha(x)$. So $\alpha(x) < \alpha(q) < \alpha(\alpha(x))$; but $\alpha(q) = q$, so that $\alpha(x) < q$, a contradiction. Similarly if $x > \alpha(x)$.   □

Theorem 8.37. $\mathcal{A}(\mathbb{R}^2) = \mathcal{C}(\mathbb{R}^2)$.

Proof.   We already know $\mathcal{A}(\mathbb{R}^2) \subseteq \mathcal{C}(\mathbb{R}^2)$ (Proposition 8.14), so let $g \in \mathcal{C}(\mathbb{R}^2)$, and suppose $g(0, 0) = P$, $g(1, 0) = Q$, and $g(0, 1) = R$. By Proposition 8.19 there is an affine transformation h with $h(0, 0) = P$, $h(1, 0) = Q$, and $h(0, 1) = R$. Since $\mathcal{A}(\mathbb{R}^2) \subseteq \mathcal{C}(\mathbb{R}^2)$, and $\mathcal{C}(\mathbb{R}^2)$ is a group, we have that h is a collineation, and so is h^{-1}, and hence so is $f = h^{-1}g$. But f fixes $(0, 0)$, $(1, 0)$, and $(0, 1)$, so $f = 1$, by Lemma 8.36, and therefore $g = h$; but $h \in \mathcal{A}(\mathbb{R}^2)$, so $g \in \mathcal{A}(\mathbb{R}^2)$, as required.   □

ANSWERS TO EXERCISES

8.1: Apply a translation to move the focus of the first parabola into coincidence with the focus of the second, and then a rotation to make the directrices parallel, and on the same side of the common focus. A central dilation, centre the focus, can now be chosen to make the two directrices coincide; but a parabola is determined by its focus and directrix, so we have finished.

8.2:   The perpendiculars from X onto BA and AC are of equal length, so $\text{area}(ABX)/\text{area}(ACX) = |BA|/|AC|$. On the other hand, $\triangle ABX$ and $\triangle ACX$ have common base-line BXC and the same other vertex, A, so that $\text{area}(ABX)/\text{area}(ACX) = |BX|/|XC|$.

8.3: Let the circles be Σ and Σ', with centres be O and O'. (Construct these as in Exercise 7.4.) Put in a diameter AB of Σ (with A not on the line OO') and a parallel diameter $A'B'$ of Σ'. Then one centre of similitude is the meet $AA' \cdot BB'$, and the other is the meet $AB' \cdot A'B$.

8.4: If the circles have centres A_j and radii r_j, $j = 1, 2, 3$, then the points O_1, O_2, O_3 lie on A_2A_3, A_3A_1, A_1A_2 respectively, and

$$\frac{A_2 O_1}{O_1 A_3} \frac{A_3 O_2}{O_2 A_1} \frac{A_1 O_3}{O_3 A_2} = \left(-\frac{r_2}{r_3}\right)\left(-\frac{r_3}{r_1}\right)\left(-\frac{r_1}{r_2}\right) = -1.$$

8.5: We have $f_1 f_2 = f_3$, as before. Then the line $O_1 O_2$ is invariant under both f_1 (since it passes through O_1), and f_2 (since it passes through O_2), and hence under $f_1 f_2$, that is, under f_3. So $O_1 O_2$ must pass through O_3.

8.6: The proof of Lemma 8.6 still works if two of the a_j are negative (so that the corresponding f_j are dilative half-turns). This means that the two *internal* centres of similitude of any two of the pairs of circles are collinear with the *external* centre of similitude of the remaining pair. (The other proofs can be adapted, also.)

8.7: It is the same as the nine-point circle of $\triangle ABC$ (and also of $\triangle HCA$ and of $\triangle HAB$).

8.8: From the fact that P' is the mid-point of HP, the reflection of the circumcircle of $\triangle ABC$ in BC passes through H. Similarly for the other reflections.

8.9: Let f, $g \in \mathcal{D}(\mathbb{R}^2)$. Given points P, Q, let $g(P) = P'$, $g(Q) = Q'$, $f(P') = P''$, and $f(Q') = Q''$. Then $PQ \parallel P'Q'$ and $P'Q' \parallel P''Q''$, so $PQ \parallel P''Q''$, whence $fg \in \mathcal{D}(\mathbb{R}^2)$. Also, the fact that $P'Q' \parallel PQ$ shows $g^{-1} \in \mathcal{D}(\mathbb{R}^2)$ (OK?), and clearly $1 \in \mathcal{D}(\mathbb{R}^2)$.

8.10: Let f be a dilatation, let P, Q, R be collinear points, and let $f(P) = P'$, $f(Q) = Q'$, and $f(R) = R'$. So $PQ \parallel QR$, and also $PQ \parallel P'Q'$ and $QR \parallel Q'R'$. Thus $P'Q' \parallel Q'R'$, from which P', Q', and R' are collinear.

8.11: For any z_1, z_2 we have $f(z_1) - f(z_2) = a(z_1 - z_2)$, and the result is immediate, given that a is real.

8.12: Composites: $\mathbf{A}_1(\mathbf{A}_2\mathbf{v} + \mathbf{b}_2) + \mathbf{b}_1 = (\mathbf{A}_1\mathbf{A}_2)\mathbf{v} + (\mathbf{A}_1\mathbf{b}_2 + \mathbf{b}_1)$. Inverses: if $\mathbf{w} = \mathbf{A}\mathbf{v} + \mathbf{b}$ then $\mathbf{v} = \mathbf{A}^{-1}\mathbf{w} - \mathbf{A}^{-1}\mathbf{b}$. Clearly the trivial map is affine.

8.13: Suppose P, Q, R are not collinear, and suppose their images under the affine transformation f are P', Q', R'. The inverse map f^{-1} is also affine (Exercise 8.12), so if P', Q', R' are collinear, then so are P, Q, R, by Proposition 8.14 applied to f^{-1}. This is a contradiction, whence the result.

8.14: Note first that A', B', C' do form a triangle, by Exercise 8.13. If D, E are the mid-points of BC, CA, then the images D', E' are the mid-points of $B'C'$, $C'A'$. But $G = AD \cdot BE$ (the meet), so that $G' = A'D' \cdot B'E'$, and the result follows.

8.15: For any P', $Q' \in f(X)$ there exist P, $Q \in X$ with $f(P) = P'$ and $f(Q) = Q'$. If R' lies on the line segment $P'Q'$ then $P'R' : R'Q' = \lambda : \mu$ for some positive λ, μ. But now if R is on the line segment PQ with $PR : RQ = \lambda : \mu$, then $f(R) = R'$ by Corollary 8.15. Then $R \in X$, since X is convex, whence $R' \in f(X)$, so that $f(X)$ is convex.

8.16: Writing $\mathbf{p}, \mathbf{q}, \mathbf{r}, \mathbf{s}$ for the four points, and $\mathbf{v} \mapsto \mathbf{A}\mathbf{v} + \mathbf{b}$ for the affine transformation, we have $\overrightarrow{P'Q'} = (\mathbf{A}\mathbf{q} + \mathbf{b}) - (\mathbf{A}\mathbf{p} + \mathbf{b}) = \mathbf{A}(\mathbf{q} - \mathbf{p})$, and similarly $\overrightarrow{R'S'} = \mathbf{A}(\mathbf{s} - \mathbf{r})$. But if $PQ \parallel RS$ and $PQ/RS = \lambda$, then $\mathbf{q} - \mathbf{p} = \lambda(\mathbf{s} - \mathbf{r})$, so that $\overrightarrow{P'Q'} = \mathbf{A}(\mathbf{q} - \mathbf{p}) = \lambda\mathbf{A}(\mathbf{s} - \mathbf{r}) = \lambda\overrightarrow{R'S'}$, and we are done.

8.17: Let $f : \mathbf{v} \mapsto \mathbf{A}\mathbf{v} + \mathbf{b}$, and suppose the three non-collinear fixed points are represented by vectors $\mathbf{p}, \mathbf{q}$, and $\mathbf{r}$. We have $\mathbf{p} = \mathbf{A}\mathbf{p} + \mathbf{b}$, $\mathbf{q} = \mathbf{A}\mathbf{q} + \mathbf{b}$, and $\mathbf{r} = \mathbf{A}\mathbf{r} + \mathbf{b}$, whence $\mathbf{q} - \mathbf{p} = \mathbf{A}(\mathbf{q} - \mathbf{p})$ and $\mathbf{r} - \mathbf{p} = \mathbf{A}(\mathbf{r} - \mathbf{p})$. Let $\mathbf{B}$ be the 2×2 matrix with $\mathbf{q} - \mathbf{p}$, $\mathbf{r} - \mathbf{p}$ as its columns: then $\mathbf{B}$ is non-singular, and $\mathbf{B} = \mathbf{A}\mathbf{B}$, whence $\mathbf{A} = \mathbf{I}_2$. So $\mathbf{p} = \mathbf{A}\mathbf{p} + \mathbf{b} = \mathbf{p} + \mathbf{b}$, and thus $\mathbf{b} = \mathbf{0}$ and $f = 1$.

8.18: The existence of a suitable f is assured by Proposition 8.19. If g is another affine transformation, with $g(P) = P'$, $g(Q) = Q'$, and $g(R) = R'$, then $g^{-1}f$ fixes P, Q, and R, so $g^{-1}f = 1$ by Exercise 8.17, and thus $g = f$.

8.19: (Is it clear that $GL_2(\mathbb{Z})$ is a group? Check.) The three given points form a triangle with centroid at $(0, 0)$, so all the required affine transformations are in fact linear transformations. For each permutation of the vertices, there is a unique affine transformation achieving this permutation, by Exercise 8.18. The six matrices are found by writing the images of $(1, 0)$ and $(0, 1)$ as the two columns, in each case:

$$\begin{pmatrix} 1 & 0 \\ 0 & 1 \end{pmatrix}, \begin{pmatrix} 0 & -1 \\ 1 & -1 \end{pmatrix}, \begin{pmatrix} -1 & 1 \\ -1 & 0 \end{pmatrix}, \begin{pmatrix} 0 & 1 \\ 1 & 0 \end{pmatrix}, \begin{pmatrix} 1 & -1 \\ 0 & -1 \end{pmatrix}, \begin{pmatrix} -1 & 0 \\ -1 & 1 \end{pmatrix},$$

and because $(0, 0)$ is fixed, the image of $(-1, -1)$ is then automatically correct. (Check.)

8.20: $ABCD$ is affine regular if there is a square $PQRS$ to which it is affinely equivalent. Since affine transformations preserve parallels (Proposition 8.16), and $PQ \parallel SR$ and $QR \parallel PS$, we must have $AB \parallel DC$ and $BC \parallel AD$, so that $ABCD$ is a parallelogram. Conversely, given a parallelogram $ABCD$ and a square $PQRS$, there is an affine transformation f with $f(A) = P$, $f(B) = Q$, and $f(C) = R$, by Proposition 8.19. But now if $f(D) = S'$, we must have $PQ \parallel S'R$ and $QR \parallel PS'$ by Proposition 8.16 again, whence $S = S'$, and $ABCD$ is affine regular.

8.21:

$$\tau = \frac{|PV|}{|VR|} = \frac{|PV|}{|PU|} = \frac{|PU| + |UV|}{|PU|} = 1 + \frac{|UV|}{|PU|},$$

so

$$\frac{|PU|}{|UV|} = (\tau - 1)^{-1} = \tau.$$

8.22: The parallels given show that area(PQR) = area(TPQ) = area(STP) = area(RST) = area(QRS), from which $PS \parallel QR$.

8.23: Let the vertices P, Q, R, S, T be represented by the vectors $\mathbf{p}$, $\mathbf{q}$, $\mathbf{r}$, $\mathbf{s}$, $\mathbf{t}$ respectively. We have $\mathbf{q} - \mathbf{t} = \alpha(\mathbf{r} - \mathbf{s})$, $\mathbf{r} - \mathbf{p} = \beta(\mathbf{s} - \mathbf{t})$, $\mathbf{s} - \mathbf{q} = \gamma(\mathbf{t} - \mathbf{p})$, and $\mathbf{t} - \mathbf{r} = \delta(\mathbf{p} - \mathbf{q})$, for some scalars α, β, γ, δ. Let us take T as the origin, so that $\mathbf{t} = \mathbf{0}$. Our equations can now be rewritten $\mathbf{q} = \alpha(\mathbf{r} - \mathbf{s})$, $\mathbf{r} = \beta\mathbf{s} + \mathbf{p}$, $\mathbf{q} = \mathbf{s} + \gamma\mathbf{p}$, and $\mathbf{r} = \delta(\mathbf{q} - \mathbf{p})$. Eliminating $\mathbf{r}$ and $\mathbf{q}$, we have $\mathbf{s} + \gamma\mathbf{p} = \alpha(\beta\mathbf{s} + \mathbf{p} - \mathbf{s})$ and $\beta\mathbf{s} + \mathbf{p} = \delta(\mathbf{s} + \gamma\mathbf{p} - \mathbf{p})$. If we assume that no three of P, Q, R, S, T are collinear, then $\mathbf{s}$, $\mathbf{p}$ are linearly independent and we can compare coefficients, obtaining $1 = \alpha(\beta - 1)$, $\gamma = \alpha$, $\beta = \delta$, and $1 = \delta(\gamma - 1) = \beta(\alpha - 1)$, whence $\alpha(\beta - 1) = \beta(\alpha - 1)$ and $\alpha = \beta$. So $\alpha = \beta = \gamma = \delta$, and $\alpha(\alpha - 1) = 1$, so $\alpha = \frac{1+\sqrt{5}}{2} = \tau$ (the convex case) or else $\alpha = \frac{1-\sqrt{5}}{2} = -\tau^{-1}$, in which case $PQRST$ is not convex, and is a *star* pentagon, with its sides crossing each other internally. Since these two cases correspond to the conditions $\alpha > 0$ and $\alpha < 0$, we could, instead of using the word 'convex', have insisted that the line segments in (8.4) are all *directly* parallel (see Exercise 3.1), the star pentagon case being given by insisting instead that they are *oppositely* parallel (see Exercise 3.1 again). We are deliberately ignoring the degenerate case when P, Q, R, S, T are collinear.

8.24: There are many ways to do this; see, for example, [9]. Here is a method based on the proof of Proposition 8.21. Draw a circle, centre Q, of unit radius, and put in a diameter RX, and a radius YQ orthogonal to RX. Construct the mid-point K of QR, so that $|KY|^2 = |YQ|^2 + |QK|^2 = 1 + \frac{1}{4}$, and $|KY| = \frac{1}{2}\sqrt{5}$. So, with centre K, draw an arc of a circle through Y to cut QX at L, and we have $|RL| = |RK| + |KL| = \frac{1}{2} + \frac{1}{2}\sqrt{5} = \tau$. With centre R, draw an arc of a circle through L to cut the original circle at P, and join PR, PQ to give a triangle PQR with $|PQ| = 1 = |QR|$ and $|QR| = \tau$. Now mark U, V on PR with $|UR| = 1 = |PV|$, and mark S, T on QV produced and QU produced, respectively, such that $|VS| = 1$ and $|UT| = 1$, and join up to give the regular pentagon $PQRST$.

8.25: If $P_0 P_1 \ldots P_{14}$ is a regular 15-gon, then $P_0 P_5 P_{10}$ is an equilateral triangle, and $P_0 P_3 P_6 P_9 P_{12}$ is a regular pentagon. So construct these two figures first, inscribed in the same circle, and then $P_5 P_6$ is one of the sides of the 15-gon, so step this length off around the circle. (If you thought you could do this construction by starting with a regular pentagon and then trisecting angles, then think again. See the remarks on page 18.)

8.26: The 15-gon has a central angle of $\frac{360}{15} = 24°$, and by bisecting we can get $12°$, then $6°$, and then $3°$. We cannot construct $2°$ or $1°$, else we could construct $20°$ also: see page 18.

8.27: In the transformed diagram, the images of ℓ and m are perpendicular diameters of the circle, so each is parallel to the tangents at the extremities of the other.

8.28: As with Proposition 8.26 and Exercise 8.27, this is just a matter of noting that reflection in the image of ℓ is a symmetry of the transformed diagram.

8.29: Proposition 8.26 works with the word 'hyperbola' in place of the word 'ellipse'. The affine transformation

$$f : \begin{pmatrix} x \\ y \end{pmatrix} \mapsto \begin{pmatrix} \cosh\alpha & ab^{-1}\sinh\alpha \\ a^{-1}b\sinh\alpha & \cosh\alpha \end{pmatrix} \begin{pmatrix} x \\ y \end{pmatrix}$$

maps Γ to itself: $f(a\cosh\theta, b\sinh\theta) = (a\cosh(\theta + \alpha), b\sinh(\theta + \alpha))$, and $f(-a\cosh\theta, b\sinh\theta) = (-a\cosh(\theta - \alpha), b\sinh(\theta - \alpha))$. Using this we can transform PQ

so that the image is parallel to one or other of the axes of Γ (depending on whether P, Q are on the same or different branches of Γ), and then, as with the ellipse, the result follows since reflection in an axis is a symmetry of a hyperbola.

8.30: Proposition 8.26 works with the word 'parabola' in place of the word 'ellipse', provided we omit any reference to the centre. The affine transformation

$$f : \begin{pmatrix} x \\ y \end{pmatrix} \mapsto \begin{pmatrix} 1 & s \\ 0 & 1 \end{pmatrix} \begin{pmatrix} x \\ y \end{pmatrix} + \begin{pmatrix} as^2 \\ 2as \end{pmatrix}$$

maps Γ to itself: $f(at^2, 2at) = (a(t+s)^2, 2a(t+s))$. Using this we can transform PQ so that the image is orthogonal to the axis of Γ, and then the result follows because reflection in the axis is a symmetry of Γ. Having lost any reference to the centre (because a parabola has no centre), we gain something in its place: f maps the line $y = k$ to the line $y = k + 2as$, from which we deduce that the equation of ℓ is $y = -2as$, that is, the line ℓ is parallel to the axis of Γ.

8.31: *Either* apply the affine transformation of Exercise 8.29 to Γ, noting that (i) since $f(\Gamma) = \Gamma$, the asymptotes of Γ are invariant under f, and (ii)

$$\begin{vmatrix} \cosh \alpha & ab^{-1} \sinh \alpha \\ a^{-1}b \sinh \alpha & \cosh \alpha \end{vmatrix} = 1,$$

so f preserves areas; *or* transform Γ to the hyperbola $xy = 1$, in which form the property is rather easily proved by coordinate geometry.

9

Infinity

We are now going to look at some geometries in which, to see what is going on and to do calculations properly, we need to add to our plane one or more 'points at infinity'. As a by-product, we shall find ourselves able to set up models for non-Euclidean geometry, where Euclid's parallel axiom (see Section 1.3) fails to hold.

9.1 MÖBIUS TRANSFORMATIONS

In Section 8.3 we wrote the equations of our similarities in terms of matrices, in order to generalize to affine transformations. Here we return to the use of complex numbers for similarities in order to generalize in a different way. So recall that a direct similarity is given by a formula of the kind $z \mapsto az + b$, $z \in \mathbb{C}$, where $a, b \in \mathbb{C}$ and $a \neq 0$. We shall now explore what happens if we map z to the *quotient* of two such expressions: let

$$f(z) = \frac{az + b}{cz + d},$$

where $a, b, c, d \in \mathbb{C}$. Here we obviously cannot allow both $c = 0$ and $d = 0$; and if $c = 0$ (so that $d \neq 0$) then $f(z) = a/dz + b/d$ is a direct similarity, and obviously our new set of maps includes all direct similarities as a subset. There is a further restriction on a, b, c, d, since if $ad = bc$ then $(a, b) = \lambda(c, d)$ for some $\lambda \in \mathbb{C}$, and now $az + b = \lambda(cz + d)$ for all z, so that $f(z) = \lambda$ for all z. We do not want this to happen, so we insist $ad \neq bc$; this implies the previous condition also, that is, c and d cannot both be zero.

There is one more difficulty to be resolved if $c \neq 0$, for then, putting $z_0 = -d/c$, we have $cz_0 + d = 0$ so that $f(z_0)$ is undefined. Since $az_0 + b \neq 0$ (OK?) it is easy to see that $|f(z)| \to \infty$ as $z \to z_0$, so we get around our difficulty by inventing a new point called ∞ (infinity), and boldly writing $f(z_0) = \infty$. This means f is now a map from $\mathbb{C}$ to $\mathbb{C} \cup \{\infty\}$. It is *not* (yet) surjective, because if $f(z) = a/c$ then $c(az + b) = a(cz + d)$, or $cb = ad$, a contradiction. But $(az + b)/(cz + d) \to a/c$ as $z \to \infty$, so the obvious thing to do is write $f(\infty) = a/c$. Thus when $c \neq 0$ we have a map $f : \mathbb{C} \cup \{\infty\} \to \mathbb{C} \cup \{\infty\}$; when $c = 0$ (so that $d \neq 0$) we have $f(z) \to \infty$ as $z \to \infty$, so we complete the picture by writing $f(\infty) = \infty$ in this case.

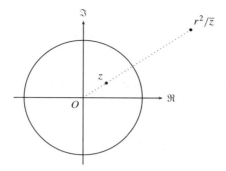

Fig. 9.1 Inversion in the circle $|z| = r$

Definition 9.1. *Let $a, b, c, d \in \mathbb{C}$ with $ad \neq bc$. The map $f : \mathbb{C} \cup \{\infty\} \to \mathbb{C} \cup \{\infty\}$ given by $f(z) = (az + b)/(cz + d)$ (with the above conventions for ∞) is a **Möbius transformation**.*

Note that f is a *transformation* of $\mathbb{C} \cup \{\infty\}$, that is, it is bijective. Indeed the reader should verify that $f^{-1}(z) = (dz - b)/(-cz + a)$, observing that if $c \neq 0$ then $f^{-1}(a/c) = \infty$ and $f^{-1}(\infty) = -d/c$.

- Ex.9.1: *Möbius transformations form a group, $\mathcal{M}^+(\mathbb{C})$, under composition of maps. The obvious map*

$$GL_2(\mathbb{C}) \to \mathcal{M}^+(\mathbb{C}), \quad \begin{pmatrix} a & b \\ c & d \end{pmatrix} \mapsto f \left(where \; f(z) = \frac{az + b}{cz + d} \right)$$

is a group homomorphism. What is its kernel?

One can also generalize *opposite* similarities $z \mapsto a\bar{z} + b$ to *conjugate* Möbius transformations, given by $z \mapsto (a\bar{z} + b)/(c\bar{z} + d)$, where $ad \neq bc$. This is the composite of the reflection $z \mapsto \bar{z}$ and the Möbius transformation $z \mapsto (az + b)/(cz + d)$. Möbius transformations and conjugate Möbius transformations together form a larger group, $\mathcal{M}(\mathbb{C})$, which has $\mathcal{M}^+(\mathbb{C})$ as a subgroup. We single out for special mention the conjugate Möbius transformation $z \mapsto r^2/\bar{z}$, where $r \in \mathbb{R}$, $r > 0$. This is called *inversion* with respect to the circle $|z| = r$, and maps P to P' where O, P, P' are collinear and $(OP)(OP') = r^2$; the points P, P' are then called *inverse points* with respect to $|z| = r$ (Fig. 9.1). It also maps O to ∞ and ∞ to O, so these are inverse points also. Points inside the circle are mapped to points outside, and vice versa; points *on* the circle are fixed, or self-inverse.

- Ex.9.2: *Write down the equation of inversion in the circle $|z - \alpha| = r$.*

- Ex.9.3: *Let AB be a diameter of the circle Σ, and let C, D lie on the line AB. Prove that C, D are inverse points with respect to Σ if and only if $\{A, B; C, D\} = -1$. (See Definition 4.35.)*

- Ex.9.4: *Let C, D be inverse points with respect to the circle Σ, and let Σ' be the circle on CD as diameter. Prove that Σ and Σ' are* orthogonal, *that is, at the points where they meet, their respective tangents are orthogonal.*

- Ex.9.5: *Prove that every Möbius transformation is either a direct similarity or else is the composite of an opposite isometry and an inversion; and every conjugate Möbius transformation is either an opposite similarity or else is the composite of a direct isometry and an inversion.*

The type of geometry we are now doing is sometimes called *inversive geometry*, and the set $\mathbb{C} \cup \{\infty\}$ is known as the *inversive plane*. In view of Exercise 9.5, it might have been preferable to call Möbius and conjugate Möbius transformations *direct* and *opposite* Möbius transformations instead. We shall continue for the most part to investigate (direct) Möbius transformations, leaving the interested reader to formulate and prove the corresponding results for the conjugate (opposite) case.

9.1.1   Lines and circles

Following Klein again, we ask: what is preserved by a Möbius transformation? Not lines, for a start: Möbius transformations that are not similarities are *not* collineations. As an example, the general point of the line $x = 1$ is $z = 1 + it, t \in \mathbb{R}$, and the image of this point under the Möbius transformation $f : z \mapsto 1/z$ is $1/(1 + it) = (1 - it)/(1 + t^2)$. Thus the image under f of the line $x = 1$ is the curve given parametrically by $x = 1/(1 + t^2)$, $y = -t/(1 + t^2)$ which, on eliminating t, turns out to be the circle $x^2 + y^2 - x = 0$. Applying f again, since $f^2 = 1$, the image of this circle is the line we started with; on the other hand it is easy to see that the x-axis and the circle $|z| = 1$ are both invariant under f. In fact:

Proposition 9.2. *The image under a Möbius transformation of a line, or a circle, is either a line or a circle.*

Proof. Any circle or line Σ can be put in the form $|z - \alpha| = k|z - \beta|$, where $k \in \mathbb{R}$, $k > 0$. If $k \neq 1$ then Σ is Apollonius' circle, with α and β as inverse points, and if $k = 1$ then Σ is a line, the perpendicular bisector of the line segment from α to β. (We shall call α and β inverse points with respect to Σ in this case also.)

Let $f(z) = (az + b)/(cz + d)$, where $ad \neq bc$. If $z' = f(z)$, then $z = (dz' - b)/(-cz' + a)$, so that $f(\Sigma) = \Sigma'$ has equation $|(dz' - b)/(-cz' + a) - \alpha| = k|(dz' - b)/(-cz' + a) - \beta|$, or $|dz' - b + \alpha(cz' - a)| = k|dz' - b + \beta(cz' - a)|$, or $|(c\alpha + d)z' - (a\alpha + b)| = k|(c\beta + d)z' - (a\beta + b)|$. Since there are infinitely many choices of α and β for the same Σ, we may as well assume that $c\alpha + d \neq 0$ and $c\beta + d \neq 0$, that is, $f(\alpha) \neq \infty$ and $f(\beta) \neq \infty$. Then Σ' is $|z' - f(\alpha)| = k'|z' - f(\beta)|$, where $k' = k|c\beta + d|/|c\alpha + d|$, which is a circle or line according as $k' \neq 1$ or $k' = 1$. $\square$

Notice that $f(\alpha)$ and $f(\beta)$ are inverse points for Σ'; and this even works if $f(\alpha) = \infty$ (when the centre of Σ' is $f(\beta)$) or if $f(\beta) = \infty$ (when the centre of Σ' is $f(\alpha)$), as the reader should verify. Note also that if O is the centre of Σ, then O and ∞ are inverse points with respect to Σ, and it is easy to show (for instance, by applying the above argument to f^{-1} in place of f) that $f(O)$ and $f(\infty)$ are inverse points with respect to Σ'. This means that $f(O)$ is *not* the centre of Σ', except when $f(\infty) = \infty$, that is, when f is a similarity.

A Möbius transformation, then, sends a line-or-circle to a line-or-circle, where a *line-or-circle* just means something that is either a line or else a circle. To decide whether a given line-or-circle gets mapped to a line, we need to look at what happens to the point ∞. Any line-or-circle Σ can be written $|z - \alpha| = k|z - \beta|$, or $|(z - \alpha)/(z - \beta)| = k$. Now $|(z - \alpha)/(z - \beta)| \to 1$ as $z \to \infty$, so we should obviously adopt the convention that $\infty \in \Sigma$ if and only if $k = 1$; that is, the point ∞ lies on every line, but not on any circle.

Let us see how this ties in with the way ∞ is treated by $f : z \mapsto (az + b)/(cz + d)$, where $c \neq 0$. We have $f(-d/c) = \infty$, so we would expect that $\Sigma' = f(\Sigma)$ is a line if and only if $-d/c$ lies on Σ. But $-d/c$ lies on Σ if and only if $|-d/c - \alpha| = k|-d/c - \beta|$, or $|c\alpha + d| = k|c\beta + d|$, and in the notation of the proof of Proposition 9.2, this means $k' = 1$, or that Σ' is a line.

In the other direction, Σ is a line if and only if $k = 1$, and this means that the equation of Σ' is $|(c\alpha + d)z' - (a\alpha + b)| = |(c\beta + d)z' - (a\beta + b)|$. This is satisfied by $z' = a/c$ (OK?), so that $f(\infty)$ lies on Σ'. To sum up, the image of a line-or-circle Σ is a line if and only if $-d/c \in \Sigma$, and the image of every line is a line-or-circle through a/c.

- Ex.9.6:   Let $f : z \to (az + b)/(cz + d)$, where $ad \neq bc$. Show that, if $c \neq 0$ and $a + d \neq 0$, then f has a unique invariant line. What happens if $a + d = 0$?

Proposition 9.3.   Let f be a Möbius transformation fixing three distinct elements of $\mathbb{C} \cup \{\infty\}$. Then $f = 1$.

Proof.   Let $f : z \mapsto (az + b)/(cz + d)$. If $z \in \mathbb{C}$ and $f(z) = z$ then $cz^2 + (d-a)z - b = 0$, which is satisfied by at most two elements of $\mathbb{C}$ provided $c \neq 0$, that is, provided $f(\infty) \neq \infty$. If on the other hand $c = 0$, then $f(\infty) = \infty$, and the equation $f(z) = z$ reduces to $(d - a)z - b = 0$, which is satisfied by only one element of $\mathbb{C}$ unless $a = d$ and $b = 0$. But if $a = d$ and $b = c = 0$, then $f = 1$.   □

Proposition 9.4.   Let a, b, c be distinct elements of $\mathbb{C} \cup \{\infty\}$. Then there is a unique Möbius transformation f with $f(a) = 0$, $f(b) = \infty$, and $f(c) = 1$.

Proof.   If $a, b, c \in \mathbb{C}$, then $f : z \mapsto ((z - a)(c - b))/((z - b)(c - a))$ will do the trick. If $a = \infty$, we use $f : z \mapsto (c - b)/(z - b)$ instead; if $b = \infty$, we use $f : z \mapsto (z - a)/(c - a)$ instead; and if $c = \infty$, we use $f : z \mapsto (z - a)/(z - b)$ instead. (Notice that these last three formulae are obtained from $((z - a)(c - b))/((z - b)(c - a))$ by letting $a \to \infty$, or $b \to \infty$, or $c \to \infty$, respectively.) If f' is a second solution, then f and f' have the same effect on a, b, and c, so that $f^{-1}f'$ fixes a, b, and c. By Proposition 9.3, $f^{-1}f' = 1$, or $f' = f$.   □

Corollary 9.5.   Let a, b, c be distinct elements of $\mathbb{C} \cup \{\infty\}$, and let a', b', c' be distinct elements of $\mathbb{C} \cup \{\infty\}$. Then there is a unique Möbius transformation f with $f(a) = a'$, $f(b) = b'$, and $f(c) = c'$.

Proof.   By Proposition 9.4, we can find Möbius transformations g and h with $g(a) = 0 = h(a')$, $g(b) = \infty = h(b')$, and $g(c) = 1 = h(c')$. Thus $f = h^{-1}g$ does what is required; and if f' is a second solution, then $f' = f$ exactly as in the proof of Proposition 9.4. $\square$

Notice that three distinct elements of $\mathbb{C} \cup \{\infty\}$ lie on a unique line-or-circle. If they are all finite (i.e., in $\mathbb{C}$) and not collinear, then they determine a triangle, on whose circumcircle they lie. If they are finite and collinear, they lie on a line! And if one of them is ∞, then they all lie on the line joining the other two, since ∞ lies on every line. As a consequence of this, if Σ is a line-or-circle, and Σ' is another line-or-circle, then there is a Möbius transformation f with $f(\Sigma) = \Sigma'$. (But f is not unique. Why?)

- **Ex.9.7:**   *Give an example of a Möbius transformation having (i) just one fixed point, (ii) exactly two fixed points, in $\mathbb{C} \cup \{\infty\}$. Can a Möbius transformation have no fixed points?*

- **Ex.9.8:**   *Give an example of a conjugate Möbius transformation having (i) just one fixed point, (ii) exactly two fixed points, in $\mathbb{C} \cup \{\infty\}$. Can a conjugate Möbius transformation have no fixed points?*

- **Ex.9.9:**   *Suppose the conjugate Möbius transformation f fixes the distinct elements $a, b, c \in \mathbb{C} \cup \{\infty\}$. Let Σ be the (unique) line-or-circle through a, b, and c. Show that f must be inversion with respect to Σ (if it is a circle), or reflection in Σ (if it is a line).*

- **Ex.9.10:**   *Let Σ be a line-or-circle, and let α, $\beta \in \mathbb{C} \cup \{\infty\}$ be inverse points with respect to Σ. Let Σ' be another line-or-circle, and let α', $\beta' \in \mathbb{C} \cup \{\infty\}$ be inverse points with respect to Σ'. Show that there is a Möbius transformation f with $f(\alpha) = \alpha'$, $f(\beta) = \beta'$, and $f(\Sigma) = \Sigma'$.*

9.1.2   Cross-ratios

We have not yet found very much that is preserved by a Möbius transformation: so far, just the property of being a line-or-circle. It is easy to see that a Möbius transformation does not preserve lengths, or even ratios of lengths (unless it is actually a similarity), and we leave the reader to construct suitable examples to show this. However, it does preserve cross-ratios, in the sense of the following definition:

Definition 9.6.   *Let a, b, c, d be distinct elements of $\mathbb{C}$. Then the **cross-ratio** $\{a, b \, ; c, d\}$ is defined by*

$$\{a, b \, ; c, d\} = \frac{(c - a)(d - b)}{(c - b)(d - a)}.$$

The definition is extended to the case where $a = \infty$ by letting $a \to \infty$ in the above formula, and similarly for b, c, and d. So

$$\{\infty, b \, ; c, d\} = \frac{d - b}{c - b},$$

and so on.

Notice that if a, b, c, d are collinear (and finite), then both $(c - a)/(c - b)$ and $(d - b)/(d - a)$ are real, and are signed ratios of lengths on the line in question; and it is easy to check that our new definition of cross-ratio reduces to the previous one (Definition 4.35) in this case.

Corollary 9.7. *Let a, b, c, d be distinct elements of $\mathbb{C} \cup \{\infty\}$, and let f be a Möbius transformation with $f(a) = 0$, $f(b) = \infty$, and $f(c) = 1$. Then $f(d) = \{a, b\,;\, d, c\}$. (Note the order.)*

Proof. This is just a restatement of Proposition 9.4: we have $f(z) = (z - a)(c - b)/((z - b)(c - a))$, and we just substitute $z = d$. (And it still works if a or b or c or d is infinite. OK?)   □

Proposition 9.8. Möbius transformations preserve cross-ratios. *Let f be a Möbius transformation, and let a, b, c, d be distinct elements of $\mathbb{C} \cup \{\infty\}$. Put $a' = f(a)$, $b' = f(b)$, $c' = f(c)$, and $d' = f(d)$. Then $\{a, b\,;\, c, d\} = \{a', b'\,;\, c', d'\}$.*

Proof. Choose a Möbius transformation g with $g(a) = 0$, $g(b) = \infty$, and $g(d) = 1$, and a Möbius transformation h with $h(a') = 0$, $h(b') = \infty$, and $h(d') = 1$. Then f and $h^{-1}g$ have the same effect on a, b and d, so that $f = h^{-1}g$, by Corollary 9.5. But $f(c) = c'$, so that $g(c) = h(c')$, or $\{a, b\,;\, c, d\} = \{a', b'\,;\, c', d'\}$, by Corollary 9.7.   □

We leave the reader to investigate whether the following suggested alternative method of proof of Proposition 9.8 would be easier: just write down the formula for f, use it to substitute for a', b', c', and d' in $\{a', b'\,;\, c', d'\}$, and simplify.

Proposition 9.9. *The distinct points a, b, c, $d \in \mathbb{C} \cup \{\infty\}$ lie on a line-or-circle if and only if $\{a, b\,;\, c, d\} \in \mathbb{R}$.*

Proof. Let Σ be the line-or-circle containing a, b, and d, and let f be the Möbius transformation with $f(a) = 0$, $f(b) = \infty$, and $f(d) = 1$. Then $f(\Sigma) = \Sigma'$ is the line-or-circle containing 0, ∞, and 1, that is, it is the real axis, $y = 0$. Then a, b, c, d lie on a line-or-circle if and only if $c \in \Sigma$, which happens if and only if $f(c) \in \Sigma'$, that is, if and only if $\{a, b\,;\, c, d\} \in \mathbb{R}$.   □

We have met this last idea before, in Section 7.4, where we proved some theorems about a cyclic tetragram $ABCD$. There, we let the points A, B, C, D be represented by the complex numbers a, b, c, d respectively, and we put

$$\frac{c - b}{a - b} = re^{i\theta}, \quad \frac{c - d}{a - d} = se^{i\varphi}, \quad \text{and} \quad \lambda = \left(\frac{c - b}{a - b}\right)\left(\frac{a - d}{c - d}\right).$$

It then followed that $\theta = \angle ABC$ and $\varphi = \angle ADC$, so that $\lambda \in \mathbb{R}$ if and only if $\theta \equiv \varphi$ modulo π. But now we see that $\lambda = \{a, c\,;\, d, b\}$, so the argument was that if $ACDB$ is cyclic, then $\{a, c\,;\, d, b\} \in \mathbb{R}$, and conversely, provided the points are not collinear.

Astute readers will have noticed that the vertices of the cyclic tetragram $ABCD$ somehow got into the 'wrong' order above, namely $ACDB$, but that this did not invalidate the argument. To clarify this, let us explore what happens to a cross-ratio $\lambda = \{a_1, a_2\,;\, a_3, a_4\}$

when we choose a permutation $\rho \in S_4$ and work out $\lambda_\rho = \{a_{\rho_1}, a_{\rho_2} ; a_{\rho_3}, a_{\rho_4}\}$ instead. Put $V = \{1, (12)(34), (13)(24), (14)(23)\}$, a subgroup of S_4 known as *Klein's four-group*, or the *Vierergruppe*. It is immediate that if $\rho \in V$, then $\lambda_\rho = \lambda$, since, rather obviously,

$$\frac{(a_3 - a_1)(a_4 - a_2)}{(a_3 - a_2)(a_4 - a_1)} = \frac{(a_4 - a_2)(a_3 - a_1)}{(a_4 - a_1)(a_3 - a_2)} = \frac{(a_1 - a_3)(a_2 - a_4)}{(a_1 - a_4)(a_2 - a_3)}$$
$$= \frac{(a_2 - a_4)(a_1 - a_3)}{(a_2 - a_3)(a_1 - a_4)}.$$

It follows that, as ρ runs through the $4! = 24$ elements of S_4, we shall *not* get 24 different values of λ_ρ, but at most $24/4 = 6$ values, one from each coset $V\rho$ of V in S_4. To see what they are, given λ, observe that by Corollary 9.5, each permutation of the set $\{0, 1, \infty\}$ extends to a unique Möbius transformation. It is easy to check that these are

$$z \mapsto z, \ (1 - z)^{-1}, \ 1 - z^{-1}, \ z^{-1}, \ 1 - z, \ (1 - z^{-1})^{-1}, \qquad (9.1)$$

corresponding to the permutations 1, $(0 \ 1 \ \infty)$, $(0 \ \infty \ 1)$, $(0 \ \infty)$, $(0 \ 1)$, and $(1 \ \infty)$ respectively. It follows that the possible values of λ_ρ are

$$\lambda, \ (1 - \lambda)^{-1}, \ 1 - \lambda^{-1}, \ \lambda^{-1}, \ 1 - \lambda, \ (1 - \lambda^{-1})^{-1}. \qquad (9.2)$$

- **Ex.9.11:** *By using Exercises 8.19 and 9.1, or otherwise, show that the six maps in (9.1) form a subgroup G of $\mathcal{M}^+(\mathbb{C})$.*

- **Ex.9.12:** *Are the six values in (9.2) necessarily distinct? Use orbit–stabilizer methods (see Section 6.7) to investigate the possibilities.*

- **Ex.9.13:** *Draw a diagram showing the real axis, the line $|z| = |z - 1|$, and the circles $|z| = 1$ and $|z - 1| = 1$. Where do these curves meet? What are their images under the six maps in (9.1)? What do the six maps do to the various regions lying between these curves?*

To get a better picture of what the group G of the last three exercises is doing, the reader is invited place our plane $\mathbb{C}$, or $\mathbb{R}^2$, inside $\mathbb{R}^3$, and consider the stereographic projection f from the sphere with the line segment from ω to $\overline{\omega}$ as diameter, to our plane; cf. Section 7.6. Every point of the sphere corresponds to a point of the plane except the 'north pole', the point of projection, which corresponds to ∞. The pairs of points $\{\omega, \overline{\omega}\}$, $\{-1, 1\}$, $\{0, 2\}$, and $\{\frac{1}{2}, \infty\}$ will now all correspond to pairs of antipodal (i.e., diametrically opposite) points on the sphere. Then G corresponds to a group of rotations of the sphere, with 3-poles at the images under f^{-1} of ω and $\overline{\omega}$, and 2-poles at the images under f^{-1} of $-1, 0, \frac{1}{2}, 1, 2$, and ∞. These last six image points will lie at the vertices of a regular hexagon, with the group being D_3, the rotation group (in $\mathbb{R}^3$) of the equilateral triangle with vertices at the images of 0, 1, and ∞.

As a first application of the theory of cross-ratios, we prove a celebrated theorem of Ptolemy:

Proposition 9.10 (Ptolemy's theorem). *Let A, B, C, D be distinct points in a plane. Then*

$$|AB||CD| + |BC||DA| \geq |AC||BD|,$$

with equality occurring if and only if either *ABCD is a convex cyclic tetragram or else A, B, C, D lie on a line in the given order or a cyclic permutation of it.*

Proof. Let A, B, C, D be represented by the complex numbers a, b, c, d respectively, and put $\lambda = \{a, b\,; c, d\}$. It follows that $\{b, c\,; a, d\} = 1 - \lambda^{-1}$ and $\{a, b\,; d, c\} = \lambda^{-1}$, so that, since $(1 - \lambda^{-1}) + \lambda^{-1} = 1$, we have $\{b, c\,; a, d\} + \{a, b\,; d, c\} = 1$. Taking absolute values and applying the triangle inequality gives

$$|\{b, c\,; a, d\}| + |\{a, b\,; d, c\}| \geq 1, \tag{9.3}$$

with equality holding if and only if $1 - \lambda^{-1}$ and λ^{-1} are real and lie between 0 and 1; and this holds if and only if $\lambda \in \mathbb{R}$ and $\lambda > 1$. Applying the Möbius transformation f with $f(a) = \infty$, $f(b) = 0$, and $f(c) = 1$, we have $f(d) = \lambda$, and A, B, C, D lie in the given cyclic order on a line-or-circle if and only if $\infty, 0, 1, \lambda$ lie on the real axis in the given order, that is, if and only if $\lambda \in \mathbb{R}$ and $\lambda > 1$ once again.

To finish the proof, we just need to write out (9.3), which says

$$\left|\frac{(a-b)(d-c)}{(a-c)(d-b)}\right| + \left|\frac{(d-a)(c-b)}{(d-b)(c-a)}\right| \geq 1, \quad \text{or} \quad \frac{|AB||CD|}{|AC||BD|} + \frac{|BC||DA|}{|AC||BD|} \geq 1,$$

and the result follows on multiplying through by $|AC||BD|$. $\qquad\square$

9.1.3 Angles

As a second application of cross-ratios, we shall show that Möbius transformations preserve *angles*. By this we mean that if, for example, the angle between two lines ℓ and m is θ, and $f \in \mathcal{M}^+(\mathbb{C})$, then the angle between $f(\ell)$ and $f(m)$ is also θ. However, either or both of $f(\ell)$, $f(m)$ might be circles, so we have to say what we mean by the angle between two circles, or between a line and a circle. But before we get to this, there is already a difficulty with two lines. If ℓ and m meet at A, and if B, B' are on ℓ, on opposite sides of A, and C, C' are on m, on opposite sides of A, and if we put $\angle BAC = \theta$, then $\angle B'AC' = \theta$ also, but $\angle CAB' = \angle C'AB = \pi - \theta$, the supplementary angle. Which do we say is the angle between ℓ and m, θ or $\pi - \theta$?

The way out of the difficulty is to use rotations: there is a rotation g, fixing A, that sends the line ℓ to the line m (Fig. 9.2), and if θ is the angle of rotation, we shall say that θ is the angle *between ℓ and m* (at A). If A is represented by $a \in \mathbb{C}$, then we have $g(z) - a = e^{i\theta}(z - a)$, for all $z \in \mathbb{C}$. There are two crucial things to notice here. First, $g^{-1}(m) = \ell$, and $g^{-1}(z) - a = e^{-i\theta}(z - a)$, for all $z \in \mathbb{C}$, so the angle between m and ℓ is not θ, but is $-\theta$, so the order of naming ℓ and m is important, and we must work here with *signed* angles. Second, if h is the composite of g with a half-turn about A,

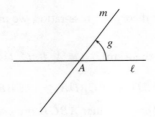

Fig. 9.2   The rotation g takes ℓ to m

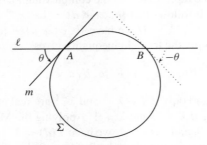

Fig. 9.3   The angle between ℓ and Σ at A and at B

then $h(\ell) = m$, and $h(z) - a = e^{i(\theta+\pi)}(z - a)$, so that the angle between ℓ and m now seems to be $\theta + \pi$. We therefore have to take values of angles modulo π, that is, we must regard angles as the same here if they differ by an integer multiple of π.

As before, let B, B' be on ℓ, on opposite sides of A, and let C, C' be on m, similarly; and to make life easier, assume $|AB| = |AB'|$ and $|AC| = |AC'|$. Let b, b', c, c' be the complex numbers representing B, B', C, C' respectively, so that $b' - a = -(b - a)$ and $c' - a = -(c - a)$. Since (with g as before) $g(\ell) = m$, we have that $g(b)$ is on m. Interchanging C and C' if necessary, assume $g(b)$ and c are on the same side of a, so that a suitable central dilation with centre a will send $g(b)$ to c. That is, $c - a = k(g(b) - a) = ke^{i\theta}(b - a)$ for some $k \in \mathbb{R}$, where in fact $k = |c-a|/|b-a|$. Note that we also have $c'-a = ke^{i\theta}(b'-a)$, and $c'-a = -ke^{i\theta}(b-a)$ (or, if you prefer, $c' - a = ke^{i(\theta+\pi)}(b - a)$), and $c - a = -ke^{i\theta}(b' - a)$ (or, if you prefer, $c - a = ke^{i(\theta+\pi)}(b' - a)$). So, with $\theta = \angle BAC$, we now have, for example, $\angle B'AC = \theta - \pi \equiv \theta$ (modulo π), and $\angle CAB' = \pi - \theta \equiv -\theta$ (modulo π).

With this in mind, we can define the angle between circles, or between a line and a circle. If the line ℓ meets the circle Σ at A and B, then let m be the tangent to Σ at A. We define the angle θ between ℓ and Σ at A to be the angle between ℓ and m (Fig. 9.3). Notice that reflection in the perpendicular bisector of AB shows that the angle between ℓ and Σ at B is $-\theta$, not θ; and of course, the angle between Σ and ℓ (changing the order) at A is $-\theta$, also. Next, if Σ, Σ' are circles meeting at A and B, then we define the angle θ between Σ and Σ' at A to be the angle between their respective tangents at A (Fig. 9.4). Once again, note that the angle between Σ and Σ' at B is then $-\theta$. Finally, if

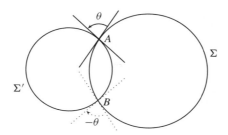

Fig. 9.4   The angle between Σ and Σ' at A and at B

two circles meet at only one point, then they touch, and their respective tangents at that point coincide, so the angle between them (in either order) is zero; and similarly for a line and circle that meet at only one point.

For completeness, if two lines ℓ and m meet at A, then they meet again at $B = \infty$, so to be consistent with the above we adopt the convention that the angle between ℓ and m at ∞ is $-\theta$, where θ is the angle between ℓ and m at A. If ℓ and m meet *only* at ∞, then they are parallel, and the angle between them (in either order) is zero.

Lemma 9.11.   *Let the circle Σ and the line ℓ meet in the two points A and B, and let $f : z \mapsto (z - a)/(z - b)$, where a, b are the complex numbers representing A, B respectively. Then $f(\Sigma)$ is a line through the origin, O, and the angle between the real axis and this line (at O) is equal to the angle between ℓ and Σ at A.*

Proof.   Since $f(b) = \infty$, the image $f(\Sigma)$ must be a line, and since $f(a) = 0$, it is a line through O. Call this line m.

Let the angle between ℓ and Σ be θ, so that if P is on the tangent to Σ at A, then $\angle BAP = \theta$. By the alternate segment theorem, if C is a point of Σ, on the opposite side of AB from P, then $\angle BCA = \theta$ also. Thus $a - c = ke^{i\theta}(b - c)$ for some $k \in \mathbb{R}$, from which $(c - a)/(c - b) = ke^{i\theta}$. But $(c - a)/(c - b) = f(c)$, which is on m, so $ke^{i\theta}$ is on m, whence θ is the angle between the real axis and m. $\square$

- **Ex.9.14:**   *What are the images under f of the line ℓ, the tangent to Σ at B, and the tangent to Σ at A?*

The next proposition shows precisely how to use the cross-ratio to measure the angle between a line-or-circle and another line-or-circle:

Proposition 9.12.   *For $i = 1, 2$ let Σ_i be a line-or-circle (with $\Sigma_1 \neq \Sigma_2$), and let Σ_1, Σ_2 meet at A and B. Let C be on Σ_1 and let D be on Σ_2, such that A, B, C, D are distinct, and let these points be represented respectively by a, b, c, $d \in \mathbb{C} \cup \{\infty\}$. Let $\{a, b\,; c, d\} = ke^{i\theta}$ where $k \in \mathbb{R}$. Then θ is the angle between Σ_2 and Σ_1 at A. (Note the order. Note also that we do not specify $k > 0$, since we are only interested in the value of θ modulo π, and replacing θ by $\theta \pm \pi$ involves replacing k by $-k$.)*

Proof.   Denote the line AB by ℓ. The proof involves many cases: a, b, c, d are either all finite or else *one* of them is infinite; and Σ_1 is either a line or a circle, and similarly for Σ_2.

Case 1: Suppose $d = \infty$, so that $\Sigma_2 = \ell$. Then $ke^{i\theta} = \{a, b\,; c, d\} = (c - a)/(c - b)$, so by Lemma 9.11 the angle between Σ_2 ($= \ell$) and Σ_1 at A is θ, as required.

Case 2: Suppose $c = \infty$. Then $\{a, b\,; d, c\} = k^{-1}e^{-i\theta}$, so by case 1 the angle between Σ_1 and Σ_2 (note the order) at A is $-\theta$, and the result follows.

Case 3: Suppose $b = \infty$. Then both of Σ_1, Σ_2 are lines, meeting at the (finite) point A. We have $ke^{i\theta} = \{a, b\,; c, d\} = (c - a)/(d - a)$, so that $\theta = \angle DAC$, the angle between Σ_2 and Σ_1 at A, as required.

Case 4: Suppose $a = \infty$. Then both of Σ_1, Σ_2 are lines, meeting at the (finite) point B. We have $\{b, a\,; c, d\} = k^{-1}e^{-i\theta}$, so by case 3 the angle between Σ_2 and Σ_1 at B is $-\theta$, and this means that, by convention, the angle between Σ_2 and Σ_1 at ∞ ($= A$) is θ, as required.

Case 5: Now suppose a, b, c, d are finite. Σ_1 and Σ_2 cannot both be lines—why?—so suppose Σ_2 is a line (so that $\Sigma_2 = \ell$), and Σ_1 is a circle. We have $\{a, b\,; c, d\} = ke^{i\theta}$, so that $(c - a)/(c - b) = k'e^{i\theta}$, where $k' = k(d - a)/(d - b)$. Since A, B, D are collinear, k' is real, and so by Lemma 9.11 the angle between Σ_2 ($= \ell$) and Σ_1 at A is θ, as required.

Case 6: Suppose a, b, c, d are finite, Σ_1 is a line, and Σ_2 is a circle. Then $\{a, b\,; d, c\} = k^{-1}e^{-i\theta}$, so by case 5 the angle between Σ_1 and Σ_2 (note the order) at A is $-\theta$, and the result follows.

Case 7: Finally, suppose a, b, c, d are finite, and that both Σ_1 and Σ_2 are circles. By Lemma 9.11 we have $(c - a)/(c - b) = k_1 e^{i\theta_1}$ and $(d - a)/(d - b) = k_2 e^{i\theta_2}$, where $k_1, k_2 \in \mathbb{R}$, the angle between ℓ and Σ_1 at A is θ_1, and the angle between ℓ and Σ_2 at A is θ_2. But this means the angle between Σ_2 and ℓ at A is $-\theta_2$, so that the angle between Σ_2 and Σ_1 at A is $(-\theta_2) + \theta_1$, or $\theta_1 - \theta_2$. But

$$ke^{i\theta} = \{a, b\,; c, d\} = \frac{c - a}{c - b} \bigg/ \frac{d - a}{d - b} = \frac{k_1}{k_2} e^{i(\theta_1 - \theta_2)},$$

from which we deduce that $\theta \equiv \theta_1 - \theta_2$ modulo π, and our proof is complete. $\square$

Corollary 9.13.   *Möbius transformations preserve angles.*

Proof.   Let f be a Möbius transformation, and let each of Σ_1, Σ_2 be a line-or-circle, meeting in A. If they have no other common point, finite or infinite, then they touch at A (or are parallel lines if $A = \infty$), and their images under f are either two circles only meeting at $f(A)$ (if $f(A) \neq \infty$), or a line and a circle only meeting at $f(A)$ (if $f(A) \neq \infty$), or two parallel lines (if $f(A) = \infty$). In each case, the angle between Σ_2 and Σ_1 at A is zero, and so is the angle between their respective images at $f(A)$.

If Σ_1 and Σ_2 meet in a second point B, then choose C on Σ_1 and D on Σ_2, so that A, B, C, D are distinct. Representing these points respectively by $a, b, c, d \in \mathbb{C} \cup \{\infty\}$, we have, by Proposition 9.12, $\{a, b\,; c, d\} = ke^{i\theta}$, where $k \in \mathbb{R}$ and θ is the angle between Σ_2 and Σ_1 at A. But $\{a, b; c, d\} = \{f(a), f(b); f(c), f(d)\}$, by Proposition 9.8, so that $\{f(a), f(b); f(c), f(d)\} = ke^{i\theta}$. Now $f(\Sigma_1)$ is a (in fact, *the*) line-or-circle through $f(a)$, $f(b)$, and $f(c)$; and $f(\Sigma_2)$ is a (*the*) line-or-circle through $f(a)$, $f(b)$, and $f(d)$, so that, by Proposition 9.12 again, the angle between $f(\Sigma_2)$ and $f(\Sigma_1)$ at $f(A)$ is also θ, as required. □

We are now going to prove a sort of converse of Corollary 9.13, namely that if a transformation of $\mathbb{C} \cup \{\infty\}$ sends each line and each circle to a line-or-circle, preserving angles, then it must be a Möbius transformation. One might object that a reflection f, being an isometry, *also* preserves angles, but it would be more accurate to say that it *reverses* angles, since if the angle between Σ_1 and Σ_2 at A is θ, then the angle between $f(\Sigma_1)$ and $f(\Sigma_2)$ at $f(A)$ is clearly $-\theta$, not θ. Since any conjugate Möbius transformation is the composite of the reflection $z \mapsto \bar{z}$ and a Möbius transformation, we see that every such map reverses angles also; but because $-\pi/2 \equiv \pi/2$ modulo π, conjugate Möbius transformations preserve *right* angles; and so do Möbius transformations, of course.

Proposition 9.14. *Let f be a transformation of $\mathbb{C} \cup \{\infty\}$ that sends each line and each circle to a line-or-circle. Then if f preserves right angles, it is either a Möbius transformation or a conjugate Möbius transformation.*

Proof. Let $f(0) = a$, $f(1) = b$, and $f(\infty) = c$. By Corollary 9.5 there is a Möbius transformation g with $g(0) = a$, $g(1) = b$, and $g(\infty) = c$. Put $h = g^{-1}f$, and h is a map that fixes 0, 1, and ∞, and sends each line and each circle to a line-or-circle; but since $h(\infty) = \infty$, then h in fact sends each line to a line, and each circle to a circle. So in particular, the restriction $h : \mathbb{C} \to \mathbb{C}$ is a collineation, and so, by Theorem 8.37, h is an affine transformation; and since f and g^{-1} both preserve right angles, so does h.

Since h fixes 0 and 1, it fixes every real point, and since it preserves right angles, every line orthogonal to the real axis and every circle with centre on the real axis is invariant under h. So the imaginary axis and the circle $|z| = 1$ are invariant under h, and hence so is their intersection, the two-element set $\{i, -i\}$.

Case 1: If $h(i) = i$, then h is an affine transformation fixing 0, 1, and i, so $h = 1$, by Exercise 8.17. Thus $f = g$, which is a Möbius transformation.

Case 2: If $h(i) = -i$, then write $\bar{h} : z \mapsto \overline{h(z)}$, and $\bar{h}$ is an affine transformation fixing 0, 1, and i. Thus $\bar{h} = 1$, by Exercise 8.17. So $\overline{g^{-1}f(z)} = z$, for all $z \in \mathbb{C}$, whence $g^{-1}f(z) = \bar{z}$, and $f(z) = g(\bar{z})$, for all $z \in \mathbb{C}$. So in this case, f is a conjugate Möbius transformation. □

We now fulfil a promise made at the end of Section 7.6.

Corollary 9.15. *Let f be stereographic projection from a sphere Σ to $\mathbb{C} \cup \{\infty\}$, and let g be a symmetry of Σ. Put $h = fgf^{-1} : \mathbb{C}\cup\{\infty\} \to \mathbb{C}\cup\{\infty\}$. Then if g is direct, h is a Möbius transformation, and if g is opposite, h is a conjugate Möbius transformation.*

Proof. We know from Proposition 7.23 that f sends circles to lines or circles, and f^{-1} sends each line-or-circle to a circle. Further, f preserves angles, or—since in Section 7.6

we were not bothering about signs of angles—maybe f reverses angles. (Or maybe it preserves some angles and reverses others?) We can be certain, though, that f preserves *right* angles. Next g, being an isometry, sends circles to circles and preserves right angles. Thus h sends each line and each circle to a line-or-circle, and preserves right angles, so, by Proposition 9.14, $h \in \mathcal{M}(\mathbb{C})$.

Now if g is a reflection, it has a fixed circle on Σ, and therefore h has a fixed line-or-circle. But this means (since certainly $h \neq 1$) that h must be a conjugate Möbius transformation, by Proposition 9.3. More generally, $g = g_1 g_2 \ldots g_r$ where each g_i is a reflection—we could insist that $r \leq 3$, but this is unimportant—and g is direct if r is even, and opposite if r is odd. But now $h = h_1 h_2 \ldots h_r$ where $h_i = f g_i f^{-1}$ is a conjugate Möbius transformation, for each i, as above. Thus h is a Möbius transformation if r is even—that is, if g is direct—and is a conjugate Möbius transformation if r is odd—that is, if g is opposite. □

- Ex.9.15: *Show that the map $g \mapsto h$ (with g, h as in Corollary 9.15) gives a group homomorphism $\alpha : O_3 \to \mathcal{M}(\mathbb{C})$. Is α injective, or surjective?*

9.1.4 Applications

First we shall look at *coaxal circles*, defined below. We need a lemma:

Lemma 9.16. *Let A, B be inverse points with respect to the line-or-circle Σ_1, and let Σ_2 be any line-or-circle through A and B. Then Σ_1 and Σ_2 are orthogonal, that is, they meet at an angle of $\pi/2$ (Fig. 9.5).*

Proof. Let f be a Möbius transformation with $f(A) = O$ and $f(B) = \infty$. Then $f(\Sigma_1)$ is a line-or-circle having O, ∞ as inverse points, that is, it is a circle with centre

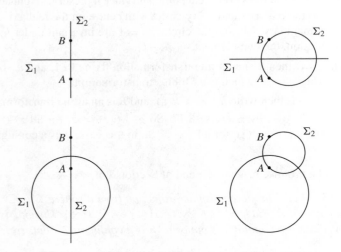

Fig. 9.5 The various cases of Lemma 9.16

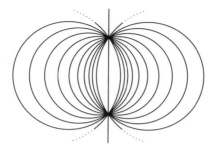

Fig. 9.6   An intersecting coaxal system

O. Next, $f(\Sigma_2)$ is a line-or-circle through O and ∞, that is, it is line through the centre of $f(\Sigma_1)$, a diameter. Thus $f(\Sigma_1)$ and $f(\Sigma_2)$ are orthogonal, and hence so are Σ_1 and Σ_2.   ☐

- Ex.9.16:   *Let Σ_1 and Σ_2 be a pair of orthogonal circles, and let a line through the centre of Σ_1 meet Σ_2 at A and B. Prove that A and B are inverse points with respect to Σ_1. What does this say about the image of Σ_2 under inversion with respect to Σ_1?*

Given A and B, consider the set of all circles through A and B. Any two of these circles have the line AB as their radical axis, and so this set of circles, together with the line AB, is called an *intersecting coaxal system* (Fig. 9.6). Note that each point of $\mathbb{C} \cup \{\infty\}$, apart from A and B, lies on exactly one member of the system.

If a circle Σ has A, B as inverse points, it is orthogonal to *every* member of the above coaxal system. Let A, B be represented by a, $b \in \mathbb{C}$, respectively. Then Σ has equation $|z - a| = k|z - b|$, for some $k \in \mathbb{R}$, $k > 0$, and each such value of k determines such a circle—except when $k = 1$, for $|z - a| = |z - b|$ is a line, the perpendicular bisector of AB. This line is the radical axis for any two circles $|z - a| = k_1|z - b|$ and $|z - a| = k_2|z - b|$, and these do not meet at all if $k_1 \neq k_2$. So we have a *non-intersecting coaxal system* (Fig. 9.7), consisting of all circles $|z - a| = k|z - b|$, together with the line $|z - a| = |z - b|$. Note that *every* line-or-circle in the one system is orthogonal to *every* line-or-circle in the other, and so they are called *orthogonal* systems (Fig. 9.9). Every point of $C \cup \{\infty\}$, apart form A and B, lies on exactly one member of the second system (OK?), so there is exactly one member of *each* of the two systems through each such point. The limit as $k \to 0$ of $|z - a| = k|z - b|$ is the *point-circle* $|z - a| = 0$—its radius is zero—and the limit as $k \to \infty$ is the point-circle $|z - b| = 0$. (OK?) So A, B are called the *limiting points* of the second system.

There is a third type of coaxal system. If two circles touch at A, then the common tangent ℓ at A is their radical axis, and so the complete set of circles through A with tangent ℓ, together with ℓ itself, is a coaxal system which we shall call *tangential* (Fig. 9.8), to distinguish it from the intersecting and non-intersecting types above. If m is the line through A orthogonal to ℓ, then the circles through A with tangent m, together

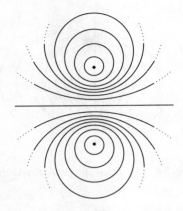

Fig. 9.7   A non-intersecting coaxal system

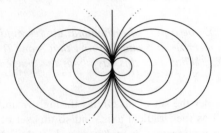

Fig. 9.8   A tangential coaxal system

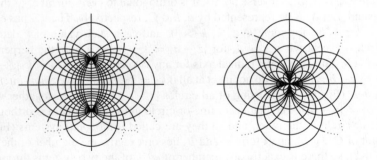

Fig. 9.9   Two pairs of orthogonal coaxal systems

with m, form another tangential system, and these two systems are orthogonal (Fig. 9.9). They can be thought of as the limiting case of the intersecting and non-intersecting systems determined by A and B, as $B \to A$.

If we apply a Möbius transformation f to the two orthogonal coaxal systems determined by A and B, we clearly get the two orthogonal coaxal systems determined by $f(A)$

and $f(B)$, provided these are finite. If on the other hand $f(B) = \infty$ (say), then the intersecting system is mapped to the set of all lines through $f(A)$, and the non-intersecting system is mapped to the set of all circles with centre $f(A)$. So clearly we should regard a set of concentric circles (i.e., circles with a common centre) as a (non-intersecting) coaxal system, and the set of all their diametral lines as the orthogonal (intersecting) coaxal system, in spite of the fact that neither system now has a radical axis. If we apply f to two orthogonal tangential systems with common point A, we get two orthogonal tangential systems with common point $f(A)$, unless $f(A) = \infty$, when we get two sets of mutually orthogonal parallel lines, and no circles at all!

There are two other ways of handling coaxal systems which we sketch here. First, linear systems. If $p_j(x, y) = 0$ is the cartesian equation of the line-or-circle Σ_j, $j = 1, 2$, then $p_1(x, y) + \lambda p_2(x, y) = 0$ is the equation of another line-or-circle, for all $\lambda \in \mathbb{R} \cup \{\infty\}$, giving what is called a *linear system* of curves. (We get Σ_1 from $\lambda = 0$, and Σ_2 by letting $\lambda \to \infty$—OK?) If Σ_1 and Σ_2 meet at A and B, then each of these new curves goes through A and B also, and we have an intersecting coaxal system. If Σ_1 and Σ_2 touch at A, then they each touch all of these new curves at A, and we have a tangential coaxal system. If Σ_1 and Σ_2 do not meet, then we leave it to the reader to show that we get a non-intersecting coaxal system, and a whole heap of imaginary circles besides. So we should now prohibit some values of λ; and at the boundary of the allowed and prohibited values will be two values that give the equations of two point-circles, the limiting points of the system.

- Ex.9.17: *Write down the cartesian equation of the general circle of the coaxal system having (α_1, β_1), (α_2, β_2) as its limiting points.*

A second approach is via stereographic projection. Let Σ be a sphere in $\mathbb{R}^3$, and let $f : \Sigma \to \mathbb{C} \cup \{\infty\}$ be stereographic projection from some point of Σ to the corresponding 'equatorial' (inversive) plane. Suppose a line ℓ in $\mathbb{R}^3$ is given. Then the set of planes containing ℓ meet Σ in a system of circles. If ℓ meets Σ in A and B, then the circles all go through A and B, and their images under f form an intersecting coaxal system, through $f(A)$ and $f(B)$. If ℓ touches Σ at A, then the circles all touch ℓ at A, and their images are a tangential coaxal system through $f(A)$. Finally, if ℓ and Σ do not meet, then no two of the system of circles on Σ meet, and their images are a non-intersecting coaxal system. In this last case, some of the planes containing ℓ do not meet Σ, or (equivalently) they meet Σ in imaginary circles. At the boundary between the planes with real and imaginary intersections with Σ are two planes that *touch* Σ, at two points A and B, and their images $f(A)$ and $f(B)$ are the limiting points of the non-intersecting system. We leave the reader to supply the details, and also to decide what happens if we take the limiting case as $\ell \to \infty$, whatever that means.

- Ex.9.18: *Given two non-intersecting circles Σ_1 and Σ_2, show that there is a Möbius transformation f such that $f(\Sigma_1)$ and $f(\Sigma_2)$ are concentric circles. Suppose now that Σ_1 lies inside Σ_2. Choose arbitrarily a circle Λ_1 touching both Σ_1 and Σ_2, and then a succession of circles Λ_j, $j \geq 2$, each touching both Σ_1 and Σ_2, but also touching Λ_{j-1}, and with $\Lambda_j \neq \Lambda_{j-2}$ for $j \geq 3$.*

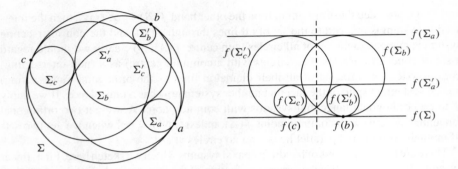

Fig. 9.10   Proposition 9.17, and its proof

Show that, if $\Lambda_n = \Lambda_1$ for some $n \geq 4$, then every other choice of Λ_1 will also give $\Lambda_n = \Lambda_1$. (This is called Steiner's porism.*)*

Coaxal circles allow us to visualize the behaviour of a Möbius transformation f. Suppose f fixes $a, b \in \mathbb{C}$, and let $g : z \mapsto (z - a)/(z - b)$, and put $h = gfg^{-1}$. Then $h \in \mathcal{M}^+(\mathbb{C})$, and h fixes 0 and ∞, so h is a direct similarity, hence a spiral similarity, and there exist $r \in \mathbb{R}$, $r > 0$, and an angle θ, such that $h(z) = re^{i\theta}z$, for all $z \in \mathbb{C}$. Then h permutes the members of the coaxal system consisting of lines through 0; and they are all invariant if $\theta = 0$, that is, if h is a central dilation. Next, h permutes the members of the orthogonal coaxal system consisting of concentric circles, with centre 0; and they are all invariant if $r = 1$, that is, if h is a rotation. Going back to f, we see that $gf(z) = hg(z)$, or $(f(z) - a)/(f(z) - b) = re^{i\theta}(z - a)/(z - b)$. Then f will permute the elements of each of the two coaxal systems determined by a and b; if $\theta = 0$ it will leave invariant all the elements of the intersecting system, and if $r = 1$ it will leave invariant all the elements of the non-intersecting system. (This last case is the only one that can arise as the image of the rotation of a sphere, under stereographic projection.) We leave the reader to produce similar arguments for the case of a Möbius transformation with only one fixed point, and also to explore what can be said about conjugate Möbius transformations.

The above is part of some Grand Scheme, trying to understand What Is Going On. Here, by way of light relief, is an application that is just for enjoyment. It is taken from [14], which contains many other beautiful theorems from inversive geometry.

Proposition 9.17.   *Let Σ be a circle, and let a, b, c be on Σ. Let Σ_a be a circle inside Σ, touching Σ at a; let Σ_b be the circle touching Σ at b, and also touching Σ_a; and let Σ_c be the circle touching Σ at c, and also touching Σ_b. Then let Σ_a' be the circle touching Σ at a, and also touching Σ_c; let Σ_b' be the circle touching Σ at b, and also touching Σ_a'; and let Σ_c' be the circle touching Σ at c, and also touching Σ_b'. Then Σ_c' touches Σ_a (Fig. 9.10).*

Proof. If we apply a Möbius transformation to the entire figure, prove the theorem for the image figure, and transform back, that will be good enough! So apply a transformation f that sends a to ∞. So now $f(\Sigma)$ is a line, with $f(b)$ and $f(c)$ on $f(\Sigma)$; then $f(\Sigma_b)$ and $f(\Sigma_b')$ are circles touching $f(\Sigma)$ at $f(b)$; and $f(\Sigma_c)$, $f(\Sigma_c')$ are circles touching $f(\Sigma)$ at $f(c)$. Next, $f(\Sigma_a)$ and $f(\Sigma_a')$ are lines, parallel to $f(\Sigma)$; also $f(\Sigma_a')$ touches $f(\Sigma_c)$ and $f(\Sigma_b')$, and $f(\Sigma_a)$ touches $f(\Sigma_b)$. We must show that $f(\Sigma_a)$ touches $f(\Sigma_c')$.

Let g be reflection in the perpendicular bisector of the line joining $f(b)$ and $f(c)$. Then g leaves $f(\Sigma)$, $f(\Sigma_a)$, and $f(\Sigma_a')$ invariant, and swaps $f(b)$ and $f(c)$, so that it swaps $f(\Sigma_b')$ and $f(\Sigma_c)$ also. Then $f(\Sigma_b)$ touches $f(\Sigma)$ at $f(b)$ and also touches $f(\Sigma_c)$, so $gf(\Sigma_b)$ is a circle touching $gf(\Sigma)$ at $gf(b)$, that is, touching $f(\Sigma)$ at $f(c)$, and also touching $gf(\Sigma_c)$, that is, $f(\Sigma_b')$. But $f(\Sigma_c')$ is the unique circle that does these things, so we deduce that $gf(\Sigma_b) = f(\Sigma_c')$. Finally, $f(\Sigma_b)$ touches $f(\Sigma_a)$, so $gf(\Sigma_b)$ touches $gf(\Sigma_a)$, that is, $f(\Sigma_c')$ touches $f(\Sigma_a)$, and we have finished: transform back and we have that Σ_c' touches Σ_a. □

- Ex.9.19: *In the proof of Proposition 9.17, what is the transformation $f^{-1}gf$, and what does it do to the original figure?*

- Ex.9.20: *Suggest (in outline) a method for constructing a circle to touch three given circles.*

- Ex.9.21: *Given two non-intersecting circles (not concentric), explain how to construct the limiting points of the coaxal system to which they belong.*

9.1.5 Hyperbolic geometry

We are now in a position to describe one type of non-Euclidean geometry. We shall do this by constructing a *model* of the *hyperbolic plane*. A model should be thought of as a map, as in cartography; in this sense an atlas consists of a collection of models of (parts of) the earth, drawn on sheets of paper and bound into a book. This is not simply a matter of choosing a scale factor and applying a similarity, because the earth is curved, being (roughly) a sphere, whereas the paper is flat, and so the maps in an atlas all involve some distortion of what they represent, in that they cannot have just one scale factor for distances that applies to all parts of the map. For example, Mercator's projection can show almost the whole earth and is good for navigation, as it gives correct compass bearings of one point from another. But it is bad at areas, showing Greenland as several times bigger than India, which it is not. Various kinds of azimuthal projection (of which stereographic projection is one type) can be made to represent areas in proportion, or to preserve angles (and hence, in some sense, shapes), or to show great circles as straight lines. Small areas, such as the United Kingdom, are usually drawn using some form of conical projection, but even in maps as small as the Ordnance Survey 1 : 50 000 sheets used by tourists there is measurable distortion. For example, the north–south parallel grid-lines on the map cannot all represent true north. Grid line 00, near Swanage, *does* point exactly north–south, but go East by 18 miles to the Needles, at the western end of the Isle of Wight, and grid north now differs from true north by about $\frac{1}{3}°$; another 23 miles to the other end of the island adds another $\frac{1}{2}^{\circ}$ to the error. At the easternmost point of the UK, Lowestoft, the error is $3°$.

Suppose we wanted to make an atlas of the plane $\mathbb{R}^2$. We could divide it up into rectangles, apply suitable similarities, and bind the resulting maps into a book. There would be no distortion of lines, proportions of lengths or areas, but unfortunately the atlas would need infinitely many pages. In fact, there would be no saving in paper in applying a similarity: one could use a scale of $1:1$ (an isometry) at no extra cost! To get a finite (and therefore affordable) map, some distortion is inevitable. We could use stereographic projection, and represent $\mathbb{R}^2$ as points on a globe; this turns lines into circles and preserves angles, but does strange things to distances and areas. Alternatively, to produce a finite map that lies flat, we could proceed as follows. For $z \in \mathbb{C}$, write $z = re^{i\theta}$, where $r \in \mathbb{R}$, $r \geq 0$, and put $f(z) = se^{i\theta}$ where $s \in \mathbb{R}$, $s \geq 0$, and $s^2 = r^2/(r^2 + 1)$. We leave it to the reader to show that f is a bijection between $\mathbb{C}$ and the unit disc $\{z \in \mathbb{C} : |z| < 1\}$. So we can squeeze a map of the whole plane into a disc.

● Ex.9.22: *Show that the image under f of the line $y = k$, $k \neq 0$, is the half-ellipse given parametrically by $x = \cos\theta$, $y = (k/\sqrt{k^2 + 1})\sin\theta$, for $0 < \theta < \pi$. What happens if $k = 0$?*

From this exercise, the image under f of every line in the plane not through the origin is half an ellipse, where the major axis of the ellipse is a diameter of the unit disc. Parallel lines give rise to ellipses with the same major axis, and meet on the circle $|z| = 1$, which is *not* part of the disc, but can be thought of as representing 'points at infinity'. So in this model of $\mathbb{C}$ we can *see* which lines are parallel, and can see parallel lines 'meeting at infinity'. It is easy to check that lines which are not parallel always meet in exactly one (proper) point.

We are well used, in fiction, to the idea that an author can invent a country from pure imagination, and even draw maps of it to bring the imaginary landscape more vividly to life. In the context of the story, the map has perfect validity and self-consistency: it simply does not matter whether or not the place described actually exists. We are about to use the same device for the hyperbolic plane, by describing a model for it, due to Poincaré. Whether or not the model describes something real or something fictional is quite unimportant. In this plane, we shall have *hyperbolic* points, lines, and so on, which we shall call h-points, h-lines, and so on, so as not to get in a muddle with the previous meanings of the terms. Write $T = \{z \in \mathbb{C} : |z| = 1\}$, the unit circle, and $\mathcal{D} = \{z \in \mathbb{C} : |z| < 1\}$, the unit disc, consisting of all points inside T. We used $\mathcal{D}$ above as the points of a model for $\mathbb{R}^2$, but now we are going to use it for a model of the hyperbolic plane: an h-point is defined to be a point $z \in \mathcal{D}$. But this time the lines are nothing to do with ellipses: an h-line is defined to be a set $\Sigma \cap \mathcal{D}$, where Σ is a line-or-circle orthogonal to T.

The reader should check that through any two h-points there is a unique h-line (Fig. 9.11), but that given an h-line ℓ and an h-point P not on ℓ, there are many— infinitely many—h-lines through P that do not meet ℓ, that is, that have no h-point in common with ℓ. We might define two h-lines to be parallel if they have no h-point in

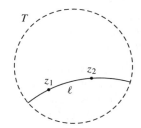

Fig. 9.11 The *h*-line ℓ through z_1 and z_2

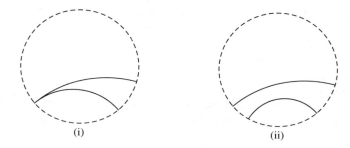

Fig. 9.12 (i) Parallel and (ii) ultra-parallel *h*-lines

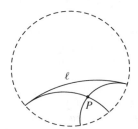

Fig. 9.13 The two *h*-lines through P, parallel to ℓ

common, but something a little more subtle works better. We say the *h*-lines $\Sigma_1 \cap \mathcal{D}$ and $\Sigma_2 \cap \mathcal{D}$ are *parallel* if the line-or-circle Σ_1 meets the line-or-circle Σ_2 on T, and *ultra-parallel* if Σ_1 and Σ_2 do not meet at all (Fig. 9.12). (So parallel *h*-lines are *not* ultra-parallel, and ultra-parallel *h*-lines are *not* parallel.) With this definition, there are *two* lines through P parallel to ℓ (Fig. 9.13). Think of the points of T as being points at infinity, just as in the model of $\mathbb{R}^2$ above. Then each *h*-line has two points at infinity, which are *not* *h*-points, of course. If Q is an *h*-point moving on ℓ, and we draw the *h*-line through P and Q, then the limit as $Q \to \infty$ of this *h*-line is an *h*-line parallel to ℓ. In Euclidean geometry, there would have been just one such limit, but in hyperbolic geometry there are two different limits as $Q \to \infty$ in one direction or the other.

• Ex.9.23: *Let Σ be the circle with cartesian equation $x^2 + y^2 + 2gx + 2fy + c = 0$.*
Give necessary and sufficient conditions on the real numbers c, f, g to
ensure that Σ is orthogonal to T.

We now need to define the *hyperbolic distance $d_h(z_1, z_2)$* between two h-points z_1, z_2.
Just as calculating distances from a map of the world is not simply a matter of applying
a scale factor, so also hyperbolic distances are a little complicated to measure; but take
heart, as we have all that we need already to hand. Once d_h is defined, we can define
an h-isometry to be a transformation of $\mathcal{D}$ that preserves hyperbolic distance, and if
this behaves at all like Euclidean distance, then an h-isometry will be an h-collineation,
that is, it will send h-lines to h-lines. But an h-line is (part of) a line-or-circle, and we
already know lots of maps that send each line-or-circle to a line-or-circle, namely the
Möbius and conjugate Möbius transformations. We shall define d_h in such a way that the
h-isometries are precisely the Möbius and conjugate Möbius transformations that leave
$\mathcal{D}$ invariant. Such a map f must also leave T invariant, and since it preserves angles, it
sends every line-or-circle orthogonal to T to a line-or-circle orthogonal to T, and thus it
is an h-collineation.

What does such an h-collineation preserve? We know it preserves or conjugates cross-
ratios (Proposition 9.8), but this depends on four points, whereas d_h has to depend on just
two h-points, z_1 and z_2. But these two h-points determine a unique h-line, which meets
T at two infinite points ω_1 and ω_2; so can we use the quantity $\lambda = \{\omega_1, \omega_2 \,; z_1, z_2\}$ to
define d_h? The *good* news is that λ is real (Proposition 9.9), and indeed $\lambda > 0$. The *bad*
news is that (i) if $z_1 = z_2$, then $\lambda = 1$, whereas we would like the distance from z_1 to z_1
to be 0; and (ii) if we interchange z_1 and z_2 (or ω_1 and ω_2), then λ is replaced by λ^{-1},
whereas we would like the distance from z_2 to z_1 to be the *same* as the distance from z_1
to z_2, and not to depend on the order of naming ω_1 and ω_2. But now logarithms come to
our rescue, since $\ln 1 = 0$ and $\ln(\lambda^{-1}) = -\ln\lambda$, so we can get around our problem as
follows:

Proposition 9.18. *Let $z_1, z_2 \in \mathcal{D}$. Put $d_h(z_1, z_2) = |\ln\{\omega_1, \omega_2 \,; z_1, z_2\}|$, the hyper-
bolic distance from z_1 to z_2. (Here ω_1 and ω_2 are the points at infinity on the h-line
through z_1 and z_2.) Then*

(i) $d_h(z_1, z_1) = 0$, *for all $z_1 \in \mathcal{D}$.*
(ii) $d_h(z_1, z_2) > 0$, *for all $z_1, z_2 \in \mathcal{D}$ with $z_1 \neq z_2$.*
(iii) *If $z_1, z_2, z_3 \in \mathcal{D}$ are h-collinear, then $d_h(z_1, z_2) = d_h(z_1, z_3) + d_h(z_3, z_2)$ if z_3
lies between z_1 and z_2, and $d_h(z_1, z_2) < d_h(z_1, z_3) + d_h(z_3, z_2)$ otherwise.*
(iv) *If $z_1, z_2, z_3 \in \mathcal{D}$ are not h-collinear, then $d_h(z_1, z_2) < d_h(z_1, z_3) + d_h(z_3, z_2)$.*

Proof. (i) and (ii) are immediate, as is the fact that $d_h(z_1, z_2)$ does not depend on the
order of naming ω_1 and ω_2. For (iii), we have

$$\{\omega_1, \omega_2 \,; z_1, z_3\} \times \{\omega_1, \omega_2 \,; z_3, z_2\} = \frac{(z_1 - \omega_1)(z_3 - \omega_2)}{(z_1 - \omega_2)(z_3 - \omega_1)} \times \frac{(z_3 - \omega_1)(z_2 - \omega_2)}{(z_3 - \omega_2)(z_2 - \omega_1)}$$

$$= \frac{(z_1 - \omega_1)(z_2 - \omega_2)}{(z_1 - \omega_2)(z_2 - \omega_1)} = \{\omega_1, \omega_2 \,; z_1, z_2\},$$

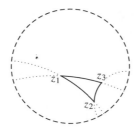

Fig. 9.14   A hyperbolic triangle

so that $\ln\{\omega_1, \omega_2 ; z_1, z_2\} = \ln\{\omega_1, \omega_2 ; z_1, z_3\} + \ln\{\omega_1, \omega_2 ; z_3, z_2\}$. If z_3 lies between z_1 and z_2, then the three logarithms have the same sign, and so $d_h(z_1, z_2) = d_h(z_1, z_3) + d_h(z_3, z_2)$. In the other cases, the two logarithms on the right have opposite signs, so $d_h(z_1, z_2) < d_h(z_1, z_3) + d_h(z_3, z_2)$. The next two exercises are needed for the proof of (iv):

- Ex.9.24: *Let z_0 be an h-point. Show that there is an h-isometry f with $f(0) = z_0$.*
- Ex.9.25: *Let z_0 be an h-point. Prove that $d_h(0, z_0) = \ln(1 + |z_0|)/(1 - |z_0|)$.*

We can replace z_1, z_2, z_3 by their images under any h-isometry f, and by Exercise 9.24 we can choose f so that $f(z_3) = 0$. (OK?) So we may as well assume $z_3 = 0$ in the first place, and prove that $d_h(z_1, z_2) < d_h(z_1, 0) + d_h(0, z_2)$. Now if ω_1, ω_2 are the points at infinity on the h-line through z_1 and z_2, we have

$$1 - |z_j| = |\omega_k| - |z_j| < |\omega_k - z_j| < |\omega_k| + |z_j| = 1 + |z_j|$$

for $j = 1, 2$ and $k = 1, 2$, by Lemma 3.16. Then

$$\frac{(z_1 - \omega_1)(z_2 - \omega_2)}{(z_1 - \omega_2)(z_2 - \omega_1)} = \left| \frac{(z_1 - \omega_1)(z_2 - \omega_2)}{(z_1 - \omega_2)(z_2 - \omega_1)} \right| < \left(\frac{1 + |z_1|}{1 - |z_1|} \right) \left(\frac{1 + |z_2|}{1 - |z_2|} \right),$$

so the result follows on taking logarithms, using Exercise 9.25.   □

Proposition 9.18(iv) is the hyperbolic version of the triangle inequality (Lemma 3.16). The h-points z_1, z_2, z_3 determine a *hyperbolic triangle* (Fig. 9.14), whose sides are the h-lines through pairs of these points, and whose side lengths are the corresponding hyperbolic distances. So the hyperbolic triangle inequality says that any side length of a hyperbolic triangle is less than the sum of the other two side lengths.

Notice that $\{\omega_1, \omega_2 ; z_1, z_2\} \to \infty$ as $z_2 \to \omega_1$, and $\{\omega_1, \omega_2 ; z_1, z_2\} \to 0$ as $z_2 \to \omega_2$, and in either case $d_h(z_1, z_2) \to \infty$; so for fixed z_1, we can say $d_h(z_1, z_2) \to \infty$ as $z_2 \to \infty$.

Hyperbolic geometry can now be developed by a careful analysis of the properties of h-isometries, rather as we dealt with isometries of $\mathbb{R}^2$ in Chapter 5. We shall provide a sketch of the beginning of this story, leaving the reader to fill in details, and to explore further. Let Σ be a line-or-circle orthogonal to T, and let the conjugate Möbius transformation f

be inversion with respect to Σ (if Σ is a circle) or reflection in Σ (if Σ is a line). Then f leaves $\mathcal{D}$ invariant and fixes every h-point of the h-line $\ell = \Sigma \cap \mathcal{D}$; and further, $f^2 = 1$. We shall call f the *h-reflection* in ℓ.

• **Ex.9.26:** *Let z_1, z_2 be distinct h-points. Then the set of h-points z with $d_h(z, z_1) = d_h(z, z_2)$ is an h-line, and reflection in this h-line swaps z_1 and z_2.*

It can now be shown that an h-isometry that fixes two h-points must fix every point of the h-line they determine (cf. Corollary 5.8), and so is trivial or is h-reflection in this line (cf. Corollary 5.9). Then it follows that every h-isometry is the composite of at most three h-reflections (cf. Corollary 5.7), and hence is a Möbius or conjugate Möbius transformation. Since these maps preserve angles, we can define angles in our hyperbolic plane by measuring them directly from our model $\mathcal{D}$, and now the right name for the line constructed in Exercise 9.26 is the *perpendicular h-bisector* of the h-line segment determined by z_1, z_2.

The next step would be to classify the h-isometries. A direct h-isometry f is the composite of h-reflections in two h-lines. There are *three* cases. If the h-lines meet in an h-point z, then z is fixed by f. (So is its inverse point with respect to T; but that is outside T.) In this case f is an h-rotation about z. If the two h-lines are parallel, then the only fixed point of f is on T, an infinite point. If the two h-lines are ultra-parallel, then f has two fixed points, both on T, and again no fixed h-points. Then, an opposite h-isometry, if not an h-reflection, must be the composite of three h-reflections, and there are rather a lot of possibilities according as each pair of the three h-lines in question meet, are parallel, or are ultra-parallel, so this seems like a good place to stop.

9.2   PROJECTIVE GEOMETRY

We have been using the idea of infinity in two quite separate ways. In common parlance, parallel lines 'meet at infinity'; but this is a fiction, and our Euclidean plane $\mathbb{R}^2$ has *no* infinite points, and parallel lines in $\mathbb{R}^2$ are precisely those pairs of lines that do *not* meet anywhere. Similarly in the hyperbolic plane, lines which are parallel do not meet (but, this time, not conversely), though to define what we mean by 'parallel' we have to consider the points on the boundary circle T of our model $\mathcal{D}$, and in this sense parallel lines in the hyperbolic plane also 'meet at infinity'. But, as in the Euclidean plane, there are *no* infinite points in the hyperbolic plane.

By contrast, in the inversive plane $\mathbb{C} \cup \{\infty\}$, the point ∞ is treated as a point like any other, and occurs in our equations and formulae. We write things like: let $f(z) = 2z/(z+1)$, so $f(-1) = \infty$, and $f(\infty) = 2$. Also, every line acquires an extra point, ∞, so that every pair of lines (parallel or not) meet at ∞, and this time we don't need those apologetic quotation marks. However, in this geometry lines which are not parallel meet in *two* points, a point of $\mathbb{C}$ and the point ∞; though this is entirely appropriate when we are considering lines as special sorts of circle, which is what inversive geometry is all about.

In projective (plane) geometry we are going to return to the idea that two lines ought to meet in no more than one point; in fact, two lines are *always* going to meet in one point. As a first attempt, start with $\mathbb{R}^2$. Every line in $\mathbb{R}^2$ has a direction (its gradient, if

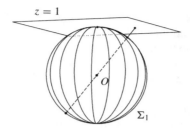

Fig. 9.15   Gnomonic projection

you like), and two distinct lines meet if they have different directions, or fail to meet if they have the same direction. We are going to add to our plane $\mathbb{R}^2$ a whole set of *new* points, namely the set of all possible directions. Thus each line acquires *one* of these new points (its own direction). Given two lines, either they have different directions, so that they still have exactly one point in common, or else they are parallel and now share the single *new* point, which is their common direction.

Of course, two points must lie on a unique line; this is fine if the two points are in $\mathbb{R}^2$, but we just invented some new points. Given a point of $\mathbb{R}^2$ and one of our new points, a direction, there is of course a unique line through the given point of $\mathbb{R}^2$ and the new point, that is, with the given direction. However, given *two* new points, there is as yet no line containing them both; so to remedy this we declare that the full set of new points constitutes a new line. Notice that this new line meets every other line in one new point, so that (suppressing the word 'new') we now have that every two points determine a unique line containing them both, and every two lines determine a unique point lying on them both.

The above rather wordy approach to projective geometry is all very well, but we need to be able to calculate with our points and lines, and so somehow, having just invented infinitely many new points and one new line, we have to have a suitable algebraic representation for them: coordinates, in fact. We shall do this by *gnomonic projection*, which is another of the azimuthal projections used by cartographers to make flat maps of the spherical earth; but we shall use it in reverse, to make a spherical map of our plane. Start by placing $\mathbb{R}^2$ inside $\mathbb{R}^3$ as the plane $z = 1$, which touches the unit sphere $\Sigma_1 : x^2 + y^2 + z^2 = 1$ at the 'north pole' $(0, 0, 1)$. So a point (x, y) of $\mathbb{R}^2$ is identified with $(x, y, 1) \in \mathbb{R}^3$. Join this point by a line to the origin O, and let this line meet Σ_1 at $\pm(x', y', z')$ (Fig. 9.15). Here $x' = \lambda x$, $y' = \lambda y$, and $z' = \lambda$, where $\lambda^2 = (x^2 + y^2 + 1)^{-1}$, so that $z' \neq 0$; and $x = x'/z'$, $y = y'/z'$. Thus every point of $\mathbb{R}^2$ corresponds to a pair of antipodal (diametrically opposite) points of Σ_1, and every such pair of points of Σ_1 (except those on the 'equator', where Σ_1 meets the plane $z = 0$) correspond to a point of $\mathbb{R}^2$. The banned pairs of points $\pm(x', y', 0) \in \Sigma_1$ are going to represent our new points.

Consider a line ℓ in $\mathbb{R}^2$, with equation $ax + by + c = 0$; identifying $\mathbb{R}^2$ with $z = 1$ in $\mathbb{R}^3$ as above, our line now lies where the plane $ax + by + cz = 0$ meets $z = 1$. The plane $ax + by + cz = 0$ passes through the origin, and so meets Σ_1 in a great circle, that is, a circle of maximum radius on Σ_1. Since a and b are not both zero, this great circle is not the equatorial circle, and indeed it meets the equatorial circle in a pair of antipodal points. These correspond to the new point on ℓ; and the equatorial circle corresponds to the new line, as described above. To find the coordinates of the new points on ℓ, we note that $ax + by + cz = 0$ meets $z = 0$ where $ax + by = 0$, so the pair of antipodal points in question are $\pm\lambda(b, -a, 0)$, where $\lambda^2 = (a^2 + b^2)^{-1}$.

Thus we get a rather curious model of the real projective plane $\mathbb{RP}^2$ ($\mathbb{R}^2$ together with its new points and new line) in which each point is represented by a pair of points, $\pm(x, y, z)$, and each line by a great circle. It would be possible to get rid of some of this peculiarity by throwing away half of Σ_1, and concentrating on the hemisphere $x^2 + y^2 + z^2 = 1$, $z \geq 0$. This would have the advantage that each point of $\mathbb{R}^2$ would be represented by a single point (x, y, z) on the hemisphere; but each *new* point would still give rise to a *pair* of points $\pm(x, y, 0)$ on the equatorial circle. Lines in $\mathbb{R}^2$ would now correspond to semicircles, but the *new* line would still correspond to a circle. This goes against what we are trying to achieve, to set up a geometry in which all points and all lines (new or not) have similar properties. We could now throw away half the points on the equatorial circle so as to leave ourselves with a model of $\mathbb{RP}^2$ in which each point (new or not) is represented by just one point of Σ_1, but the way of choosing which points to keep will be somewhat arbitrary. There are other disadvantages, too: for example, when θ is small and positive, the two points $(\pm\cos\theta, 0, \sin\theta)$ belong to our chosen hemisphere, but their limits as $\theta \to 0$ are $(\pm 1, 0, 0)$, only one of which is allowed to belong to our model. So notions of limit and continuity are troublesome with such a model: it is a discontinuous model. We have done violence to $\mathbb{RP}^2$, by slicing it open along a line (the new line) in order to place it inside $\mathbb{R}^3$.

It is a theorem from topology that it is impossible to form a model of $\mathbb{RP}^2$ inside $\mathbb{R}^3$ (with points standing for points) without some such violence. To get a 'nice' model, we need to use $\mathbb{R}^4$, and we now give a brief sketch of one way this can be done. We start with the map $f : \mathbb{RP}^2 \to \mathbb{R}^5$ given by $(x, y, z) \mapsto (x^2, y^2, z^2 + xy, yz, zx)$; notice that this is well-defined, because $f(x, y, z) = f(-x, -y, -z)$.

- Ex.9.27: *Suppose* (x, y, z), $(x_1, y_1, z_1) \in \Sigma_1$ *are such that* $f(x, y, z) = \lambda f(x_1, y_1, z_1)$ *for some* $\lambda \in \mathbb{R}$. *Prove that* $(x, y, z) = \pm(x_1, y_1, z_1)$.

Now let $g : \mathbb{R}^5 \setminus \{\mathbf{0}\} \to S^4$ be the map $\mathbf{v} \mapsto \mathbf{v}/|\mathbf{v}|$, where the length $|\mathbf{v}|$ of $\mathbf{v}$ is given by $|(v_1, v_2, v_3, v_4, v_5)|^2 = v_1^2 + v_2^2 + v_3^2 + v_4^2 + v_5^2$, and the 4-sphere S^4 consists of all vectors of length 1 in $\mathbb{R}^5$. By Exercise 9.27, the composite map $gf : \mathbb{RP}^2 \to S^4$ is an injection. It is easy to see that the point $(0, 0, 0, 0, 1)$ is not in the image of gf, so, taking this as 'north pole', let $h : S^4 \setminus \{(0, 0, 0, 0, 1)\} \to \mathbb{R}^4$ be stereographic projection to the 'equatorial hyperplane' $v_5 = 0$, which we identify with $\mathbb{R}^4$ in the obvious way: $h(v_1, v_2, v_3, v_4, v_5) = (v_1/(1 - v_5), v_2/(1 - v_5), v_3/(1 - v_5), v_4/(1 - v_5))$. We thus have a continuous injection $hgf : \mathbb{RP}^2 \to \mathbb{R}^4$, and if we wished we could take the image as a model for $\mathbb{RP}^2$ inside $\mathbb{R}^4$. For more details, see [24]. We take this no further, but instead proceed in the opposite direction, and make a model in which *more* than two points represent a point of our projective plane.

9.2.1 Homogeneous coordinates

Recall that we join a point of the plane $z = 1$ (representing $\mathbb{R}^2$) to the origin, and make this correspond to the two points in which the line meets the sphere Σ_1. If instead of this sphere we choose $r > 0$ and use the sphere $\Sigma_r : x^2 + y^2 + z^2 = r^2$, then we still get two antipodal points (but on Σ_r instead of Σ_1), the only difference being that if $(x, y) \in \mathbb{R}^2$ corresponds to $\pm(x', y', z') \in \Sigma_1$, then it corresponds to $\pm r(x', y', z') \in \Sigma_r$. (We are still identifying $(x, y) \in \mathbb{R}^2$ with $(x, y, 1) \in \mathbb{R}^3$; the fact that $z = 1$ does not touch Σ_r unless $r = 1$ does not affect the argument.) As it really does not matter what value r takes, we decide *not* to choose a value of r, but to allow the points $\pm r(x', y', z')$ to represent (x, y), for *all* $r > 0$; equivalently, (x', y', z') and (x'', y'', z'') represent the same point of our projective plane if and only if $x' = \lambda x''$, $y' = \lambda y''$ and $z' = \lambda z''$ for some $\lambda \neq 0$. This should induce a feeling of *déjà vu*: we have used triples of coordinates to represent points in a plane once before, and a glance back at Section 4.8 reminds us that barycentric coordinates have a similar property, namely that (ξ, η, ζ) and $(\lambda\xi, \lambda\eta, \lambda\zeta)$ represent the same point of $\mathbb{R}^2$ for all $\lambda \neq 0$. Further examination of that section reveals the frequent appearance of the triple $(x, y, 1)$, and other points on $z = 1$, and the reader should become convinced that there are connections here that are more than superficial.

Let us summarize. Each point $(x, y, z) \in \mathbb{R}^3$, except $(0, 0, 0)$, lies on a unique line through $(0, 0, 0)$, and all the points on this line, except $(0, 0, 0)$, are of the form $(\lambda x, \lambda y, \lambda z)$ for some $\lambda \neq 0$, and all represent the *same* point of our real projective plane $\mathbb{RP}^2$. In effect, the whole line in $\mathbb{R}^3$ stands for a single point in $\mathbb{RP}^2$, and likewise a line $ax + by + c = 0$ in $\mathbb{R}^2$ is represented by the plane $ax + by + cz = 0$, being the union of all the lines joining the origin to points of the given line $ax + by + c = 0, z = 1$. So the *points* of $\mathbb{RP}^2$ are *lines* (through O) in $\mathbb{R}^3$, and the *lines* of $\mathbb{RP}^2$ are the *planes* (through O) in $\mathbb{R}^3$.

The triples (x, y, z) are called *homogeneous coordinates* of the points of $\mathbb{RP}^2$ they represent. Given $(x, y) \in \mathbb{R}^2$, this is identified with $(x, y, 1) \in \mathbb{R}^3$, and $(x, y, 1)$ will do perfectly well as the homogeneous coordinates of the corresponding point of $\mathbb{RP}^2$, the other possibilities being $(\lambda x, \lambda y, \lambda)$ for $\lambda \neq 0$. Conversely, the homogeneous coordinates (x, y, z), where $z \neq 0$, represent the same point of $\mathbb{RP}^2$ as $(x/z, y/z, 1)$ (by choosing $\lambda = 1/z$), and this corresponds to $(x/z, y/z) \in \mathbb{R}^2$. The line $px + qy + r = 0$ in $\mathbb{R}^2$, a linear equation, yields the homogeneous linear equation $px + qy + rz = 0$ in $\mathbb{RP}^2$ (which is why the coordinates are called homogeneous coordinates); think of this as being done by replacing x, y by x/z, y/z respectively, and multiplying through by z, a process known as *homogenization*. The reverse process, *dehomogenization*, is easier: in $px + qy + rz = 0$, just put $z = 1$ to get the cartesian equation $px + qy + r = 0$. There is one homogeneous linear equation for which this fails, namely the equation $z = 0$; this is our new line, and the points $(x, y, 0)$ which lie on it are the new points. Let us investigate these a little further.

Choose a point (x_0, y_0) on the line $px + qy + r = 0$ in $\mathbb{R}^2$. Then $(x_0 + tq, y_0 - tp)$ also lies on the same line, for all t. Going over to homogeneous coordinates, $(x_0 + tq, y_0 - tp, 1)$ lies on $px + qy + rz = 0$, for all t, or, multiplying through by t^{-1}, $(x_0/t + q, y_0/t - p, 1/t)$ lies on $px + qy + rz = 0$ for all $t \neq 0$. Letting $t \to \infty$, we see that $(q, -p, 0)$ lies on $px + qy + rz = 0$, which is obvious anyway; but as t becomes large the cartesian point $(x_0 + tq, y_0 - tp)$ moves arbitrarily far from the origin, so that we might say it goes to infinity as $t \to \infty$. So our new points $(x, y, 0)$ can now be renamed *points at infinity*, and our new line $z = 0$ is the *line at infinity*. Note that the limit $(q, -p, 0)$ just obtained is the same whether $t \to +\infty$ or $t \to -\infty$: each line has just *one* point at infinity. We have now, as promised, many points at infinity, and, as promised, a perfectly good way of handling them algebraically. As an example, the cartesian lines $px + qy + r = 0$ and $px + qy + r' = 0$ are parallel. The corresponding homogeneous equations $px + qy + rz = 0$ and $px + qy + r'z = 0$ have the non-trivial common solution $(q, -p, 0)$, i.e., parallel lines meet at infinity.

Just as with barycentric coordinates (fig. 4.14), the points $(1, 0, 0)$, $(0, 1, 0)$ and $(0, 0, 1)$ are the vertices of the *triangle of reference*. The first two are at infinity on the x-axis and y-axis of $\mathbb{R}^2$, and the third is the origin of $\mathbb{R}^2$. The sides of the triangle of reference are $x = 0$, $y = 0$, and $z = 0$: the cartesian y-axis and x-axis, and the line at infinity. We shall see how to change to a different triangle of reference in Section 9.2.3.

Suppose $\mathbf{a} = (a_1, a_2, a_3)$, $\mathbf{b} = (b_1, b_2, b_3)$ are the homogeneous coordinates of two points of $\mathbb{RP}^2$, so that $\mathbf{a}$, $\mathbf{b}$ are linearly independent vectors in $\mathbb{R}^3$. Every vector of the plane through $\mathbf{0}$, $\mathbf{a}$ and $\mathbf{b}$ can be written $\mathbf{c} = \lambda \mathbf{a} + \mu \mathbf{b}$ for some $\lambda, \mu \in \mathbb{R}$, and this means that every point of the line ℓ through $\mathbf{a}$ and $\mathbf{b}$ in $\mathbb{RP}^2$ has homogeneous coordinates $\lambda \mathbf{a} + \mu \mathbf{b}$ for some $\lambda, \mu \in \mathbb{R}$, not both zero. If this were Euclidean geometry, we would have had to divide this expression by $\lambda + \mu$, because we would have had to insist that the scalars multiplying $\mathbf{a}$ and $\mathbf{b}$ add up to 1; but these are homogeneous coordinates, and multiplying or dividing by a scalar just gives an alternative set of coordinates for the *same* point of $\mathbb{RP}^2$. This means that each point of ℓ is represented by $(\lambda, \mu) \in \mathbb{R}^2$, with $(k\lambda, k\mu)$ representing the same point for all $k \in \mathbb{R}$, $k \neq 0$. So we have a system of one-dimensional homogeneous coordinates on ℓ, with reference points $\mathbf{a}$, $\mathbf{b}$ being represented by $(1, 0)$ and $(0, 1)$ respectively. If we put $\theta = \mu/\lambda$, then every point of ℓ, except $\mathbf{b}$, has homogeneous coordinates $\mathbf{a} + \theta \mathbf{b}$ for some $\theta \in \mathbb{R}$; if we write $\theta = \infty$ when $\lambda = 0$, then every point of ℓ corresponds to a unique $\theta \in \mathbb{R} \cup \{\infty\}$. Such a θ is called a *projective parameter* for ℓ.

Let us use these techniques to prove a famous theorem:

Proposition 9.19 (Desargues' theorem).   *Given $\triangle ABC$ and $\triangle A'B'C'$, label the meets of corresponding sides as follows: $L = BC \cdot B'C'$, $M = CA \cdot C'A'$, and $N = AB \cdot A'B'$. If AA', BB', CC' are concurrent, then L, M, N are collinear.*

Proof.   Let AA', BB', CC' meet at P (Fig. 9.16). If any two of A, A', P coincide, the proof is rather easy (check!), so assume A, A', P are distinct, and similarly for B, B', and P, and for C, C', and P. Let A, B, C, P have homogeneous coordinates $\mathbf{a}$, $\mathbf{b}$, $\mathbf{c}$, $\mathbf{p}$ respectively. Then the coordinates of A', B', C' are $\mathbf{a}' = \mathbf{p} + \lambda \mathbf{a}$, $\mathbf{b}' = \mathbf{p} + \mu \mathbf{b}$, and $\mathbf{c}' = \mathbf{p} + \nu \mathbf{c}$, respectively, for some $\lambda, \mu, \nu \in \mathbb{R}$. We now have $\mathbf{b}' - \mathbf{c}' = \mu \mathbf{b} - \nu \mathbf{c}$; here the left-hand side represents some point of the line $B'C'$, while the right-hand side represents some point of the line BC. So the point represented must be $BC \cdot B'C' = L$, and the coordinates of L may be taken as $\mathbf{l} = \mathbf{b}' - \mathbf{c}'$. Similarly the coordinates of M may

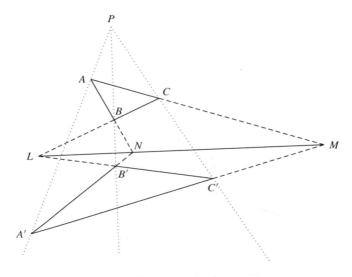

Fig. 9.16   Desargues' theorem

be taken as $\mathbf{m} = \mathbf{c}' - \mathbf{a}'$, and the coordinates of N as $\mathbf{n} = \mathbf{b}' - \mathbf{a}'$. But now $\mathbf{n} = \mathbf{l} + \mathbf{m}$, which means that N lies on LM, as required.   ☐

● Ex.9.28:   *The Desargues diagram can be relabelled in many ways, using exactly the same points and lines. For example, the triangles PBC and $A'NM$ already lie in the diagram, and PA', BN, CM meet at A. Then $BC \cdot NM = L$, $CP \cdot MA' = C'$, $PB \cdot A'N = B'$, and the line through these three points is already in the diagram. Thus the permutation $(AP)(B'N)(C'M)$ of the ten vertices 'preserves' Desargues' theorem. So how many such permutations are there, and (harder) can you recognize the subgroup of S_{10} so obtained?*

9.2.2   Duality

In $\mathbb{RP}^2$, two points determine a line, and two lines determine a point. This symmetry of properties is called *duality*, and allows us, under suitable circumstances, to interchange the notions of line and point, join and meet, collinearity and concurrency, and obtain a diagram or theorem *dual* to another. A triangle is a figure determined by three points (its vertices) which are not collinear; the dual figure is a *trilateral*, which consists of three lines (its sides) which are not concurrent.

The line $\ell : px + qy + rz = 0$ can be represented by the triple $(p, q, r) \in \mathbb{R}^3 \setminus \{(0, 0, 0)\}$, called *line coordinates* of ℓ. Clearly (kp, kq, kr) represents the same line for any $k \in \mathbb{R} \setminus \{0\}$, and so line coordinates behave very much like point (homogeneous) coordinates. Any piece of algebra performed with point coordinates can be repeated, interpreting the triples as line coordinates instead, and this is what is meant by the dual result. The one piece of strangeness that occurs is how to represent a point in line

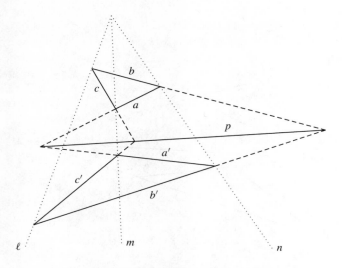

Fig. 9.17   The dual of Desargues' theorem

coordinates. In point coordinates, a point is given by its point coordinates (naturally!) but a line by an equation such as $px + qy + cz = 0$. This means that (p, q, r) have been chosen, and the line consists of all (x, y, z) satisfying the equation. Dually, in line coordinates, a line is given by its line coordinates, but this time a point has to be given by an equation: choose (x, y, z), and the line equation of this point is $px + qy + rz = 0$, where it is x, y, z that are fixed, and p, q, r that are varying. The line equation thus gives all lines (p, q, r) through the chosen point.

As an example, we now write down the dual of Desargues' theorem:

Proposition 9.20.   *Given trilaterals abc and a′b′c′, let $\ell = bc \cdot b'c'$, $m = ca \cdot c'a'$ and $n = ab \cdot a'b'$. (Here bc means the meet of the lines b and c, and bc · b′c′ is the join of the points bc and b′c′, and so on.) If the points aa′, bb′, and cc′ are collinear, then ℓ, m and n are concurrent.*

Proof.   Let the line through aa', bb', cc' be p (Fig. 9.17), choose line coordinates for a, b, c, p, and repeat all the algebra from the proof of Desargues' theorem.   □

We can tie Desargues' theorem and its dual together, by writing $a = BC$, $b = CA$, $c = AB$, $a' = B'C'$, $b' = C'A'$, and $c' = A'B'$. This makes the meet bc the meet of CA and AB, i.e., A, and so on; and thus $\ell = AA'$, $m = BB'$, and $n = CC'$; and also $L = aa'$, $M = bb'$, and $N = cc'$. So the dual of Desargues' theorem says that if L, M, N are collinear, then AA', BB', CC' are concurrent: in other words, the *dual* of Desargues' theorem is nothing more nor less than the *converse* of Desargues' theorem. In general, however, the dual of a theorem is not its converse.

It should be noted that Desargues' theorem is a perfectly good theorem in $\mathbb{R}^2$. However, to state it as a theorem in $\mathbb{R}^2$, one needs a good deal of caution about when various lines might happen to be parallel. For example, if $AA' \parallel BB' \parallel CC'$, then the theorem works just as well, but P is at infinity. If, instead, $BC \parallel B'C'$, then L is at infinity, which means that $MN \parallel BC$; unless we have $BC \parallel B'C'$ and $CA \parallel C'A'$ as well, so that both L and M are at infinity. In this case LM is the line at infinity, so N must be at infinity also, and we deduce that $AB \parallel A'B'$. It must be plain that to prove all of this in $\mathbb{R}^2$, with all the special cases, would be a great deal more work than our projective proof.

9.2.3   Projective transformations and cross-ratios

It is time to introduce transformations into $\mathbb{RP}^2$. Let $\mathbf{A} \in GL_3(\mathbb{R})$. We define the *projective transformation* $f : \mathbb{RP}^2 \to \mathbb{RP}^2$ (corresponding to $\mathbf{A}$) by mapping the point with coordinates $\mathbf{u}$ to the point with coordinates $\mathbf{Au}$, where

$$\mathbf{u} = \begin{pmatrix} x \\ y \\ z \end{pmatrix}.$$

Notice that if $k \in \mathbb{R} \setminus \{0\}$, then $k\mathbf{A} \in GL_3(\mathbb{R})$, and also $(k\mathbf{A})\mathbf{u} = \mathbf{A}(k\mathbf{u})$, which means that (i) f is well defined: it does not matter whether we use $\mathbf{u}$ or $k\mathbf{u}$ to represent the chosen point in $\mathbb{RP}^2$; and (ii) the matrices $\mathbf{A}$ and $k\mathbf{A}$ give rise to the same map f. In particular, the scalar matrix $k\mathbf{I}_3$ gives rise to the trivial map $1 : \mathbb{RP}^2 \to \mathbb{RP}^2$. Write H for the subgroup of $GL_3(\mathbb{R})$ consisting of all such scalar matrices. The map f, corresponding to or represented by $\mathbf{A}$, is also represented by every matrix of the coset $\mathbf{A}H$. To compose two such maps, f_1 and f_2, one multiplies the corresponding matrices $\mathbf{A}_1$ and $\mathbf{A}_2$ or (equivalently) the corresponding cosets: $(\mathbf{A}_1 H)(\mathbf{A}_2 H) = (\mathbf{A}_1 \mathbf{A}_2)(HH) = (\mathbf{A}_1 \mathbf{A}_2)H$, because scalar matrices commute with everything (so $H\mathbf{A}_2 = \mathbf{A}_2 H$), and H is a subgroup (so $HH = H$). This group, whose elements are cosets of H, is the *quotient* of $GL_3(\mathbb{R})$ by H, or $GL_3(\mathbb{R})/H$. It is called $PGL_3(\mathbb{R})$, the *projective general linear group*. In fact, because each real number has a real cube root, we can pick out of each coset $\mathbf{A}H$ a unique matrix of determinant 1, whence, writing $SL_3(\mathbb{R}) = \{\mathbf{A} \in GL_3(\mathbb{R}): \det \mathbf{A} = 1\}$, the *special linear* group, we have $PGL_3(\mathbb{R}) \cong SL_3(\mathbb{R})$; but this is a fluke to do with the fact that 3 is odd and we are using $\mathbb{R}$. For one-dimensional or three-dimensional real projective geometry ($\mathbb{RP}^1$, $\mathbb{RP}^3$), the corresponding groups of transformations are *not* isomorphic to $SL_2(\mathbb{R})$ or $SL_4(\mathbb{R})$, and neither is there any such isomorphism in complex projective geometry: we have to work in the quotient group.

• Ex.9.29: *Prove that the affine transformations of $\mathbb{R}^2$, thought of as projective transformations, are precisely those projective transformations that leave the line at infinity invariant.*

Lemma 9.21.   *Let A, B, C, P be points of $\mathbb{RP}^2$, no three being collinear. Then there is a projective transformation f with $f(1, 0, 0) = A$, $f(0, 1, 0) = B$, $f(0, 0, 1) = C$, and $f(1, 1, 1) = P$.*

Proof. Let A, B, C, P have homogeneous coordinates $\mathbf{a}$, $\mathbf{b}$, $\mathbf{c}$, $\mathbf{p}$ respectively. Since A, B, C are not collinear, the vectors $\mathbf{a}$, $\mathbf{b}$, $\mathbf{c}$ are linearly independent in $\mathbb{R}^3$, and so form a basis of $\mathbb{R}^3$. In particular, $\mathbf{p} = \lambda\mathbf{a} + \mu\mathbf{b} + \nu\mathbf{c}$ for some λ, μ, $\nu \in \mathbb{R}$; and $\lambda \neq 0$ (else P, B, C are collinear), and similarly $\mu \neq 0$ and $\nu \neq 0$. The result now follows on putting

$$\mathbf{A} = \begin{pmatrix} \lambda a_1 & \mu b_1 & \nu c_1 \\ \lambda a_2 & \mu b_2 & \nu c_2 \\ \lambda a_3 & \mu b_3 & \nu c_3 \end{pmatrix},$$

where $\mathbf{a} = (a_1, a_2, a_3)$, $\mathbf{b} = (b_1, b_2, b_3)$, and $\mathbf{c} = (c_1, c_2, c_3)$.   □

This map f allows us to set up a change of coordinates, so that the new triangle of reference is $\triangle ABC$. Since $f^{-1}(A) = (1, 0, 0)$, and so on, the new coordinates of any point Q are defined to be the previous (homogeneous) coordinates of $f^{-1}(Q)$. The point P, whose new coordinates are now $(1, 1, 1)$, is the *unit point* of this system. If we choose A, B, $C \in \mathbb{R}^2$ and P to be the centroid of $\triangle ABC$, then the new coordinates we get are just barycentric coordinates with respect to $\triangle ABC$.

A projective transformation is a collineation of $\mathbb{RP}^2$: it sends lines to lines. In fact the converse is true: every collineation of $\mathbb{RP}^2$ is a projective transformation. Here is a sketch of the proof. If f is a collineation of $\mathbb{RP}^2$, then the image under f of the line at infinity, $z = 0$, is some line $px + qy + rz = 0$. Find a projective transformation g that sends $px + qy + rz = 0$ to $z = 0$. Then gf leaves $z = 0$ invariant, and so leaves the complement of $z = 0$ invariant also, that is, it induces a collineation h of $\mathbb{R}^2$. But this means h is an affine transformation, which is just a projective transformation leaving $z = 0$ invariant. Then $gf = h$, so $f = g^{-1}h$, the composite of two projective transformations, and hence itself a projective transformation.

Let A, B, C, D be four distinct collinear points of $\mathbb{RP}^2$, represented by coordinates $\mathbf{a}$, $\mathbf{b}$, $\mathbf{c}$, $\mathbf{d}$ respectively. Then $\mathbf{c} = \lambda_1\mathbf{a} + \lambda_2\mathbf{b}$ and $\mathbf{d} = \mu_1\mathbf{a} + \mu_2\mathbf{b}$ for some non-zero λ_1, λ_2, μ_1, $\mu_2 \in \mathbb{R}$. We define the *cross-ratio* $\{A, B\,;\, C, D\}$ by

$$\{A, B\,;\, C, D\} = \frac{\lambda_2}{\lambda_1} \bigg/ \frac{\mu_2}{\mu_1} = \frac{\lambda_2\mu_1}{\lambda_1\mu_2}.$$

- Ex.9.30: *Show that $\{A, B\,;\, C, D\}$ is well defined: let $\mathbf{a}' = k_1\mathbf{a}$, $\mathbf{b}' = k_2\mathbf{b}$, $\mathbf{c}' = k_3\mathbf{c}$, and $\mathbf{d}' = k_4\mathbf{d}$ for some non-zero $k_i \in \mathbb{R}$. Find λ_1', λ_2', μ_1', μ_2' such that $\mathbf{c}' = \lambda_1'\mathbf{a}' + \lambda_2'\mathbf{b}'$ and $\mathbf{d}' = \mu_1'\mathbf{a}' + \mu_2'\mathbf{b}'$, and show that $(\lambda_2\mu_1)/(\lambda_1\mu_2) = (\lambda_2'\mu_1')/(\lambda_1'\mu_2')$.*

- Ex.9.31: *Prove that this definition coincides with Definition 4.35 when A, B, C, D are collinear points of $\mathbb{R}^2$.*

- Ex.9.32: *Let θ be a projective parameter for the line ℓ, and let A_i be a point of ℓ with parameter θ_i, for $i = 1, 2, 3, 4$. Show that $\{A_1, A_2\,;\, A_3, A_4\} = (\theta_3 - \theta_1)(\theta_4 - \theta_2)/((\theta_3 - \theta_2)(\theta_4 - \theta_1))$.*

In the notation of Definition 9.6, this exercise says $\{A_1, A_2\,;\, A_3, A_4\} = \{\theta_1, \theta_2\,;\, \theta_3, \theta_4\}$.

Proposition 9.22. *Projective transformations preserve cross-ratios.*

Proof. Let f be a projective transformation, let A, B, C, D be collinear points, and let $f(A) = A'$, $f(B) = B'$, $f(C) = C'$, and $f(D) = D'$. We have to show that $\{A, B \,;\, C, D\} = \{A', B' \,;\, C', D'\}$. Let f be given by the matrix $\mathbf{A}$, and let $\mathbf{a}, \mathbf{b}, \mathbf{c}, \mathbf{d}$ be coordinates for A, B, C, D respectively. Let $\mathbf{c} = \lambda_1 \mathbf{a} + \lambda_2 \mathbf{b}$ and $\mathbf{d} = \mu_1 \mathbf{a} + \mu_2 \mathbf{b}$, so that $\{A, B \,;\, C, D\} = (\lambda_2 \mu_1)/(\lambda_1 \mu_2)$. But now A', B', C', D' have coordinates $\mathbf{Aa}$, $\mathbf{Ab}$, $\mathbf{Ac}$, and $\mathbf{Ad}$ respectively. Further, $\mathbf{Ac} = \mathbf{A}(\lambda_1 \mathbf{a} + \lambda_2 \mathbf{b}) = \lambda_1(\mathbf{Aa}) + \lambda_2(\mathbf{Ab})$, and similarly $\mathbf{Ad} = \mu_1(\mathbf{Aa}) + \mu_2(\mathbf{Ab})$, from which $\{A', B' \,;\, C', D'\} = (\lambda_2 \mu_1)/(\lambda_1 \mu_2)$ also. $\qquad\square$

The dual of the cross-ratio of four collinear points is the cross-ratio of four concurrent lines. If p, q, r, s are four concurrent lines, with corresponding line coordinates $\mathbf{p}$, $\mathbf{q}$, $\mathbf{r}$, $\mathbf{s}$, then $\mathbf{r} = \lambda_1 \mathbf{p} + \lambda_2 \mathbf{q}$ and $\mathbf{s} = \mu_1 \mathbf{p} + \mu_2 \mathbf{q}$ for some non-zero $\lambda_1, \lambda_2, \mu_1, \mu_2 \in \mathbb{R}$. (OK?) So we define

$$\{p, q \,;\, r, s\} = \frac{\lambda_2}{\lambda_1} \bigg/ \frac{\mu_2}{\mu_1} = \frac{\lambda_2 \mu_1}{\lambda_1 \mu_2}. \tag{9.4}$$

The reader should compare this with Definition 4.36, and we shall now show that the two definitions agree.

Let $\mathbf{u}_1$, $\mathbf{u}_2$ be homogeneous coordinates of two points. What are the line coordinates of their join? If we let $\mathbf{u}_i = (x_i, y_i, z_i)$, then we are trying to solve the simultaneous equations $px_1 + qy_1 + rz_1 = 0$, $px_2 + qy_2 + rz_2 = 0$ for p, q, r. A solution is $p = y_1 z_2 - y_2 z_1$, $q = z_1 x_2 - z_2 x_1$, $r = x_1 y_2 - x_2 y_1$, which is the *cross product* or *vector product* $\mathbf{u}_1 \times \mathbf{u}_2$ of $\mathbf{u}_1$ and $\mathbf{u}_2$: explicitly,

$$\mathbf{u}_1 \times \mathbf{u}_2 = (y_1 z_2 - y_2 z_1, z_1 x_2 - z_2 x_1, x_1 y_2 - x_2 y_1).$$

So the line coordinates of the join of $\mathbf{u}_1$ and $\mathbf{u}_2$ are $\mathbf{u}_1 \times \mathbf{u}_2$. (And dually, as the reader should verify, if $\mathbf{u}_1$ and $\mathbf{u}_2$ are the line coordinates of two lines, then $\mathbf{u}_1 \times \mathbf{u}_2$ are the point coordinates of their meet.) The reader should verify the following properties of the vector product, for all $\mathbf{u}_i \in \mathbb{R}^3$ and $\lambda_i \in \mathbb{R}$:

$$\mathbf{u}_1 \times \mathbf{u}_1 = \mathbf{0};$$

$$\mathbf{u}_1 \times \mathbf{u}_2 = -\mathbf{u}_2 \times \mathbf{u}_1;$$

$$(\lambda_1 \mathbf{u}_1 + \lambda_2 \mathbf{u}_2) \times \mathbf{u}_3 = \lambda_1(\mathbf{u}_1 \times \mathbf{u}_3) + \lambda_2(\mathbf{u}_2 \times \mathbf{u}_3);$$

$$\mathbf{u}_1 \times (\lambda_2 \mathbf{u}_2 + \lambda_3 \mathbf{u}_3) = \lambda_2(\mathbf{u}_1 \times \mathbf{u}_2) + \lambda_3(\mathbf{u}_1 \times \mathbf{u}_3).$$

So now let A, B, C, D be four collinear points, and let P be a point not on the line AB. Let $\mathbf{a}, \mathbf{b}, \mathbf{c}, \mathbf{d}, \mathbf{p}$ be homogeneous coordinates for A, B, C, D, P respectively. Let $\mathbf{c} = \lambda_1 \mathbf{a} + \lambda_2 \mathbf{b}$ and $\mathbf{d} = \mu_1 \mathbf{a} + \mu_2 \mathbf{b}$, so that $\{A, B \,;\, C, D\} = (\lambda_2 \mu_1)/(\lambda_1 \mu_2)$. The line coordinates of PA, PB, PC, PD are $\mathbf{p} \times \mathbf{a}$, $\mathbf{p} \times \mathbf{b}$, $\mathbf{p} \times \mathbf{c}$, and $\mathbf{p} \times \mathbf{d}$ respectively, and

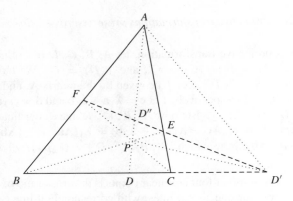

Fig. 9.18 $\{B, C\,;\, D, D'\} = -1$

according to Definition 4.36 the cross-ratio of these lines is $\{A, B\,;\, C, D\}$. But

$$\mathbf{p} \times \mathbf{c} = \mathbf{p} \times (\lambda_1 \mathbf{a} + \lambda_2 \mathbf{b}) = \lambda_1 (\mathbf{p} \times \mathbf{a}) + \lambda_2 (\mathbf{p} \times \mathbf{b}), \quad \text{and}$$

$$\mathbf{p} \times \mathbf{d} = \mathbf{p} \times (\mu_1 \mathbf{a} + \mu_2 \mathbf{b}) = \mu_1 (\mathbf{p} \times \mathbf{a}) + \mu_2 (\mathbf{p} \times \mathbf{b}),$$

and thus according to (9.4) the cross-ratio of the four lines is $(\lambda_2 \mu_1)/(\lambda_1 \mu_2)$, and so the two definitions agree.

Here is an application. Suppose we are given $\triangle ABC$ with D, E, F on BC, CA, AB respectively, and with AD, BE, CF concurrent at P. Let EF meet BC, AD at D', D'' respectively (Fig. 9.18). Then because the lines FB, EC, $D''D$, AD' meet at A, we have $\{B, C\,;\, D, D'\} = \{F, E\,;\, D'', D'\}$, each being the cross-ratio of the four lines. Then again, because the lines FC, EB, $D''D$, PD' meet at P, we have $\{F, E\,;\, D'', D'\} = \{C, B\,;\, D, D'\}$, each being the cross-ratio of *these* four lines. But now $\{B, C\,;\, D, D'\} = \{C, B\,;\, D, D'\} = 1/\{B, C\,;\, D, D'\}$, from which $\{B, C\,;\, D, D'\} = -1$. Compare this with Exercises 4.19 and 4.20.

Here is another way, using coordinates. Take $\triangle ABC$ as the triangle of reference, and P as the unit point. So $A = (1, 0, 0)$, $P = (1, 1, 1)$, and AP has line coordinates $(1, 0, 0) \times (1, 1, 1) = (0, -1, 1)$. (So its equation is $-y + z = 0$.) Then BC has line coordinates $(0, 1, 0) \times (0, 0, 1) = (1, 0, 0)$ (so its equation is $x = 0$) and the lines AP and BC meet at D, which thus has coordinates $(0, -1, 1) \times (1, 0, 0) = (0, 1, 1)$. Similarly, E has coordinates $(1, 0, 1)$ and F has coordinates $(1, 1, 0)$, so EF has line coordinates $(1, 0, 1) \times (1, 1, 0) = (-1, 1, 1)$; so D' must have coordinates $(-1, 1, 1) \times (1, 0, 0) = (0, 1, -1)$. But now $(0, 1, 1) = (0, 1, 0) + (0, 0, 1)$ and $(0, 1, -1) = (0, 1, 0) - (0, 0, 1)$ so the projective parameters of D, D' with respect to B, C are 1 and -1, and $\{B, C\,;\, D, D'\} = 1/(-1) = -1$ again.

- **Ex.9.33:** *Let ℓ_i be the line $y = m_i x + c_i$, in $\mathbb{R}^2$, for $i = 1, 2, 3, 4$.* (i) *Prove that if ℓ_1, ℓ_2, ℓ_3, ℓ_4 are concurrent, then $\{\ell_1, \ell_2\,;\, \ell_3, \ell_4\} = (m_3 - m_1)(m_4 - m_2)/((m_3 - m_2)(m_4 - m_1))$, and express this in terms of the angles between the lines.*

(ii) *Find a formula for* $\{\ell_1, \ell_2 ; \ell_3, \ell_4\}$ *when the four lines are parallel.*

9.2.4 Classification of conics (for the last time)

A conic in $\mathbb{R}^2$ is given by an equation of the form $ax^2 + 2hxy + by^2 + 2gx + 2fy + c = 0$, or

$$
\begin{pmatrix} x & y & 1 \end{pmatrix}
\begin{pmatrix} a & h & g \\ h & b & f \\ g & f & c \end{pmatrix}
\begin{pmatrix} x \\ y \\ 1 \end{pmatrix} = 0.
$$

To get the corresponding equation in $\mathbb{RP}^2$, we homogenize by replacing x, y by x/z, y/z respectively, and then multiplying through by z^2. This yields $ax^2 + by^2 + cz^2 + 2fyz + 2gzx + 2hxy = 0$, or

$$
\begin{pmatrix} x & y & z \end{pmatrix}
\begin{pmatrix} a & h & g \\ h & b & f \\ g & f & c \end{pmatrix}
\begin{pmatrix} x \\ y \\ z \end{pmatrix} = 0,
$$

or $\mathbf{u}^T \mathbf{M} \mathbf{u} = 0$, where

$$
\mathbf{u} = \begin{pmatrix} x \\ y \\ z \end{pmatrix} \quad \text{and} \quad \mathbf{M} = \begin{pmatrix} a & h & g \\ h & b & f \\ g & f & c \end{pmatrix}.
$$

Notice that, because this equation is homogeneous in x, y, z, it is satisfied by (kx, ky, kz) whenever it is satisfied by (x, y, z). Then again, applying a projective transformation to our conic amounts to replacing $\mathbf{u}$ by $\mathbf{Au}$ for some $\mathbf{A} \in GL_3(\mathbb{R})$, and so the transformed equation is $(\mathbf{Au})^T \mathbf{M}(\mathbf{Au}) = 0$, or $\mathbf{u}^T \mathbf{N} \mathbf{u} = 0$ (where $\mathbf{N} = \mathbf{A}^T \mathbf{M} \mathbf{A}$), another conic. So projective transformations take conics to conics, and indeed they take proper conics to proper conics, since $\det \mathbf{M} \neq 0$ if and only if $\det \mathbf{N} \neq 0$. Two conics so related are said to be *projectively equivalent*. So when are two proper real conics in $\mathbb{RP}^2$ projectively equivalent? The answer is *always*, as we now show:

Theorem 9.23 (Classification of conics (IV)). *All proper real conics in $\mathbb{RP}^2$ are projectively equivalent.*

Proof. Conics that are affinely equivalent are necessarily projectively equivalent. By Theorem 8.23, every proper real conic in $\mathbb{R}^2$ is affinely equivalent to one of $x^2 + y^2 - 1 = 0$ or $x^2 - y^2 - 1 = 0$ or $y^2 - 4x = 0$, so on homogenizing, we have that every proper real conic in $\mathbb{RP}^2$ is projectively equivalent to $x^2 + y^2 - z^2 = 0$ or $x^2 - y^2 - z^2 = 0$ or $y^2 - 4xz = 0$. All we have to do is show that these three conics are projectively equivalent to each other. But this is easy: putting $x = z_1$, $y = y_1$, and $z = x_1$ turns $x^2 - y^2 - z^2 = 0$ into $z_1^2 - y_1^2 - x_1^2 = 0$, or $x_1^2 + y_1^2 - z_1^2 = 0$; and putting $x = \frac{1}{2}(z_1 - x_1)$, $y = y_1$, and $z = \frac{1}{2}(z_1 + x_1)$ turns $y^2 - 4zx = 0$ into $y_1^2 - (z_1^2 - x_1^2) = 0$, or $x_1^2 + y_1^2 - z_1^2 = 0$, again. The changes of variable are given by

$$\begin{pmatrix} x \\ y \\ z \end{pmatrix} = \begin{pmatrix} 0 & 0 & 1 \\ 0 & 1 & 0 \\ 1 & 0 & 0 \end{pmatrix} \begin{pmatrix} x_1 \\ y_1 \\ z_1 \end{pmatrix} \quad \text{and} \quad \begin{pmatrix} x \\ y \\ z \end{pmatrix} = \begin{pmatrix} -\frac{1}{2} & 0 & \frac{1}{2} \\ 0 & 1 & 0 \\ \frac{1}{2} & 0 & \frac{1}{2} \end{pmatrix} \begin{pmatrix} x_1 \\ y_1 \\ z_1 \end{pmatrix},$$

respectively, and the 3×3 matrices are both invertible, so the substitutions do correspond to projective equivalences.   □

We now have four classification theorems for conics, under isometry (Theorem 7.11), under similarity (Theorem 8.5), under affine transformation (Theorem 8.23), and under projective transformation (Theorem 9.23), the proof of the last three each depending on the previous one. Could we have gone straight for Theorem 9.23, without bothering about the others? The answer is yes, and we now outline two possible methods.

Two 3×3 real matrices $\mathbf{M}$, $\mathbf{N}$ satisfying $\mathbf{N} = \mathbf{A}^T \mathbf{M} \mathbf{A}$ for some $\mathbf{A} \in GL_3(\mathbb{R})$ are said to be *congruent*, and there is a theorem in linear algebra that every real symmetric matrix is congruent to a diagonal matrix with entries ± 1, 0 on the diagonal; and two of these diagonal matrices are congruent if and only if they have the same number of 1's and -1's on the diagonal. (This is *Sylvester's law of inertia*, and it can be proved by the familiar process of completing the square: see [6] for details.) Thus every real symmetric *invertible* 3×3 matrix is congruent to one of the four diagonal matrices

$$\pm \begin{pmatrix} 1 & 0 & 0 \\ 0 & 1 & 0 \\ 0 & 0 & 1 \end{pmatrix}, \quad \pm \begin{pmatrix} 1 & 0 & 0 \\ 0 & 1 & 0 \\ 0 & 0 & -1 \end{pmatrix}.$$

Note that, since $\mathbf{M}$ and $-\mathbf{M}$ give the same conic, we obtain *two* conics from these four matrices, namely $x^2 + y^2 \pm z^2 = 0$, or (dehomogenizing) $x^2 + y^2 = \mp 1$, an imaginary circle and a real circle. Thus every proper real conic (which of course excludes imaginary conics) is projectively equivalent to $x^2 + y^2 = z^2$, which gives Theorem 9.23 again.

• Ex.9.34: *What is the corresponding theorem in* $\mathbb{RP}^3$?

The other way to obtain Theorem 9.23 is to use Lemma 9.21 to change the triangle of reference and put the equation of a given proper real conic Γ in a special form. Specifically, we choose X and Z on Γ, and let the tangents to Γ at X and Z meet at Y, and take $\triangle XYZ$ as triangle of reference, with another point W on Γ chosen as unit point (Fig. 9.19). Let Γ have equation $\mathbf{u}^T \mathbf{A} \mathbf{u} = 0$, and let $\mathbf{u}_1$ be a point of Γ. Let the line ℓ have line coordinates $\mathbf{A}\mathbf{u}_1$, and note that $\mathbf{u}_1$ lies on ℓ, since $\mathbf{u}_1^T \mathbf{A} \mathbf{u}_1 = 0$. We claim ℓ is the tangent to Γ at $\mathbf{u}_1$. For suppose $\mathbf{u}_2$ is another point of ℓ, so that $\mathbf{u}_2^T \mathbf{A} \mathbf{u}_1 = 0 = \mathbf{u}_1^T \mathbf{A} \mathbf{u}_2$. If $\mathbf{u}_2$ also lies on Γ, then $\mathbf{u}_i^T \mathbf{A} \mathbf{u}_j = 0$ for $i = 1, 2$ and $j = 1, 2$, and it follows that $(\mathbf{u}_1 + \theta \mathbf{u}_2)^T \mathbf{A} (\mathbf{u}_1 + \theta \mathbf{u}_2) = 0$ for all θ. Thus ℓ is a subset of Γ, which is impossible if Γ is proper. Consequently ℓ meets Γ *only* at $\mathbf{u}_1$, so it is the tangent to Γ at $\mathbf{u}_1$.

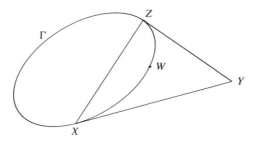

Fig. 9.19 Putting Γ in the form $y^2 = xz$

Now we want the tangent at $(1, 0, 0)$ to be $z = 0$, and the tangent at $(0, 0, 1)$ to be $x = 0$, so we want

$$\begin{pmatrix} a & h & g \\ h & b & f \\ g & f & c \end{pmatrix} \begin{pmatrix} 1 \\ 0 \\ 0 \end{pmatrix} = \begin{pmatrix} 0 \\ 0 \\ p \end{pmatrix}, \quad \text{and} \quad \begin{pmatrix} a & h & g \\ h & b & f \\ g & f & c \end{pmatrix} \begin{pmatrix} 0 \\ 0 \\ 1 \end{pmatrix} = \begin{pmatrix} q \\ 0 \\ 0 \end{pmatrix}$$

for some p, q. This gives $a = h = 0$ (and $g = p$), and $f = c = 0$ (and $g = q$), so our conic Γ is $by^2 + 2gxz = 0$. Since it also passes through $(1, 1, 1)$, we have $b + 2g = 0$, and the equation of Γ is $y^2 = xz$. If we have a *second* conic Γ', we can choose a second triangle of reference $\triangle X'Y'Z'$ and unit point W' so that Γ' has equation $(y')^2 = x'y'$ in the second system. Lemma 9.21 gives us a projective transformation f with $f(X) = X'$, $f(Y) = Y'$, $f(Z) = Z'$, and $f(W) = W'$, and this f will also send Γ to Γ', giving yet another proof of Theorem 9.23.

- Ex.9.35: *In the above notation, suppose YW meets Γ again at W_1, and meets XZ at Y_1. Show that $\{W, W_1 ; Y, Y_1\} = -1$. Show also that the tangents to Γ at W and W_1 meet on XZ.*

- Ex.9.36: *Use Exercise 9.35 to give a ruler-only construction for the tangents from a point to a conic.*

Suppose $\mathbf{u}$ is a general point of the proper conic $\Gamma : \mathbf{u}^T \mathbf{M} \mathbf{u} = 0$. Then, as we have just seen, the line coordinates of the tangent to Γ at $\mathbf{u}$ are $\mathbf{v} = \mathbf{Mu}$. Putting $\mathbf{N} = \mathbf{M}^{-1}$, we have $\mathbf{v}^T \mathbf{N} \mathbf{v} = (\mathbf{u}^T \mathbf{M}) \mathbf{M}^{-1} (\mathbf{Mu}) = \mathbf{u}^T \mathbf{Mu} = 0$. This gives Γ as an *envelope*, which is the dual of a curve: a set of lines satisfying a polynomial equation in the line coordinates. If the curve has degree 2, it is called a *conic envelope*. Since we have obtained a second-degree equation $\mathbf{v}^T \mathbf{N} \mathbf{v} = 0$ for the tangents to Γ, we now know that a conic can be regarded equally as a conic envelope. Said loosely, the dual of a conic is a conic.

We can now give an example of a dual theorem that looks quite different from the original theorem. Recall that Pascal's theorem says that if the six vertices of a hexagram lie on a circle Γ, then the opposite sides of the hexagram meet pairwise in three points that

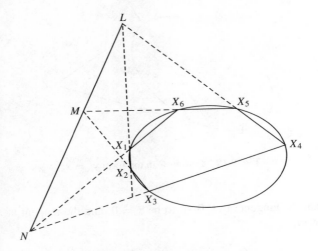

Fig. 9.20 Pascal's theorem

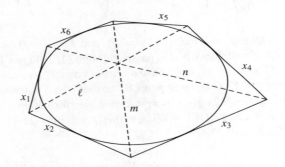

Fig. 9.21 Brianchon's theorem

are collinear. Since we know that all proper conics are projectively equivalent, Pascal's theorem still works if Γ is any proper (real) conic (Fig. 9.20). (There is a complex version of Pascal's theorem still to come, in Theorem 10.13.) If we take the dual of Pascal's theorem, then the six points on Γ become six tangents to the dual conic, and we obtain *Brianchon's theorem*: if the six sides of a hexagram touch a proper conic, then the three joins of opposite vertices of the hexagram are concurrent (Fig. 9.21).

Returning to our conic $\Gamma : y^2 = xz$, notice that the point $(\theta^2, \theta, 1)$ always lies on Γ, and indeed every point of Γ can be so represented, except $(1, 0, 0)$. But $(\theta^2, \theta, 1)$ and $(1, \theta^{-1}, \theta^{-2})$ represent the same point, and $(1, \theta^{-1}, \theta^{-2}) \rightarrow (1, 0, 0)$ as $\theta \rightarrow \infty$, so we obtain all the points of Γ as θ runs through $\mathbb{R} \cup \{\infty\}$. This θ is called a *projective parameter* for Γ, by analogy with the projective parameter for a line that we met above. We finish this section with a theorem that should be compared with Exercise 9.32, and shows that the analogy is more than superficial:

Proposition 9.24. *Let X_i be fixed points of $\Gamma : y^2 = xz$, with projective parameters θ_i, for $i = 1, 2, 3, 4$. Let P be a fifth point of Γ, and let $\ell_i = PX_i$, all i. Then*

$$\{\ell_1, \ell_2 ; \ell_3, \ell_4\} = \frac{(\theta_3 - \theta_1)(\theta_4 - \theta_2)}{(\theta_3 - \theta_2)(\theta_4 - \theta_1)}.$$

Proof. Let P have coordinates $(\varphi^2, \varphi, 1)$. The line ℓ_i then has line coordinates

$$(\varphi^2, \varphi, 1) \times (\theta_i^2, \theta_i, 1) = (\varphi - \theta_i, \theta_i^2 - \varphi^2, \varphi^2\theta_i - \theta_i^2\varphi)$$

or, on dividing through by $\theta_i - \varphi$, it has line coordinates $(-1, \theta_i + \varphi, \theta_i\varphi)$. Now find $\lambda_1, \lambda_2, \mu_1, \mu_2$ so that

$$\lambda_1(-1, \theta_1 + \varphi, \theta_1\varphi) + \lambda_2(-1, \theta_2 + \varphi, \theta_2\varphi) = (-1, \theta_3 + \varphi, \theta_3\varphi), \quad \text{and}$$

$$\mu_1(-1, \theta_1 + \varphi, \theta_1\varphi) + \mu_2(-1, \theta_2 + \varphi, \theta_2\varphi) = (-1, \theta_4 + \varphi, \theta_4\varphi).$$

From these, $\lambda_1 + \lambda_2 = 1$ and $\lambda_1\theta_1 + \lambda_2\theta_2 = \theta_3$ (OK?), whence $\lambda_1(\theta_1 - \theta_3) + \lambda_2(\theta_2 - \theta_3) = 0$. Similarly, $\mu_1(\theta_1 - \theta_4) + \mu_2(\theta_2 - \theta_4) = 0$, and so

$$\{\ell_1, \ell_2 ; \ell_3, \ell_4\} = \frac{\lambda_2\mu_1}{\lambda_1\mu_2} = \left(-\frac{\theta_1 - \theta_3}{\theta_2 - \theta_3}\right)\left(-\frac{\theta_2 - \theta_4}{\theta_1 - \theta_4}\right) = \frac{(\theta_3 - \theta_1)(\theta_4 - \theta_2)}{(\theta_3 - \theta_2)(\theta_4 - \theta_1)}. \quad \square$$

Notice that the cross-ratio $\{\ell_1, \ell_1 ; \ell_3, \ell_4\}$ does not depend on the position of P. Dually, the tangents at X_1, X_2, X_3, X_4 meet the tangent at P in four points L_1, L_2, L_3, L_4 whose cross-ratio does not depend on the position of P; and in fact $\{\ell_1, \ell_1 ; \ell_3, \ell_4\} = \{L_1, L_2 ; L_3, L_4\}$, a result known as *Chasles' theorem*. For more details, see [29].

9.2.5 Elliptic geometry

We are now going to look briefly at a second type of non-Euclidean geometry called *elliptic* geometry. We have our model for this already: it is the real projective plane, $\mathbb{RP}^2$, with its points and lines, noting that since every pair of lines in $\mathbb{RP}^2$ meet in a point, there can be no such thing as a pair of parallel lines in $\mathbb{RP}^2$. However, to compare this geometry with Euclidean and hyperbolic geometries, we need a sensible way of measuring distances and angles in $\mathbb{RP}^2$, and then a group of elliptic 'isometries', or distance-preserving maps. To see how to do this, we return to our unit sphere $S^2 = \{(x, y, z) \in \mathbb{R}^3 : x^2 + y^2 + z^2 = 1\}$, centre O, as the model for $\mathbb{RP}^2$, with pairs of antipodal points representing points of $\mathbb{RP}^2$. Given points $A, B \in S^2$, we can replace B by the antipodal point if necessary to ensure that $\angle AOB \leq \pi/2$, and we then define the elliptic distance from A to B in $\mathbb{RP}^2$ to be $d_e(A, B) = \angle AOB$. This is the same as the arc length from A to B along the circle in which the plane AOB cuts S^2. Notice that no length is greater than $\pi/2$. Two lines of $\mathbb{RP}^2$ are represented by two great circles on S^2, and the angle between the lines is defined to be the angle between these great circles, that is, between their tangents at their common point. The isometries of this geometry are given by the Euclidean symmetries of S^2, that is, by $O_3(\mathbb{R})$, the 3×3 orthogonal matrices.

It is obvious that $d_e(A, A) = 0$ and $d_e(A, B) > 0$ if $A \neq B$; and also $d_e(A, B) = d_e(B, A)$. We shall now prove the elliptic version of the triangle inequality (cf. Lemma 3.16 and Proposition 9.18(iii),(iv)):

Proposition 9.25.   *Let $A, B, C \in \mathbb{RP}^2$. Then $d_e(A, B) \leq d_e(A, C) + d_e(C, B)$.*

Proof.   Let A, B, C be represented by the unit vectors $\mathbf{a}, \mathbf{b}, \mathbf{c}$ on S^2. So $d_e(A, B) = \cos^{-1} |\mathbf{a} \cdot \mathbf{b}|$, and so on. We have to show that $\cos^{-1} |\mathbf{a} \cdot \mathbf{b}| \leq \cos^{-1} |\mathbf{a} \cdot \mathbf{c}| + \cos^{-1} |\mathbf{c} \cdot \mathbf{b}|$ or, taking the cosine of each side (and remembering that $\cos x$ is a decreasing function of x for $0 \leq x \leq \pi$), that

$$|\mathbf{a} \cdot \mathbf{b}| \geq \cos(\cos^{-1} |\mathbf{a} \cdot \mathbf{c}| + \cos^{-1} |\mathbf{c} \cdot \mathbf{b}|)$$

$$= |\mathbf{a} \cdot \mathbf{c}||\mathbf{c} \cdot \mathbf{b}| - \sqrt{1 - |\mathbf{a} \cdot \mathbf{c}|^2}\sqrt{1 - |\mathbf{c} \cdot \mathbf{b}|^2},$$

(Here we have used the formula $\cos(x + y) = \cos x \cos y - \sin x \sin y$.) Moving the square roots to the left-hand side and everything else to the right-hand side and then squaring both sides, we must show $(1 - |\mathbf{a} \cdot \mathbf{c}|^2)(1 - |\mathbf{c} \cdot \mathbf{b}|^2) \geq (|\mathbf{a} \cdot \mathbf{c}||\mathbf{c} \cdot \mathbf{b}| - |\mathbf{a} \cdot \mathbf{b}|)^2$, or

$$1 - |\mathbf{b} \cdot \mathbf{c}|^2 - |\mathbf{c} \cdot \mathbf{a}|^2 - |\mathbf{a} \cdot \mathbf{b}|^2 + 2|\mathbf{b} \cdot \mathbf{c}||\mathbf{c} \cdot \mathbf{a}||\mathbf{a} \cdot \mathbf{b}| \geq 0.$$

This last inequality would follow if we knew

$$1 - (\mathbf{b} \cdot \mathbf{c})^2 - (\mathbf{c} \cdot \mathbf{a})^2 - (\mathbf{a} \cdot \mathbf{b})^2 + 2(\mathbf{b} \cdot \mathbf{c})(\mathbf{c} \cdot \mathbf{a})(\mathbf{a} \cdot \mathbf{b}) \geq 0.$$

Here we recognize the left-hand side as

$$\det \begin{pmatrix} 1 & \mathbf{a} \cdot \mathbf{b} & \mathbf{a} \cdot \mathbf{c} \\ \mathbf{b} \cdot \mathbf{a} & 1 & \mathbf{b} \cdot \mathbf{c} \\ \mathbf{c} \cdot \mathbf{a} & \mathbf{c} \cdot \mathbf{b} & 1 \end{pmatrix} = \det(\mathbf{A}^T \mathbf{A}),$$

where $\mathbf{A} = (\mathbf{a}, \mathbf{b}, \mathbf{c})$ is the 3×3 matrix whose columns are $\mathbf{a}, \mathbf{b}$, and $\mathbf{c}$. But $\det(\mathbf{A}^T \mathbf{A}) = (\det \mathbf{A}^T)(\det \mathbf{A}) = (\det \mathbf{A})^2$, which cannot be negative; so we have finished. (This proof was done back-to-front. Please check that the argument does indeed reverse.)   □

We leave the reader to investigate precisely when $d_e(A, B) = d_e(A, C) + d_e(B, C)$.

Once lengths are defined, we can measure areas, and since lengths are expressed in terms of angles, it is perhaps no surprise that areas can be expressed in terms of angles also; the surprise is that the angles of a triangle determine its area, unlike the situation in Euclidean geometry where triangles with the same angles do not have the same area, in general. Suppose A, B, C on S^2 represent vertices of a triangle in $\mathbb{RP}^2$, with angles α, β, γ at A, B, C. The sides AB, BC, CA correspond to three great circles on S^2. Now the unit sphere S^2 has surface area 4π, and the great circle BC divides this into two hemispheres of area 2π. The second great circle CA meets the first at an angle γ, and divides each of these hemispheres into two *lunes* of area proportional to their respective angles, that is, of area 2γ and $2\pi - 2\gamma$. The third great circle AB cuts the lune of area 2γ into two triangular regions of area $\Delta = \text{area}(ABC)$ and (say) Δ_γ, so that $\Delta + \Delta_\gamma = 2\gamma$. In a similar way, $\Delta + \Delta_\alpha = 2\alpha$ and $\Delta + \Delta_\beta = 2\beta$, and adding these three equations gives $3\Delta + \Delta_\alpha + \Delta_\beta + \Delta_\gamma = 2(\alpha + \beta + \gamma)$. But every point of S^2 either lies in the region represented by $\Delta + \Delta_\alpha + \Delta_\beta + \Delta_\gamma$ (i.e., the union of the three lunes of angle α, β, and γ), or its antipodal point lies in this region. So $\Delta + \Delta_\alpha + \Delta_\beta + \Delta_\gamma = 2\pi$, and on substituting back, $\Delta = \alpha + \beta + \gamma - \pi$, the *spherical excess* formula for area.

There is a similar formula in hyperbolic geometry, with the sign changed: in the hyperbolic plane, a triangle with angles α, β, γ has area (proportional to) $\pi - \alpha - \beta - \gamma$. (For a proof, see [9].) As a consequence of these results, the angle sum of a triangle is *greater* than π in elliptic geometry, *less* than π in hyperbolic geometry, and of course it is *equal* to π in Euclidean geometry. So can we now decide whether we live in an elliptic, hyperbolic or Euclidean world, by drawing a triangle, measuring the angles, and comparing the sum with π? Of course, the answers *will* be different, because all measurement is approximate. If the answers came out substantially different there might be some conclusion to be drawn, but this is not what happens. The fact that the angle sum is π, as nearly as we can measure it, means that either our world is Euclidean, or else it is non-Euclidean and we are measuring a triangle of rather small area, so small that we cannot distinguish its angle sum from π. To get a meaningful measurement, we would probably need a triangle of intergalactic size, and now we are into cosmology and relativity, and have to face the fact that our question is not answerable as easily as we might at first have hoped.

ANSWERS TO EXERCISES

9.1: We already have that if $f \in \mathcal{M}^+(\mathbb{C})$ then $f^{-1} \in \mathcal{M}^+(\mathbb{C})$; and of course $1 \in \mathcal{M}^+(\mathbb{C})$, by taking $a = d = 1$ and $b = c = 0$. Now let f, $g \in \mathcal{M}^+(\mathbb{C})$, where $f(z) = (az + b)/(cz + d)$ and $g(z) = (a'z + b')/(c'z + d')$. We have

$$fg(z) = f(g(z)) = \frac{ag(z) + b}{cg(z) + d} = \frac{a(a'z + b') + b(c'z + d')}{c(a'z + b') + d(c'z + d')} = \frac{(aa' + bc')z + (ab' + bd')}{(ca' + dc')z + (cb' + dd')}.$$

Note that

$$\begin{pmatrix} a & b \\ c & d \end{pmatrix}\begin{pmatrix} a' & b' \\ c' & d' \end{pmatrix} = \begin{pmatrix} aa' + bc' & ab' + bd' \\ ca' + dc' & cb' + dd' \end{pmatrix},$$

and also $ad - bc$ and $a'd' - b'c'$ are non-zero. Taking determinants shows $(aa' + bc')(cb' + dd') - (ab' + bd')(ca' + dc') \neq 0$, whence $fg \in \mathcal{M}^+(\mathbb{C})$ and we have a group. On the way we have also shown that the map $GL_2(\mathbb{C}) \to \mathcal{M}^+(\mathbb{C})$ is a homomorphism; its kernel consists of the *scalar* matrices: all $a\mathbf{I}_2$ where $a \in \mathbb{C}$, $a \neq 0$. (Check.)

9.2:
$$f(z) - \alpha = \frac{r^2}{\bar{z} - \bar{\alpha}}, \quad \text{or} \quad z \mapsto \frac{r^2}{\bar{z} - \bar{\alpha}} + \alpha.$$

9.3: Let O be the centre of Σ. Then C, D are inverse points if and only if $(OA)^2 = (OC)(OD)$, which happens if and only if $\{A, B; C, D\} = -1$, by Exercise 4.33.

9.4: Let CD meet Σ at A, B; let O be the centre of Σ, and let Σ, Σ' meet at P and Q. Then $(OP)^2 = (OA)^2 = (OC)(OD)$, so that OP is the tangent to Σ' at P, by Proposition 7.21. But of course the radius OP of Σ is orthogonal to the tangent to Σ at P, and we are done. Similarly for Q.

9.5: Let the given Möbius transformation be $f : z \mapsto (az + b)/(cz + d)$, where $ad \neq bc$. Then f is a similarity if $c = 0$, so assume $c \neq 0$. We try to write f as the composite of $z \mapsto \varepsilon\overline{z} + \delta$, where $|\varepsilon| = 1$, and $z \mapsto r^2/(\overline{z} - \overline{\alpha}) + \alpha$, where $r \in \mathbb{R}$, $r > 0$. Thus we want

$$\frac{az + b}{cz + d} = \frac{r^2}{\overline{\varepsilon z + \delta} - \overline{\alpha}} + \alpha,$$

for all $z \in \mathbb{C} \cup \{\infty\}$. Putting $z = \infty$ shows $\alpha = a/c$, so that now $f(z) = (bc - ad)/(c^2(z + d/c)) + \alpha$. So, using polar form, let $(bc - ad)/c^2 = r^2\varepsilon$, where $r \in \mathbb{R}$, $r > 0$, and $|\varepsilon| = 1$. With these values, $f(z) = r^2/(\overline{\varepsilon}z + \overline{\varepsilon}d/c) + \alpha$, so we finish the job by making $\overline{\delta} - \overline{\alpha} = \overline{\varepsilon}d/c$, that is, by putting $\delta = \alpha + \varepsilon \overline{d}/\overline{c}$. The proof for conjugate Möbius transformations is much the same, conjugating or 'unconjugating' where necessary.

9.6: By the preceding remarks, any invariant line must pass through $f(\infty) = a/c$ and through $f^{-1}(\infty) = -d/c$, so there will be exactly one such line provided $a/c \neq -d/c$, or $a + d \neq 0$. When $a + d = 0$ (and $c \neq 0$), all lines through a/c get mapped to lines through a/c. Are any of them invariant? A direct calculation is rather messy, so we'll use some low cunning instead. We have $z' - a/c = ((bc + a^2)/c^2)(z - a/c)^{-1}$, where $z' = f(z)$. Find $\alpha \in \mathbb{C}$ with $\alpha^2 = c^2/(bc + a^2)$, so that $\alpha(z' - a/c) = (\alpha(z - a/c))^{-1}$. Write $g : z \mapsto \alpha(z - a/c)$ (a similarity), and $h : z \mapsto z^{-1}$ (a Möbius transformation). Thus $g(z') = hg(z)$, or $f(z) = z' = g^{-1}hg(z)$, so that $f = g^{-1}hg$. If ℓ is a line, then $f(\ell) = \ell \Leftrightarrow g^{-1}hg(\ell) = \ell \Leftrightarrow hg(\ell) = g(\ell)$. So ℓ is an invariant line of $f \Leftrightarrow g(\ell)$ is an invariant line of h. But it is easy to see that h has just two invariant lines: the real axis, $y = 0$, and the imaginary axis, $x = 0$—check this!—and they are orthogonal. Since g is a similarity, and so is an angle-preserving collineation, we deduce that f also has just two invariant lines (meeting at a/c), and they are orthogonal. (One might hope for a shorter proof like this: the join of a fixed point of f to the point a/c must be an invariant line—this is true—and the equation $z = f(z)$ reduces to a quadratic equation, giving two fixed points—also true—so surely there are two invariant lines? Unfortunately for this argument, *both* the fixed points lie on one of the invariant lines, and *neither* on the other—e.g., for h the fixed points are ± 1, both on $y = 0$—so the method fails.)

9.7: (i) Try $z \mapsto z/(z + 1)$, which fixes only 0; or the translation $z \mapsto z + 1$, which fixes only ∞. (ii) For example, $z \mapsto 2z/(z + 1)$ fixes only 0 and 1; or $z \mapsto -z$ fixes only 0 and ∞.

Last part: no. Let $f : z \mapsto (az + b)/(cz + d)$, where $ad \neq bc$. If $c = 0$, then $f(\infty) = \infty$, and if $c \neq 0$ then the quadratic equation $z(cz + d) = az + b$ is satisfied by at least one $z \in \mathbb{C}$. So f has at least one fixed point.

9.8: Try $z \mapsto \overline{z}/(\overline{z} + 1)$, which fixes only 0; or the glide-reflection $z \mapsto \overline{z} + 1$, which fixes only ∞. (ii) For example, $z \mapsto 2\overline{z}/(\overline{z} + 1)$ fixes only 0 and 1; or the dilative reflection $z \mapsto 2\overline{z}$ fixes only 0 and ∞. (To make the calculations easier, note that if f fixes z, then f^2 also fixes z; but of course the converse is false, in general.)

Last part: yes. For example, the map $z \mapsto i/\overline{z}$ does not fix ∞, and if it fixes the finite z, then $z\overline{z} = i$, which is impossible!

9.9: Let g be inversion with respect to Σ or reflection in Σ, as appropriate. Then g is a conjugate Möbius transformation fixing a, b, and c, so that fg^{-1} is a Möbius transformation fixing a, b, and c. (OK?) Thus $fg^{-1} = 1$, by Proposition 9.3, and so $f = g$.

9.10: Choose γ on Σ, and a Möbius transformation g with $g(\alpha) = 1$, $g(\beta) = -1$, and $g(\gamma) = \infty$. Then $g(\Sigma)$ is a line-or-circle containing $g(\gamma) = \infty$, so it is a line, and it has $g(\alpha)$, $g(\beta)$, or ± 1, as inverse points, so it is the line $x = 0$. Likewise we can find a Möbius transformation h with $h(\alpha') = 1$, $h(\beta') = -1$, and with $h(\Sigma')$ being the line $x = 0$. Put $f = h^{-1}g$. (It is not unique. Why?)

9.11: Written in the more usual $(az + b)/(cz + d)$ form, the six maps correspond to the six matrices given in the solution to Exercise 8.19, which form a group. Alternatively, if we write $f : z \mapsto (1 - z)^{-1}$ and $g : z \mapsto z^{-1}$, then it is easy to see that $G = \{1, f, f^2, g, gf, gf^2\}$ with $f^3 = 1 = g^2$ and $fg = gf^2$, so that not only is G a group, but $G \cong D_3 \cong S_3$.

9.12: Using the notation of the solution to Exercise 9.11, the fixed points of f (and f^2, which is just f^{-1}) are ω and $\overline{\omega}$ where $\omega = e^{i\pi/3}$; the fixed points of g are 1 and -1; the fixed points of gf are $\frac{1}{2}$ and ∞; and the fixed points of gf^2 are 0 and 2. So the stabilizer of ω is $G_\omega = \{1, f, f^2\}$, whence its orbit ω^G has size $|G|/|G_\omega| = 6/3 = 2$, and similarly for $\overline{\omega}$. It is easy to see now that $\omega^G = \{\omega, \overline{\omega}\}$. Next, $G_{-1} = \{1, g\}$, from which $|(-1)^G| = 6/2 = 3$, and in fact $(-1)^G = \{-1, \frac{1}{2}, 2\}$. Finally, $G_1 = \{1, g\}$ also, so $|1^G| = 3$, and in fact $1^G = \{0, 1, \infty\}$; however, this is a degenerate case for the cross-ratios, as these three values are impossible if the four points z_1, z_2, z_3, z_4 are distinct.

For all other values of λ, that is, for $\lambda \notin \{-1, 0, \frac{1}{2}, 1, 2, \omega, \overline{\omega}, \infty\}$, we have $G_\lambda = \{1\}$, so that $|\lambda^G| = 6$ and the six values in (9.2) are distinct.

9.13: The real axis (which can alternatively be described as the line $|z - \omega| = |z - \overline{\omega}|$, where $\omega = e^{i\pi/3}$) contains $-1, 0, \frac{1}{2}, 1, 2$, and (being a line) ∞. The line $|z| = |z - 1|$ contains $\omega, \frac{1}{2}, \overline{\omega}$, and ∞. The circle $|z| = 1$ contains $1, \omega, -1$, and $\overline{\omega}$; and the circle $|z - 1| = 1$ contains $0, \omega, 2$, and $\overline{\omega}$.

It is clear that all six maps send the real axis to itself. Then to find, for example, the image of the line $|z| = |z - 1|$ under $f : z \mapsto (1 - z)^{-1}$, note that $f(\infty) = 0$, $f(\frac{1}{2}) = 2$, and $f(\omega) = \omega$, so that the image must be the circle $|z - 1| = 1$. Alternatively, the image is obtained by replacing z (in the equation $|z| = |z - 1|$) by $f^{-1}(z) = 1 - z^{-1}$, giving $|1 - z^{-1}| = |(1 - z^{-1}) - 1|$. Multiplying each side by $|z|$, we get $|z - 1| = 1$ again. Then to find, for example, the image under $g : z \mapsto z^{-1}$ of the region with 'corners' at $\frac{1}{2}$, 1, and ω, note that $g(\frac{1}{2}) = 2$, $g(1) = 1$, and $g(\omega) = \overline{\omega}$, so the image must be the region with 'corners' at $2, 1$, and $\overline{\omega}$. Alternatively, the region is the set of all z with $|z - 1| < |z| < 1$ and $|z - \omega| < |z - \overline{\omega}|$, so its image is obtained by replacing z by $g^{-1}(z) = z^{-1}$, giving $|z^{-1} - 1| < |z^{-1}| < 1$ and $|z^{-1} - \omega| < |z^{-1} - \overline{\omega}|$, that is, $|z - 1| < 1 < |z|$ and $|z - \overline{\omega}| < |z - \omega|$. Other cases are done similarly.

9.14: $f(b) = \infty$, so the line ℓ is sent to a line, through $f(a)$ and $f(\infty)$, that is, through 0 and 1, so it is the real axis. The tangent to Σ at B meets Σ only at B, so its image is a line through $f(\infty)$, meeting m only at $f(b)$. So it is the line through 1, parallel to m. Finally, the tangent to Σ at A meets Σ only at A, so its image is a circle meeting m only at $f(a)$, that is, it touches m at O. It also passes through $f(\infty)$, so its centre is where the line $|z| = |z - 1|$ meets the perpendicular to m through O.

9.15: We have $\alpha(g) = fgf^{-1}$, so $\alpha(g_1g_2) = fg_1g_2f^{-1} = (fg_1f^{-1})(fg_2f^{-1}) = \alpha(g_1)\alpha(g_2)$, and we have a homomorphism. (Did you spot that this was assumed in the proof of Corollary 9.15?) If $\alpha(g) = 1$, then $fgf^{-1} = 1$, so $g = f^{-1}f = 1$. Thus α is injective. Every rotation of a sphere has a pair of poles, i.e., diametrically opposite fixed points, so its image under α will have two fixed points; however, there are Möbius transformations with only one fixed point (see Exercise 9.7), and these cannot be in the image of α. So α is not surjective.

9.16: Again, let f be a Möbius transformation with $f(A) = O$ and $f(B) = \infty$. Then $f(\Sigma_2)$ is a line through O, and $f(\Sigma_1)$ is *not* a line, but is a circle orthogonal to $f(\Sigma_2)$, that is, with its centre on $f(\Sigma_2)$. Let ℓ be the line AB. Then $f(\ell)$ is another line (not a circle) through O, and since ℓ is orthogonal to Σ_1, the centre of $f(\Sigma_1)$ lies on $f(\ell)$. But $f(\ell)$ and $f(\Sigma_2)$ meet at O, so O is the centre of $f(\Sigma_1)$. Since $f(A)$ ($= O$) and $f(B)$ ($= \infty$) are inverse points with respect to $f(\Sigma_1)$, it follows that A and B are inverse points with respect to Σ_1.

Inversion with respect to Σ_1 sends A to B and B to A, and it does this for every choice of ℓ. So inversion with respect to Σ_1 maps every point of Σ_2 to a point of Σ_2, and therefore Σ_2 is invariant under this inversion.

9.17: $(x - \alpha_1)^2 + (y - \beta_1)^2 + \lambda((x - \alpha_2)^2 + (y - \beta_2)^2) = 0$.

9.18: Σ_1 and Σ_2 determine a non-intersecting coaxal system; just choose f to send one of its limiting points to ∞. Steiner's porism is obvious if the given circles are concentric, and the general case is done by finding f to map the given circles Σ_1, Σ_2 to concentric circles, as above, and then applying the result to the circles $f(\Lambda_j)$.

9.19: Put $h = f^{-1}gf$. Since g is a conjugate Möbius transformation, so is h, and since g has a fixed line, h has a fixed line-or-circle, the image under f^{-1} of the line of reflection of g. Since g leaves $f(\Sigma)$ invariant and swaps $f(b)$ with $f(c)$, then h must leave Σ invariant and swap b with c. If the line joining b and c is parallel to the tangent to Σ at a, then h is reflection in the perpendicular bisector of this line, i.e., the diameter of Σ through a. If on the other hand the line joining b and c meets the tangent to Σ at d, then h is inversion with respect to the circle, centre d, through a. In either case, h leaves Σ, Σ_a, and Σ'_a invariant, and swaps b with c, Σ_b with Σ'_c, and Σ_c with Σ'_b. (It is now possible to rewrite the proof of Proposition 9.17 without using f, but just by applying h to the original diagram. Try it!)

9.20: Let the three circles be $|z - a| = r$, $|z - b| = s$, and $|z - c| = t$, where $r \le s \le t$. For the sake of ease, suppose these are mutually external to each other, and that we are trying to construct a circle $|z - d| = u$, touching them all externally; other cases (there are up to eight!) are done somewhat similarly. Then the circle $|z - d| = u + r$ passes through a, and touches the circles $|z - b| = s - r$ and $|z - c| = t - r$. So it is enough to be able to deal with the case where one of the circles is a point-circle, i.e., a point. But inversion with respect to any circle centred on this point sends the other circles into circles, and the required circle into one of their common tangents, and inverting again sends this tangent to the required circle. Inverse points can be constructed via Exercises 9.3 and 4.35, so it is easy to construct the inverse of a line-or-circle with respect to a circle. Common tangents can also be constructed: see Exercises 8.3 and 7.16; or use the above trick again: note that the external common tangent to $|z - a| = r$ and $|z - b| = s$, for example, is parallel to the tangent from a to $|z - b| = s - r$, which can be constructed as in Exercise 7.16.

9.21: Let the circles be Σ, Σ', with centres O, O' which, if not given, can be constructed. (OK?) Construct one of the common tangents, AA' as in the solution to Exercise 9.20, where A is on Σ and A' is on Σ'. Then the mid-point X of AA' is on the radical axis, and the circle centre X through A and A' is orthogonal to both Σ and Σ' (and to their radical axis), and so belongs to the orthogonal system. The required limiting points are where this circle meets OO'.

9.22: A typical point of the line is (t, k). Let $f(t, k) = (x, y)$. Then $t = kx/y$, and $x^2 + y^2 = (t^2 + k^2)/(t^2 + k^2 + 1)$, which on substituting for t gives $x^2 + (k^2 + 1)y^2/k^2 = 1$, and the result follows. If $k = 0$ we get $x = \cos\theta$, $y = 0$, $0 < \theta < \pi$, which is still the right answer, but is a diameter of the disc, part of the given line. (Think of it as a degenerate case, half of an ellipse whose minor axis is of length zero!)

9.23: We just need $c = 1$. (This is the power of $(0, 0)$ with respect to Σ; see Exercise 7.20.)

9.24: Put $a = z_0/|z_0|$, so that $z_0 = \lambda a$, where $\lambda = |z_0|$. The Möbius transformation $f : z \mapsto a(z + \lambda a)/(\lambda z + a)$ will do the trick. (Check that $f(a) = a$, $f(-a) = -a$, $f(0) = z_0$, and $f(\mathcal{D}) = \mathcal{D}$.)

9.25: Again, put $a = z_0/|z_0|$, and $\lambda = |z_0|$. Then $\pm a$ are the points at infinity on the h-line through 0 and z_0, and

$$\{a, -a \,;\, 0, z_0\} = \frac{(0-a)(\lambda a + a)}{(0+a)(\lambda a - a)} = \frac{1+\lambda}{1-\lambda},$$

and the result follows.

9.26: (Sketch) The coaxal system with limiting points z_1, z_2 contains a unique line-or-circle Σ, orthogonal to T, and z_1, z_2 are inverse points with respect to Σ. So if f is h-reflection in $\ell = \Sigma \cap \mathcal{D}$, then f is an h-isometry swapping z_1 and z_2, and fixing every point of ℓ. So certainly $d_h(z, z_1) = d_h(z, z_2)$ for every $z \in \ell$. Now suppose z is an h-point not on ℓ: we must show $d_h(z, z_1) \neq d_h(z, z_2)$. Since z_1, z_2 are on opposite sides of ℓ, suppose z and z_1 are on the same side of ℓ, so that z and z_2 are on opposite sides of ℓ. So the h-line through z and z_2 meets ℓ at a point z_3 between z and z_2. Then $d_h(z, z_1) < d_h(z, z_3) + d_h(z_3, z_1) = d_h(z, z_3) + d_h(z_3, z_2) = d_h(z, z_2)$.

9.27: We have $(x^2, y^2, z^2 + xy, yz, zx) = \lambda(x_1^2, y_1^2, z_1^2 + x_1 y_1, y_1 z_1, z_1, x_1)$. If $x \neq 0$, then $x_1 \neq 0$, so put $k = x/x_1$, so that $x = kx_1$ and $\lambda = k^2$. Then $kzx_1 = zx = \lambda z_1 x_1 = k^2 z_1 x_1$, whence $z = kz_1$. Next, $z^2 + kx_1 y = z^2 + xy = \lambda(z_1^2 + x_1 y_1) = k^2 z_1^2 + k^2 x_1 y_1 = z^2 + k^2 x_1 y_1$, whence $y = ky_1$ and we have $(x, y, z) = k(x_1, y_1, z_1)$, as required. Similarly if $y \neq 0$; and the case where $x = y = 0$ is easy.

9.28: There are ten points in the figure, and ten lines. Any of the ten points can be chosen as P. There are three lines through P, and six points on these lines other than P, namely A, A', B, B', C, C'. Any of these six points can be chosen as A. Then B has to be on a line through P and on a line through A (but not on PA), and there are two such points, namely B and C. Either can be chosen as B. The rest of the labelling is now determined: C is the 'other choice' for B, and A', B', C' are the 'other points' on PA, PB, PC respectively; and finally, $L = BC \cdot B'C'$, $M = CA \cdot C'A'$, and $N = AB \cdot A'B'$. So the size of the 'relabelling group' $G \subset S_{10}$ is $10 \times 6 \times 2 = 120$. This is 5!, so surely $G \cong S_5$? Consider the subset of points $\{P, A, B, C\}$. This has the property that the six lines PA, PB, PC, AB, AC, BC are all in the diagram, i.e., the complete quadrangle $PABC$ is part of the diagram. There are exactly five such quadrangles in the diagram: they are $PABC$, $PA'B'C'$, $AA'MN$, $BB'LN$, and $CC'LM$. (Please check.) These five subsets of the set of ten points are obviously permuted by the relabellings, and so we obtain a homomorphism $G \to S_5$, which clearly has trivial kernel (OK?) and so, since $|G| = |S_5|$, we have an isomorphism.

9.29: An affine transformation is given by a matrix $\mathbf{A} \in GL_3(\mathbb{R})$ with last row $(0, 0, 1)$, and such a matrix (though of as a projective transformation) sends each point of the form $(x, y, 0)$ to a point of the form $(x', y', 0)$; so the line $z = 0$ is invariant. Conversely, if $\mathbf{A} \in GL_3(\mathbb{R})$ sends each point of the form $(x, y, 0)$ to a point of the form $(x', y', 0)$, then its last row must be $(0, 0, k)$ for some $k \neq 0$. Then $k^{-1}\mathbf{A}$ gives the same projective transformation, and has last row $(0, 0, 1)$, so is affine.

9.30: $\lambda_1' = k_3 \lambda_1 / k_1$, $\lambda_2' = k_3 \lambda_2 / k_2$, $\mu_1' = k_4 \mu_1 / k_1$, and $\mu_2' = k_4 \mu_2 / k_2$.

9.31: We now assume $\mathbf{a}$, $\mathbf{b}$, $\mathbf{c}$ and $\mathbf{d}$ lie on the plane $z = 1$ in $\mathbb{R}^3$, so that $\lambda_1 + \lambda_2 = 1 = \mu_1 + \mu_2$. Then $\overrightarrow{CA} = \mathbf{a} - (\lambda_1 \mathbf{a} + \lambda_2 \mathbf{b}) = (1 - \lambda_1)\mathbf{a} - \lambda_2 \mathbf{b} = \lambda_2(\mathbf{a} - \mathbf{b})$, and similarly $\overrightarrow{CB} = \lambda_1(\mathbf{b} - \mathbf{a})$. So $CA/CB = -\lambda_2/\lambda_1$, and similarly $DA/DB = -\mu_2/\mu_1$, and the result follows.

9.32: For each i, the coordinates of A_i are $\mathbf{a}_i = \mathbf{a} + \theta_i \mathbf{b}$, for some $\mathbf{a}$, $\mathbf{b}$. Let $\mathbf{a}_3 = \lambda_1 \mathbf{a}_1 + \lambda_2 \mathbf{a}_2$ and $\mathbf{a}_4 = \mu_1 \mathbf{a}_1 + \mu_2 \mathbf{a}_2$. Thus $\mathbf{a} + \theta_3 \mathbf{b} = \lambda_1(\mathbf{a} + \theta_1 \mathbf{b}) + \lambda_2(\mathbf{a} + \theta_2 \mathbf{b})$, from which (since $\mathbf{a}$ and $\mathbf{b}$ are linearly independent) we have $\lambda_1 + \lambda_2 = 1$ and $\lambda_1 \theta_1 + \lambda_2 \theta_2 = \theta_3$, and thus $\lambda_1(\theta_1 - \theta_3) + \lambda_2(\theta_2 - \theta_3) = 0$.

Similarly, $\mu_1(\theta_1 - \theta_4) + \mu_2(\theta_2 - \theta_4) = 0$, and so

$$\{A_1, A_2 \,; A_3, A_4\} = \frac{\lambda_2 \mu_1}{\lambda_1 \mu_2} = \left(-\frac{\theta_1 - \theta_3}{\theta_2 - \theta_3}\right)\left(-\frac{\theta_2 - \theta_4}{\theta_1 - \theta_4}\right) = \frac{(\theta_3 - \theta_1)(\theta_4 - \theta_2)}{(\theta_3 - \theta_2)(\theta_4 - \theta_1)}.$$

(We have assumed the θ_i are all finite. What happens if one of them is infinite?)

9.33: (i) Let the common point be (x_0, y_0) so that the equation of ℓ_i is $y - y_0 = m_i(x - x_0)$. Homogenizing, ℓ_i has line coordinates $(m_i, -1, y_0 - m_i x_0) = (0, -1, y_0) + m_i(1, 0, -x_0)$. The formula for $\{\ell_1, \ell_2 \,; \ell_3, \ell_4\}$ now follows by a calculation like that in Exercise 9.32. Now let α_i be the angle between ℓ_i and the x-axis, so that the angle between ℓ_i and ℓ_j is $\alpha_i - \alpha_j$, and $m_i = \tan \alpha_i$. So

$$m_i - m_j = \tan \alpha_i - \tan \alpha_j = \frac{\sin(\alpha_i - \alpha_j)}{\cos \alpha_i \cos \alpha_j}.$$

Thus

$$\{\ell_1, \ell_2 \,; \ell_3, \ell_4\} = \frac{\sin(\alpha_3 - \alpha_1)\sin(\alpha_4 - \alpha_2)}{\sin(\alpha_3 - \alpha_2)\sin(\alpha_4 - \alpha_1)}.$$

(ii) If the lines are parallel, then $m_1 = m_2 = m_3 = m_4 = m$, say, and the line coordinates of ℓ_i are $(m, -1, c_i) = (m, -1, 0) + c_i(0, 0, 1)$. So by another calculation like that in Exercise 9.32, we have $\{\ell_1, \ell_2 \,; \ell_3, \ell_4\} = (c_3 - c_1)(c_4 - c_2)/((c_3 - c_2)(c_4 - c_1))$.

9.34: Every real symmetric invertible 4×4 matrix is congruent to one of the five diagonal matrices

$$\pm \begin{pmatrix} 1 & 0 & 0 & 0 \\ 0 & 1 & 0 & 0 \\ 0 & 0 & 1 & 0 \\ 0 & 0 & 0 & 1 \end{pmatrix}, \quad \pm \begin{pmatrix} 1 & 0 & 0 & 0 \\ 0 & 1 & 0 & 0 \\ 0 & 0 & 1 & 0 \\ 0 & 0 & 0 & -1 \end{pmatrix}, \quad \begin{pmatrix} 1 & 0 & 0 & 0 \\ 0 & 1 & 0 & 0 \\ 0 & 0 & -1 & 0 \\ 0 & 0 & 0 & -1 \end{pmatrix}.$$

Using (x, y, z, w) as homogeneous coordinates, the corresponding quadrics are $x^2 + y^2 + z^2 + w^2 = 0$, $x^2 + y^2 + z^2 - w^2 = 0$ and $x^2 + y^2 - z^2 - w^2 = 0$. Dehomogenizing by putting $w = 1$, we get $x^2 + y^2 + z^2 = -1$, an imaginary sphere; $x^2 + y^2 + z^2 = 1$, a sphere; and $x^2 + y^2 - z^2 = 1$, a *hyperboloid of one sheet*. This latter is the surface of revolution formed by rotating the hyperbola $x^2 - z^2 = 1$, $y = 0$ about the z-axis. Every proper real quadric is thus projectively equivalent to a sphere or a hyperboloid of one sheet. For example, the hyperboloid of *two* sheets, $x^2 - y^2 - z^2 = 1$, formed by rotating the same hyperbola about the x-axis instead, is projectively equivalent to a sphere. (OK?)

9.35: YW has line coordinates $(0, 1, 0) \times (1, 1, 1) = (1, 0, -1)$, and XZ has line coordinates $(0, 1, 0)$, so Y_1 is $(1, 0, -1) \times (0, 1, 0) = (1, 0, 1)$. So YW has equation $x = z$, and meets $y^2 = xz$ where $y^2 = x^2$, i.e., at $(1, \pm 1, 1) = (1, 0, 1) \pm (0, 1, 0)$, whence the result. For the second part, the tangents at W and W_1 are $x + z = \pm 2y$ (OK?) and these meet at $(1, 2, 1) \times (1, -2, 1) = (4, 0, -4)$, which is on $y = 0$.

9.36: Let the point be Y and the conic Γ. Let a line through Y meet Γ in W and W_1, and construct the harmonic conjugate Y_1 of Y with respect to W, W_1. (For this part of the construction, refer to Exercise 4.35.) Repeat, using a different line through Y, meeting Γ in W' and W_1', to obtain a second point Y_1'. Let $Y_1 Y_1'$ meet Γ at X and Z. Then YX and YZ are the required tangents. (Cf. Exercise 7.16.)

10

Complex geometry

In elementary school algebra, before we meet complex numbers, we are faced with the necessity to distinguish those quadratic equations like $x^2 - 1 = 0$ that have solutions, from those like $x^2 + 1 = 0$ which do not. The first encounter with the number i as an 'invented' solution to $x^2 + 1 = 0$ probably raised a range of feelings from scepticism and hilarity through to doubts about the sanity of our teachers. (The Greeks were equally sceptical about $\sqrt{2}$, as they thought all numbers had to be rational. When they discovered that $\sqrt{2}$ is not rational, they kept quiet about it!) Of course, soon we begin to see the power of complex numbers and the unification of seemingly unconnected facts that they bring to light, and we start to treat them as old and trusted friends.

In geometry, the same sort of difficulties arise. In $\mathbb{R}^2$, the line $y = 1$ meets the parabola $y = x^2$ where $x^2 = 1$, that is, at $(1, 1)$ and $(-1, 1)$. But the line $y = -1$ misses the same parabola, since to find a common point we need to solve $x^2 = -1$. Algebraically, this is no problem, and the common solutions are $x = \pm i$, $y = -1$; the bold step we now propose is to say that this means that the line $y = -1$ and the parabola $y = x^2$ meet at the points $(i, -1)$ and $(-i, 1)$. Of course, these are not in $\mathbb{R}^2$; but they *are* in $\mathbb{C}^2$, so we simply regard $\mathbb{R}^2$ as sitting inside $\mathbb{C}^2$, exactly as (or because) $\mathbb{R}$ sits inside $\mathbb{C}$.

Complex numbers allow the algebraist to say that, just as every linear equation has a solution, so too every quadratic equation has two solutions, possibly coincident. As geometers, we want to be able to say that, just as every pair of lines meet in a point, so too every line meets every conic in two points, possibly coincident. Of course, two lines might be parallel: the lines $x + y = 0$ and $x + y = 1$ do not meet in $\mathbb{C}^2$ any more than they meet in $\mathbb{R}^2$, so we must use projective geometry if we want every pair of lines to meet. Again, the asymptote $x = 0$ does not meet the hyperbola $xy = 1$ in $\mathbb{C}^2$ any more than it meets it in $\mathbb{R}^2$, so we shall not make much progress without points at infinity *as well as* complex points.

10.1   COMPLEX PROJECTIVE GEOMETRY

We thus define the *complex projective plane* $\mathbb{CP}^2$, as follows. Its points are given by homogeneous coordinates (x, y, z) where x, y, $z \in \mathbb{C}$, not all zero, and (x', y', z') represents the same point if and only if $(x, y, z) = \lambda(x', y', z')$ for some $\lambda \in \mathbb{C}$, $\lambda \neq 0$.

Everything works much as in $\mathbb{RP}^2$: lines are given by homogeneous linear equations, conics by homogeneous quadratic equations, cubics by homogeneous cubic equations, and so on; but the coefficients of the polynomials involved are now allowed to be complex. And a complex projective transformation is given by an invertible 3×3 *complex* matrix, an element of $GL_3(\mathbb{C})$.

In a similar way, the points of complex projective n-space $\mathbb{CP}^n$ are given by homogeneous coordinates from $\mathbb{C}^{n+1} \setminus \{\mathbf{0}\}$, $(n + 1)$-tuples of complex numbers, and the transformations by matrices in $GL_{n+1}(\mathbb{C})$. For example, a typical point of the complex projective line $\mathbb{CP}^1$ is given by $(z, w) \in \mathbb{C}^2$ (z, w not both zero), and a typical transformation by

$$ f : \begin{pmatrix} z \\ w \end{pmatrix} \mapsto \begin{pmatrix} a & b \\ c & d \end{pmatrix} \begin{pmatrix} z \\ w \end{pmatrix} = \begin{pmatrix} az + bw \\ cz + dw \end{pmatrix}, $$

where $ad \neq bc$. Now $(az + bw, cz + dw)$ and $((az + bw)/(cz + dw), 1)$ represent the same point of $\mathbb{CP}^1$, and if we dehomogenize by putting $w = 1$, we obtain a map $f : \mathbb{C} \cup \{\infty\} \to \mathbb{C} \cup \{\infty\}$ given by $z \mapsto (az+b)/(cz+d)$ (with the usual conventions for ∞), in other words, a Möbius transformation. So the inversive plane can be thought of as the complex projective line, and Möbius transformations as projective transformations of this line. In this way complex projective geometry includes within it *all* the other geometries we have studied: Euclidean, affine, real projective, inversive, hyperbolic, and elliptic. (The ambiguity over whether a plane is also a line is familiar in linear algebra, where $\mathbb{C} = \mathbb{C}^1$ is a one-dimensional complex space but also a two-dimensional real space: $\mathbb{C} \cong \mathbb{R}^2$.)

We are not going to take complex projective geometry very far. We shall content ourselves with a few sample theorems and proofs. First, as an example of how a complicated Euclidean theorem can be 'cleaned up' by more advanced methods, we sketch another proof of Theorem 7.10. Suppose the conic Γ is a line-pair, so that its equation is $(px + qy + rz)(p'x + q'y + r'z) = 0$, say. (We do not now need to distinguish cases like $x^2 - y^2 = 0$, a real line-pair, from $x^2 + y^2 = 0$, an imaginary line-pair.) So Γ is

$$ \begin{pmatrix} x & y & z \end{pmatrix} \begin{pmatrix} p \\ q \\ r \end{pmatrix} \begin{pmatrix} p' & q' & r' \end{pmatrix} \begin{pmatrix} x \\ y \\ z \end{pmatrix} = 0, $$

or $\mathbf{u}^T \mathbf{N} \mathbf{u} = 0$, where

$$ \mathbf{u} = \begin{pmatrix} x \\ y \\ z \end{pmatrix} \quad \text{and} \quad \mathbf{N} = \begin{pmatrix} p \\ q \\ r \end{pmatrix} \begin{pmatrix} p' & q' & r' \end{pmatrix} = \begin{pmatrix} pp' & pq' & pr' \\ qp' & qq' & qr' \\ rp' & rq' & rr' \end{pmatrix}. $$

This is not quite the standard matrix form for Γ, as $\mathbf{N}$ may not be symmetric; however, $\mathbf{M} = \mathbf{N} + \mathbf{N}^T$ is symmetric, and $\mathbf{u}^T \mathbf{N} \mathbf{u} = (\mathbf{u}^T \mathbf{N} \mathbf{u})^T = \mathbf{u}^T \mathbf{N}^T \mathbf{u}$, so $\mathbf{u}^T \mathbf{M} \mathbf{u} = 0$ is the matrix equation of Γ. Notice that, because $\mathbf{N}$ is the product of two matrices of rank 1, its rank is also 1, so that $\mathrm{rank}(\mathbf{M}) = \mathrm{rank}(\mathbf{N} + \mathbf{N}^T) \leq 1 + 1 = 2$, and $\det \mathbf{M} = 0$. (The relevant facts about rank can be found in [6].)

For the converse, suppose Γ has equation $\mathbf{u}^T \mathbf{M} \mathbf{u} = 0$, where $\mathbf{M}$ is symmetric. If $\det \mathbf{M} = 0$, then we can find a point $\mathbf{u}_1 \in \mathbb{CP}^2$ with $\mathbf{M} \mathbf{u}_1 = \mathbf{0}$; notice also that $\mathbf{u}_1^T \mathbf{M} = \mathbf{0}$. Let $\mathbf{u}_2$ be any point of Γ, so that $\mathbf{u}_2^T \mathbf{M} \mathbf{u}_2 = 0$, and let $\mathbf{u} = \mathbf{u}_1 + \theta \mathbf{u}_2$ be a point of the line joining $\mathbf{u}_1$ to $\mathbf{u}_2$. Then

$$\mathbf{u}^T \mathbf{M} \mathbf{u} = (\mathbf{u}_1 + \theta \mathbf{u}_2)^T \mathbf{M}(\mathbf{u}_1 + \theta \mathbf{u}_2)$$
$$= \mathbf{u}_1^T \mathbf{M} \mathbf{u}_1 + \theta \mathbf{u}_1^T \mathbf{M} \mathbf{u}_2 + \theta \mathbf{u}_2^T \mathbf{M} \mathbf{u}_1 + \theta^2 \mathbf{u}_2^T \mathbf{M} \mathbf{u}_2,$$

and here all four terms vanish, so that $\mathbf{u}$ is on Γ. Thus, for every $\mathbf{u}_2$ on Γ, the whole of the line joining $\mathbf{u}_2$ to $\mathbf{u}_1$ is part of Γ, which must therefore be two lines, or a repeated line. (Its equation is quadratic, so there are no other possibilities.)

Looking back at Theorem 7.10 again, what about all that business about $ab > h^2$, $ab = h^2$, and $ab < h^2$? The real conic $ax^2 + by^2 + cz^2 + 2fyz + 2gzx + 2hxy = 0$ meets the line $z = 0$ where $ax^2 + 2hxy + by^2 = 0$, which will give two real points if $ab < h^2$, two imaginary points if $ab > h^2$, and coincident real points if $ab = h^2$. So a proper conic is a hyperbola if it cuts the line at infinity in two real points, an ellipse if it cuts the line at infinity in two imaginary points, and a parabola if it touches the line at infinity.

As an example, suppose we want to find the centre and asymptotes of the hyperbola $x^2 + 2xy - 15y^2 - 2x + 1 = 0$. Homogenize to obtain $x^2 + 2xy - 15y^2 - 2xz + z^2 = 0$, which meets $z = 0$ where $x^2 + 2xy - 15y^2 = 0$, or $(x - 3y)(x + 5y) = 0$. So the two points at infinity on the hyperbola are $P = (3, 1, 0)$ and $Q = (-5, 1, 0)$. (Since they are real, it *is* a hyperbola.) The matrix of the conic is

$$\begin{pmatrix} 1 & 1 & -1 \\ 1 & -15 & 0 \\ -1 & 0 & 1 \end{pmatrix},$$

so the line coordinates of the tangents at P and Q are

$$\begin{pmatrix} 1 & 1 & -1 \\ 1 & -15 & 0 \\ -1 & 0 & 1 \end{pmatrix} \begin{pmatrix} 3 \\ 1 \\ 0 \end{pmatrix} = \begin{pmatrix} 4 \\ -12 \\ -3 \end{pmatrix}$$

and

$$\begin{pmatrix} 1 & 1 & -1 \\ 1 & -15 & 0 \\ -1 & 0 & 1 \end{pmatrix} \begin{pmatrix} -5 \\ 1 \\ 0 \end{pmatrix} = \begin{pmatrix} -4 \\ -20 \\ 5 \end{pmatrix}.$$

Dehomogenizing, the asymptotes are $4x - 12y - 3 = 0$ and $-4x - 20y + 5 = 0$, and these meet at $\left(\frac{15}{16}, \frac{1}{16} \right)$, the centre.

An equivalent calculation with a real ellipse will yield two imaginary points at infinity, and therefore two *imaginary* asymptotes; but these still meet at the real centre! As an example, take the circle with centre (α, β) and radius r, which has equation $(x - \alpha)^2 + (y - \beta)^2 = r^2$, or (homogenizing) $x^2 + y^2 - 2\alpha xz - 2\beta yz + cz^2 = 0$, where $c = \alpha^2 + \beta^2 - r^2$. This meets $z = 0$ where $x^2 + y^2 = 0$, that is, at $I = (1, i, 0)$ and $J = (1, -i, 0)$, and the tangents there have line coordinates

$$\begin{pmatrix} 1 & 0 & -\alpha \\ 0 & 1 & -\beta \\ -\alpha & -\beta & c \end{pmatrix} \begin{pmatrix} 1 \\ \pm i \\ 0 \end{pmatrix} = \begin{pmatrix} 1 \\ \pm i \\ -\alpha \mp \beta i \end{pmatrix},$$

so that the imaginary asymptotes are $x \pm iy = (\alpha \pm \beta i)$, meeting at (α, β).

The strange thing about the last calculation is that the points $I = (1, i, 0)$ and $J = (1, -i, 0)$ do not depend on which circle we choose: they lie on *every* circle. They are known as the *circular points at infinity*. Conversely, if the real conic $ax^2 + by^2 + cz^2 + 2fyz + 2gzx + 2hxy = 0$ passes through I and J, then $a - b \pm 2ih = 0$, so that $a = b$ and $h = 0$, and we have a circle. Notice that the imaginary asymptotes, the tangents at I and J, depend on α and β but not on r, so that concentric circles *touch* each other at I and J. The circular points have many other strange properties, for instance:

- Ex.10.1: *Let f be a projective transformation of $\mathbb{RP}^2$. Show that, as a projective transformation of $\mathbb{CP}^2$, f leaves the pair $\{I, J\}$ invariant if and only if f restricts to a similarity on $\mathbb{R}^2$.*

- Ex.10.2: *Let $\triangle ABC$ be a triangle in $\mathbb{R}^2$. Show that the cross-ratio $\{AB, AC; AI, AJ\}$ depends only on $\angle BAC$, and that $\{AB, AC; AI, AJ\} = -1$ if and only if $BA \perp AC$.*

Most of the theory we developed in $\mathbb{RP}^2$ generalizes easily to $\mathbb{CP}^2$, and indeed most of the proofs are identical. This opens the door to some highfalutin proofs of elementary theorems. For example, Chasles' theorem says that if A, B, C, D, P lie on a conic Γ, then $\{PA, PB; PC, PD\}$ depends on A, B, C, D (and Γ, of course), but not on P. If we take the case where $C = I$ and $D = J$, then Γ is a circle, and, with the aid of Exercise 10.2, Chasles' theorem just says that $\angle APB$ is constant as P moves around the circle, which is the theorem about angles in the same segment, Proposition 7.14. Again, suppose a conic Γ is put in its standard form $y^2 = xz$, so that the sides YX, YZ of the triangle of reference touch Γ at X and Z, and also the unit point W lies on Γ. We saw in Exercise 9.35 that if YW meets Γ again at W_1, and meets XZ at Y_1, then $\{W, W_1; Y, Y_1\} = -1$. If we take $X = I$ and $Z = J$, then the asymptotes are XY and ZY, so that Y is the centre of Γ, and WW_1 must be a diameter; since XZ is now the line joining the circular points, it is the line at infinity, so that Y_1 is at infinity, and the statement $\{W, W_1; Y, Y_1\} = -1$ now says that Y is the mid-point of WW_1, i.e., the centre of a circle is at the mid-point of every diameter! For P on the circle, Chasles' theorem says that $\{PW, PW_1; PI, PJ\}$ is independent of P; if we let $P \to I$ then PI becomes the tangent at I, that is, XY, and so $\{PW, PW_1; PI, PJ\} = \{XW, XW_1; XY, XY_1\} = -1$. By Exercise 10.2, $WP \perp PW_1$, or the angle in a semicircle is a right angle.

The possibility arises of trying some of the above tricks in the reverse direction; for example, in [29] Chasles' theorem for an arbitrary conic is deduced from the theorem about angles in the same segment of a circle. This kind of approach can be dangerous, and the literature is littered with attempts at quick painless proofs where not all is quite as easy (or correct) as it may seem at a first glance. For example, in [12] one of the proofs of Pascal's theorem proceeds as follows: suppose X_1, X_2, X_3, X_4, X_5, X_6 lie on a conic Γ, and let $L = X_1X_2 \cdot X_4X_5$ (the meet), $M = X_2X_3 \cdot X_5X_6$, and $N = X_3X_4 \cdot X_6X_1$. We have to show that L, M, N are collinear. Suppose LM meets Γ in P and Q. Take a projective transformation that sends these two points to the circular points, I and J. Then Γ (or, rather, its image, but we'll keep the same names) is a circle, and L, M are on the line at infinity. Thus $X_1X_2 \parallel X_4X_5$ and $X_2X_3 \parallel X_5X_6$, since the lines meet at infinity. But by Exercise 7.18, this means that $X_3X_4 \parallel X_6X_1$ also, so that N is also on the line at infinity, whence L, M, N are collinear.

There are at least two major flaws in this otherwise beautiful argument. First, the line LM might cut Γ (assumed real) in two real points, or two imaginary points (both of which cause no problem), or in two coincident points, when the proof breaks down. Secondly, even if P and Q are distinct, there is no reason why the projective transformation that takes them to I and J should be a *real* transformation, given by a real matrix; and if it is not, then although Γ is now (after transformation) a perfectly nice circle, there is no guarantee that any of the six points X_k will transform to real points on this circle. (It might even be an imaginary circle.) If the circle theorem (Exercise 7.18) had an algebraic proof that worked equally well for real or imaginary points, then the proof would hold; but the proof of Exercise 7.18 depended on the six points being real, so that we could chase angles around in two cyclic tetragrams. So the proof is a nice idea, but a swindle. We shall give a proper proof of Pascal's theorem in the next section.

10.2 ALGEBRAIC GEOMETRY

Recall that a *curve* is the set of points whose coordinates are the zeros of a polynomial equation. In $\mathbb{R}^2$ or $\mathbb{C}^2$ the polynomial will be any polynomial in two indeterminates, x and y; but in $\mathbb{RP}^2$ or $\mathbb{CP}^2$ the same curve is given by a *homogeneous* polynomial in three indeterminates, x, y, and z. Thus, for example, the line $x + y + 1 = 0$ in $\mathbb{R}^2$ becomes the line $x + y + z = 0$ in $\mathbb{RP}^2$. In general, if $p(x, y) = 0$ is the equation of a curve in $\mathbb{R}^2$, and if p has degree n in x and y together (which means that there is a non-zero term $ax^r y^s$ with $r + s = n$, and all other terms have $r + s \leq n$), then the homogenized polynomial is $z^n p(x/z, y/z)$, where *every* term has degree n in x, y, and z together: each term $ax^r y^s$ gets replaced by $ax^r y^s z^{n-r-s}$.

Plane algebraic geometry is the study of these curves. In dimensions higher than 2, one also needs to consider the common zeros of more than one polynomial: for example, in $\mathbb{R}^3$ a single linear equation gives a plane, not a line, and to get a line one needs to write down not one equation, but two. We have also seen that a conic can be obtained by intersecting a quadric surface (a cone, in fact) with a plane, so it is given by the common zeros of a quadratic and a linear equation; and more complicated curves can be obtained by intersecting two quadrics, that is, from two simultaneous quadratic polynomial equations. For example, the two quadrics $x = yz$ and $y = z^2$ in $\mathbb{R}^3$ meet in the set of points given parametrically by $x = \theta^3$, $y = \theta^2$, $z = \theta$, a 'curve' called a *twisted cubic*.

- Ex.10.3:   *Using homogeneous coordinates (x, y, z, w) in $\mathbb{RP}^3$, these quadrics homogenize to $xw = yz$ and $yw = z^2$. Show that their intersection in $\mathbb{RP}^3$ consists of the twisted cubic together with a certain line.*

We shall now confine our attention to dimension 2.

Definition 10.1.   *Let the curve Γ be given by the homogeneous equation $p(x, y, z) = 0$. Then the **order** of Γ is the degree of the polynomial p (in x, y, z together).*

It should be clear that the order of Γ is unaffected by a change of coordinates, since this involves replacing x, y, z by three homogeneous linear expressions in x, y, z in the polynomial p, and the new polynomial obtained will have the same degree as p.

A non-constant polynomial p is *irreducible* if any factorization $p = p_1 p_2$ involves one or other of the polynomials p_1, p_2 being a constant. We shall use without proof the fact that every non-constant polynomial can be factorized as a product of (one or more) irreducible polynomials, uniquely up to the order of the factors and constant multiples. (A proof can be found in [7].)

- Ex.10.4:   *For example, the polynomial $2xy$ factorizes as $(2x)(y)$ or $(y)(2x)$ or $(x)(2y)$ or $(2y)(x)$. Is that all?*

- Ex.10.5:   *Prove that if p, p_1, p_2 are polynomials with $p = p_1 p_2$, then p is homogeneous if and only if both p_1 and p_2 are homogeneous.*

Definition 10.2.   *A curve is **irreducible** if it is given by $p = 0$, where p is an irreducible polynomial; otherwise it is **reducible**.*

Definition 10.3.   *Let Γ be a curve with equation $p = 0$, and let $p = p_1 p_2 \cdots p_r$ be a factorization of p into irreducible polynomials. Let Γ_i be the (irreducible) curve with equation $p_i = 0$. Then $\Gamma_1, \Gamma_2, \ldots, \Gamma_r$ are the **components** of Γ. If, for example, $\Gamma_1 = \Gamma_2$, then Γ_1 is a **repeated** component of Γ.*

It should be clear that, as a set, $\Gamma = \Gamma_1 \cup \Gamma_2 \cup \ldots \cup \Gamma_r$, whether or not Γ has repeated components. By the uniqueness of polynomial factorization, the components are well defined; also it should be clear that a change of coordinate system does not affect the way a curve splits up into components, or whether or not a curve is irreducible.

Examples.   A line is an irreducible curve of order 1. A conic is a curve of order 2, and for conics, *irreducible* means the same as *proper*, and *reducible* means the same as *improper*. An improper conic either has two components, both lines, or a single repeated component, also a line. A *cubic* is a curve of order 3. If a cubic is reducible, it consists of either a proper conic and a line, such as $(y^2 - xz)x = 0$, or three lines, such as $xyz = 0$, or a repeated line with a second line, such as $x^2 y = 0$, or a line three times, such as $x^3 = 0$.

10.2.1   Intersections of curves

We are now going to investigate the intersections of certain pairs of curves. We start with a line and an arbitrary curve. We shall need the *fundamental theorem of algebra*, which states that over $\mathbb{C}$, every non-constant polynomial $p(x)$ in one indeterminate x has a zero $a \in \mathbb{C}$; or (equivalently, via the *remainder theorem*), $p(x)$ is divisible by $x - a$. Proofs of these theorems can be found in many places, for example [3], [30], and [6], [7], [8]. Repeating the argument, $p(x)$ can be written as a product of linear factors. If we homogenize this statement, then the polynomial becomes a homogeneous polynomial in two indeterminates, and its linear factors become homogeneous linear factors. Explicitly:

Lemma 10.4. *Let $p(x, y)$ be a non-constant homogeneous complex polynomial in the two indeterminates x and y. If the degree of p in x and y together is n, then $p(x, y)$ factorizes as a product of n homogeneous linear factors over $\mathbb{C}$.*

Proof. Write $p(x, y) = y^r q(x, y)$, where $r \geq 0$ and $q(x, y)$ is homogeneous of degree $n - r$, and $q(x, y)$ does not divide by y. If $r = n$, we have finished; so assume $r < n$. So $q(x, y)$ contains a non-zero term in x^{n-r}, and $q(x, 1)$ is a non-constant polynomial in x. By the fundamental theorem quoted above, we can write

$$q(x, 1) = a_0(x - a_1)(x - a_2) \cdots (x - a_{n-r})$$

for some $a_0, a_1, \ldots, a_{n-r} \in \mathbb{C}$, with $a_0 \neq 0$. From this,

$$q(x, y) = y^{n-r} q(x/y, 1) = a_0(x - a_1 y)(x - a_2 y) \cdots (x - a_{n-r} y),$$

and so

$$p(x, y) = a_0 y^r (x - a_1 y)(x - a_2 y) \cdots (x - a_{n-r} y).$$

□

Proposition 10.5. *In $\mathbb{CP}^2$, let ℓ be a line, and let Γ be a curve of order n. If ℓ is not a component of Γ, then ℓ and Γ meet in at least one and at most n points.*

Proof. The properties being investigated do not depend on the choice of coordinate system, so choose the triangle of reference XYZ with X, Y on ℓ, so that ℓ has equation $z = 0$. Let Γ have equation $p(x, y, z) = 0$, so that $p(x, y, z)$ is homogeneous of degree n in x, y, and z. If ℓ is not a component of Γ, then $p(x, y, z)$ is not divisible by z, so that $p(x, y, 0)$ is homogeneous of degree n in x and y. By Lemma 10.4,

$$p(x, y, 0) = (a_1 x - b_1 y)(a_2 x - b_2 y) \ldots (a_n x - b_n y)$$

for some $a_i, b_i \in \mathbb{C}$, where for each i the constants a_i and b_i are not both zero. So the point $(b_i, a_i, 0)$ lies on both Γ and ℓ, for $1 \leq i \leq n$; though of course different values of i will not necessarily give different points. On the other hand, if (a, b, c) lies on both ℓ and Γ, then $c = 0$ and $p(a, b, 0) = 0$, so that $a_i a - b_i b = 0$ for some i. But this means that $a = \lambda b_i$ and $b = \lambda a_i$ for some $\lambda \neq 0$. Thus $(a, b, c) = \lambda(b_i, a_i, 0)$, and both (a, b, c) and $(b_i, a_i, 0)$ represent the same point, so we have already found all the points lying on both ℓ and Γ.

□

Note that when $n = 1$, this says that a line meets another (different) line in exactly one point.

- **Ex.10.6:**   When $n = 2$, Proposition 10.5 says a line ℓ and a conic Γ (assumed not to contain ℓ as a component) meet in either one or two points. Precisely when do they meet in one point, rather than two?

Corollary 10.6.   *Let Γ be a curve in $\mathbb{CP}^2$. There are infinitely many points of $\mathbb{CP}^2$ on Γ, and infinitely many points not on Γ.*

Proof.   Let Γ have order n. For $n = 1$, the result is obvious, so suppose $n > 1$. Then Γ has at most n components, so we can find a line ℓ which is not a component of Γ, and, by Proposition 10.5, ℓ contains infinitely many points not on Γ. Next, if Γ contained only r points, say, then it is easy to find a line ℓ that does not contain any of these r points (OK?); but ℓ meets Γ, by Proposition 10.5, a contradiction.   □

Having investigated the intersection of a line with an arbitrary curve, we now look at the intersection of a conic with an arbitrary curve. First, we need a technical lemma. For any polynomial $p(x, y, z)$ in x, y, and z, we can write $p(x, y, z)$ as a polynomial in y whose coefficients are polynomials in x and z. The highest power of y that occurs is the *y-degree* of $p(x, y, z)$. We can then divide $p(x, y, z)$ by the polynomial $y^2 - xz$, which has y-degree 2, leaving a remainder whose y-degree is at most 1. Explicitly:

Lemma 10.7.   *Let $p(x, y, z)$ be a homogeneous polynomial in x, y, and z. Then*

$$p(x, y, z) = (y^2 - xz)q(x, y, z) + r(x, y, z),$$

where $q(x, y, z)$, $r(x, y, z)$ are homogeneous and the y-degree of $r(x, y, z)$ is at most 1.
□

Proposition 10.8.   *In $\mathbb{CP}^2$, let Λ be a conic, and let Γ be a curve of order n. If Λ and Γ do not have a common component, then Λ and Γ meet in at least one and at most $2n$ points.*

Proof.   Suppose first that Λ is reducible (improper), so that it has two components, both lines (and not necessarily distinct). Either of these lines meets Γ in at least one point, by Proposition 10.5, so Λ and Γ have at least one point in common. On the other hand, if Λ and Γ have $2n + 1$ (or more) points in common, then $n + 1$ or more of these lie on one of the two components of Λ, which is then also a component of Γ, by Proposition 10.5, and we have finished.

So now suppose Λ is irreducible (proper). Choose a coordinate system in which Λ takes its standard form $y^2 = xz$ (this is the $\mathbb{CP}^2$ version of Theorem 9.23), and suppose that in this system Γ has equation $p(x, y, z) = 0$. Now every point of Λ has coordinates $(\theta^2, \theta, 1)$ for a unique $\theta \in \mathbb{C}$, except the point $(1, 0, 0)$, which corresponds to $\theta = \infty$. From our present point of view, it is better to homogenize this last statement, and say that every point of Λ has coordinates $(\theta^2, \theta\varphi, \varphi^2)$ for some $\theta, \varphi \in \mathbb{C}$, not both zero; and here (θ, φ) and $(\lambda\theta, \lambda\varphi)$ are parameters for the same point, for any $\lambda \neq 0$. (In effect, (θ, φ) are homogeneous coordinates for Λ, which thus becomes rather like $\mathbb{CP}^1$, the projective line. This point of view can be taken much further; see [29] for details.)

We thus have that every common point of Λ and Γ is of the form $(\theta^2, \theta\varphi, \varphi^2)$, where $p(\theta^2, \theta\varphi, \varphi^2) = 0$. Now $p(\theta^2, \theta\varphi, \varphi^2)$ is a polynomial in θ and φ, regarded as indeterminates, and if as such it is not the zero polynomial, then it is homogeneous of degree $2n$ in θ and φ. In this case, by Lemma 10.4, we have

$$p(\theta^2, \theta\varphi, \varphi^2) = (a_1\theta - b_1\varphi)(a_2\theta - b_2\varphi)\ldots(a_{2n}\theta - b_{2n}\varphi)$$

where for each i, the constants a_i and b_i are not both zero. So $p(\theta^2, \theta\varphi, \varphi^2) = 0$ if and only if $a_i\theta - b_i\varphi = 0$ for some i, that is, $\theta = \lambda b_i$ and $\varphi = \lambda a_i$ for some i and some $\lambda \neq 0$. So the points $(\lambda^2 b_i^2, \lambda^2 a_i b_i, \lambda^2 a_i^2)$, or (equivalently) $(b_i^2, a_i b_i, a_i^2)$ lie on both Λ and Γ, and no others. Of course, these points may not be distinct, but there are not more than $2n$ of them, which is what we wanted.

It remains only to look at the case where $p(\theta^2, \theta\varphi, \varphi^2)$ is the zero polynomial in θ and φ. This is where we use Lemma 10.7: we can write

$$p(x, y, z) = (y^2 - xz)q(x, y, z) + r(x, y, z),$$

where $q(x, y, z), r(x, y, z)$ are homogeneous and the y-degree of $r(x, y, z)$ is at most 1. We want to show that $y^2 - xz$ divides $p(x, y, z)$, so it is enough to show that $r(x, y, z)$ is identically zero. Now on substituting $x = \theta^2$, $y = \theta\varphi$, and $z = \varphi^2$, both $p(x, y, z)$ and $y^2 - xz$ vanish identically, and hence so does $r(x, y, z)$. But since $r(x, y, z)$ is homogeneous of degree n in x, y, and z together, and has y-degree at most 1, we can write

$$r(x, y, z) = (a_0 x^n + a_2 x^{n-1}z + \ldots + a_{2n}z^n)$$
$$+ y(a_1 x^{n-1} + a_3 x^{n-2}z + \ldots + a_{2n-1}z^{n-1})$$

for some $a_i \in \mathbb{C}, 0 \leq i \leq 2n$. But now

$$r(\theta^2, \theta\varphi, \varphi^2) = a_0\theta^{2n} + a_1\theta^{2n-1}\varphi + a_2\theta^{2n-2}\varphi^2 + \ldots + a_{2n}\varphi^{2n},$$

and since this vanishes identically, we deduce that $a_i = 0$, for all i. This means that $r(x, y, z)$ is the zero polynomial, and we have finished. □

Propositions 10.5 and 10.8 are part of *Bezout's theorem*, which says that two curves Λ and Γ, of orders m and n respectively, meet in at least one and at most mn points (Fig. 10.1). Proposition 10.5 is the case $m = 1$ and Proposition 10.8 is the case $m = 2$, and this is all we need at present. Bezout's theorem actually says more: it assigns a positive integer, the *intersection multiplicity*, to each common point of Λ and Γ, in a manner that does not depend on the choice of coordinate system, and in such a way that the intersection multiplicities of the common points add up to exactly mn. For a line meeting a curve, this is easy: the intersection multiplicity is just the algebraic multiplicity, that is, the number of times the corresponding factor appears when we solve the appropriate equation. For example, the line $x = 0$ is a tangent to the conic $y^2 = xz$ at $(0, 0, 1)$, where the intersection multiplicity is 2; this is because putting

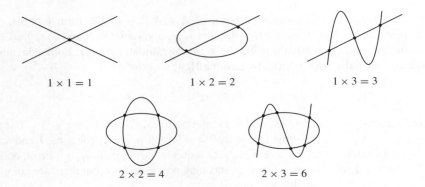

Fig. 10.1 Easy cases of Bezout's theorem

$x = 0$ in $y^2 = xz$ gives $y^2 = 0$, a double root at $y = 0$. On the other hand $y = 0$ meets $y^2 = xz$ where $xz = 0$, which has single roots at $x = 0$ and $z = 0$, so the intersection multiplicities at $(0, 0, 1)$ and $(1, 0, 0)$ are both 1. Again, the line $x = 0$ is a tangent to the cubic $y^3 = xz^2$ at $(0, 0, 1)$, where, since putting $x = 0$ gives $y^3 = 0$, a triple root, the intersection multiplicity is 3. The method we used to prove Bezout's theorem for $m = 2$ depended on being able to write x, y, and z, for an arbitrary point of Λ, as polynomials in θ and φ; a curve for which this can be done is called a *rational* curve. Every line and every irreducible conic is rational, but *not* every irreducible cubic or higher-order curve, so that different techniques are needed when $m \geq 3$. The classical approach is via resultants, for which we do not have space here: for details, see [32] or [28]. The more modern approach uses even more abstract machinery (ring theory), which while being the 'right' way to do things, makes little sense without the appropriate background knowledge. A good introduction to the modern approach can be found in [15], but try the classical approach first if you want to understand why such an enormous amount of fuss is needed to get such a seemingly simple idea on a sound footing.

10.2.2 Nine associated points

We turn to a new question: how many points are needed to determine a curve of order n? When $n = 1$, the answer is 2: a (unique) line can be drawn through any two points. Algebraically, if the points are (x_1, y_1, z_1) and (x_2, y_2, z_2), then we have to solve

$$px_1 + qy_1 + rz_1 = 0,$$
$$px_2 + qy_2 + rz_2 = 0$$

for p, q, r: two homogeneous linear equations in three unknowns, which always has a non-trivial solution. Indeed, if the vectors (x_1, y_1, z_1) and (x_2, y_2, z_2) are linearly independent (which they are, if the points are distinct—OK?), then the solution is unique up to a scalar multiple, so that the line is unique.

Now for $n = 2$: what if we try to put a conic Γ through the points $P_i = (x_i, y_i, z_i)$, $i = 1, 2, \ldots, r$? We have to solve r simultaneous equations

$$ax_i^2 + by_i^2 + cz_i^2 + 2fy_iz_i + 2gz_ix_i + 2hx_iy_i = 0 \tag{10.1}$$

for the six unknowns a, b, c, f, g, h. Notice that x_i, y_i, z_i are known, and the equations are *homogeneous linear* in the unknowns a, b, c, f, g, h. Now a set of simultaneous homogeneous linear equations always has a non-trivial solution if the number of unknowns exceeds the number of equations. More precisely, a set of r linearly independent simultaneous homogeneous linear equations in s unknowns has $s - r$ linearly independent solutions if $r < s$; and any $s - r + 1$ solutions are linearly dependent. This theorem, which we shall not prove here, can be found in almost any text on linear algebra, for example, in [6] or [3]. So, taking $s = 6$, we can find a non-trivial solution of our equations provided $r \leq 5$, that is, we can always find a conic through any 5 points. If the five equations are linearly independent, that is, no one of them is a linear combination of the others, then the solution is unique up to a scalar multiple, and so in this case the conic is unique. Does this always happen? The answer is no, since if, for example, P_1, P_2, P_3, P_4 are collinear, then the line containing them together with *any* line through P_5 is a (reducible) conic through our five points. However, if no four of our five points are collinear, then the conic through them is unique. To prove this, we need to show that each of the five equations (10.1) is linearly independent of the other four, and this follows if we can find a solution to any four that does *not* solve the fifth, that is, if we can find a conic through any four of the points that does not go through the fifth.

So let us find, for example, a conic through P_1, P_2, P_3, P_4 that does not go through P_5. The line-pair $P_1P_2 \cup P_3P_4$ is a suitable choice—no one said the conic had to be irreducible!—provided P_5 is not collinear with P_1 and P_2 or with P_3 and P_4. If P_1, P_2, P_5 are collinear, then P_1, P_3, P_5 are not collinear (else the four points P_1, P_2, P_3, P_5 are collinear, which we have banned), and similarly P_2, P_4, P_5 are not collinear. In this case the line-pair $P_1P_3 \cup P_2P_4$ does the trick. The other case, when P_3, P_4, and P_5 are collinear, is done similarly. We have proved:

Proposition 10.9. *There is a conic through any five points, and it is unique if and only if no four of the points are collinear. It is irreducible if and only if no three of the points are collinear.*   □

A special case of this is when we take two of the five points to be I and J, the circular points at infinity. This makes the conic a circle, if it is irreducible, so the proposition now says that there is a circle through any three points, provided they are not collinear. (If three points lie on a line ℓ, the conic through I, J and these three points is the line-pair consisting of ℓ together with the line at infinity.) If instead we take two of the five points to be real points at infinity, then the proposition says we can draw a hyperbola through three given points (not collinear, and with no two collinear with either point at infinity) with its asymptotes parallel to two given lines (i.e., the lines joining an arbitrary point to the given points at infinity).

The dual of a conic is a conic envelope, so the dual of Proposition 10.9 says there is a conic touching five lines. If one of these lines is the line at infinity, then the conic, if irreducible, is a parabola, so in general a parabola can be found to touch four given lines. The reducible case occurs if three or more of the given lines are concurrent.

So now we turn to the case $n = 3$: how many points determine a cubic?

Proposition 10.10.   *There is a cubic through any nine points.*

Proof.   Let the points be $P_i = (x_i, y_i, z_i)$, $1 \leq i \leq 9$. We must solve the nine simultaneous homogeneous linear equations

$$ax_i^3 + by_i^3 + cz_i^3 + dx_i^2 y_i + ey_i^2 z_i + fz_i^2 x_i$$
$$+ gx_i y_i^2 + hy_i z_i^2 + kz_i x_i^2 + \ell x_i y_i z_i = 0$$

for the ten unknowns $a, b, c, d, e, f, g, h, k, \ell$, and because there are more unknowns than equations, there will be a non-trivial solution.   □

The uniqueness question is more complicated for cubics than it was for conics. Two conics meet in $2 \times 2 = 4$ points, at most, and five points determine a conic. But if Bezout's theorem is to be believed, two cubics can meet in up to $3 \times 3 = 9$ points, and so we should *not* expect any nine points so obtained to determine a *unique* cubic! To see exactly what goes on, we first prove a lemma about cubics through eight points:

Lemma 10.11.   *Let $P_i = (x_i, y_i, z_i)$, $1 \leq i \leq 8$, be eight points, no four of which are collinear and no seven of which lie on a conic. Then the eight equations*

$$ax_i^3 + by_i^3 + cz_i^3 + dx_i^2 y_i + ey_i^2 z_i + fz_i^2 x_i$$
$$+ gx_i y_i^2 + hy_i z_i^2 + kz_i x_i^2 + \ell x_i y_i z_i = 0$$

in the ten unknowns $a, b, c, d, e, f, g, h, k, \ell$ are linearly independent.

Proof.   We need to show that there is a solution to any seven of the equations which is not a solution of the eighth, that is, we need to find a cubic that passes through any seven of the points but not through the eighth. So, for example, let us find a cubic through $P_1, \ldots, P_7$ that does not pass through P_8.

By Proposition 10.9, we can find a conic Λ_1 through P_1, P_4, P_5, P_6, P_7; a conic Λ_2 through P_2, P_4, P_5, P_6, P_7; and a conic Λ_3 through P_3, P_4, P_5, P_6, P_7. We claim that P_8 cannot lie on any two of these three conics. For suppose, for example, that P_8 lies on both Λ_1 and Λ_2. Then each of Λ_1, Λ_2 contains P_4, P_5, P_6, P_7, P_8. Since no four of these points are collinear, we deduce from Proposition 10.9 that $\Lambda_1 = \Lambda_2$. But now the seven points P_1, P_2, P_4, P_5, P_6, P_7, P_8 lie on Λ_1 $(= \Lambda_2)$, which contradicts the given conditions. So since P_8 does not lie on any two of the three cubics, let us suppose (renumbering if necessary) that it does not lie on Λ_1 or Λ_2 (but might possibly lie on Λ_3).

We now define two reducible cubics, each consisting of a conic and a line. (We did not say 'whose components are a conic and a line', because the conics involved might themselves be reducible.) The cubics are $\Gamma_1 = \Lambda_1 \cup P_2 P_3$ and $\Gamma_2 = \Lambda_2 \cup P_1 P_3$. We claim that one or other of these cubics solves our problem; both contain P_1, P_2, P_3, P_4,

P_5, P_6, P_7; and we claim that they cannot both contain P_8. For if they did, then since neither of the conics Λ_1, Λ_2 contains P_8, it follows that both the lines P_1P_2 and P_1P_3 contain P_8. But this means that P_1, P_2, P_3, P_8 are collinear, which again contradicts the given conditions, and our proof is complete. □

Theorem 10.12 (Nine associated points). *Let Γ_1 and Γ_2 be two cubic curves, meeting in precisely nine distinct points. If Γ_3 is any cubic curve containing eight of these points, then it contains the ninth point also.*

Proof. First suppose that four of the nine points lie on a line, ℓ. Then by Proposition 10.5, ℓ is a component of both Γ_1 and Γ_2, which therefore have infinitely many common points, a contradiction. So no four of the nine points are collinear.

Next suppose that seven of the nine points lie on a conic, Λ. If Λ is reducible, then $\Lambda = \ell_1 \cup \ell_2$ for some lines ℓ_1, ℓ_2 (not necessarily distinct), and now either ℓ_1 or ℓ_2 must contain at least four of the seven points, another contradiction. So Λ is irreducible, and by Proposition 10.8, Λ is a component of both Γ_1 and Γ_2, which once again therefore have infinitely many common points, a contradiction. So no seven of the nine points lie on a conic.

We are thus in the situation of Lemma 10.11. Let the nine points be $P_i = (x_i, y_i, z_i)$ for $1 \leq i \leq 9$, and suppose Γ_3 contains P_i for $1 \leq i \leq 8$. Suppose also that, for $j = 1, 2, 3$, Γ_j has equation $p_j(x, y, z) = 0$, where

$$p_j(x, y, z) = a_j x^3 + b_j y^3 + c_j z^3 + d_j x^2 y + e_j y^2 z$$
$$+ f_j z^2 x + g_j x y^2 + h_j y z^2 + k_j z x^2 + \ell_j xyz.$$

We know that, for $j = 1, 2, 3$, $(a_j, b_j, c_j, d_j, e_j, f_j, g_j, h_j, k_j, \ell_j)$ is a solution of the eight simultaneous equations

$$ax_i^3 + by_i^3 + cz_i^3 + dx_i^2 y_i + ey_i^2 z_i + fz_i^2 x_i$$
$$+ gx_i y_i^2 + hy_i z_i^2 + kz_i x_i^2 + \ell x_i y_i z_i = 0$$

($1 \leq i \leq 8$), and by Lemma 10.11 these eight equations are independent. So we are in the case $s = 10$, $r = 8$ of the theorem on simultaneous equations quoted above (page 277), and the three solutions we have must be linearly dependent, that is,

$$\lambda_1 p_1(x, y, z) + \lambda_2 p_2(x, y, z) + \lambda_3 p_3(x, y, z)$$

is the zero polynomial for some choice of scalars λ_1, λ_2, λ_3, not all zero. Now we cannot have $\lambda_3 = 0$, else $\lambda_1 p_1(x, y, z) = -\lambda_2 p_2(x, y, z)$, which would mean $\Gamma_1 = \Gamma_2$, which is impossible. But now

$$\lambda_1 p_1(x_9, y_9, z_9) + \lambda_2 p_2(x_9, y_9, z_9) + \lambda_3 p_3(x_9, y_9, z_9) = 0,$$

and since $p_1(x_9, y_9, z_9) = 0$, $p_2(x_9, y_9, z_9) = 0$, and $\lambda_3 \neq 0$, we deduce that $p_3(x_9, y_9, z_9) = 0$, in other words, Γ_3 contains P_9. □

There is no immediately obvious dual to the above result, because the dual of a cubic need not be a cubic envelope, that is, the line equation of the corresponding envelope need not have degree 3. In more detail, the *order* of a curve was defined as the degree of the defining polynomial, but by courtesy of Bezout's theorem (plus a little more work), it has a more geometrical definition: it is the maximum number of distinct points in which a line meets the curve (provided the curve does not have a repeated component). The dual concept is the *class* of the curve, which is the maximum number of distinct tangents to the curve that pass through a single point. For a conic, order and class are both 2: a line meets it in two points, and from any point, two tangents can be drawn. For a cubic, things are more complicated: the order is 3 (that is what is meant by a *cubic!*), but the class can be 3 (if the cubic has a cusp, like $y^2 = x^3$), or 4 (if the cubic has a node, like $y^2 = x^2(x+1)$), or 6 (if the cubic has neither node nor cusp, like $y^2 = (x+4)(x^2+1)$). For details, see [32].

10.2.3 Applications

We shall apply the nine associated points theorem to various situations, mostly involving conics and lines, so that our cubics will mostly be reducible, consisting of a conic and a line, or of three lines. First, the promised proof of Pascal's theorem:

Proposition 10.13 (Pascal's mystic hexagram). *Let X_1, X_2, X_3, X_4, X_5, X_6 be six distinct points on a proper conic Λ. Let $L = X_1X_2 \cdot X_4X_5$ (the meet of the joins X_1X_2 and X_4X_5); let $M = X_2X_3 \cdot X_5X_6$, and let $N = X_3X_4 \cdot X_6X_1$ (similarly). Then L, M, N are collinear. In words: if a hexagram has its vertices on a proper conic, then the meets of opposite sides are collinear.*

Proof. First note that L cannot lie on Λ, for if it did then either Λ contains three collinear points X_1, X_2, L (which is impossible by Proposition 10.5, since Λ is irreducible), or else L coincides with one of X_1 and X_2 and also with one of X_4 and X_5 (which means that two of X_1, X_2, X_4, X_5 coincide, another contradiction). Similarly for M and N.

Next, L and M are distinct, else X_1X_2 and X_2X_3 both pass through L, which means that X_1, X_2, X_3 are collinear. This again is ruled out by Proposition 10.5 and the fact that Λ is irreducible. Similarly for L and N, and for M and N. (Of course, the conclusion of the proposition holds trivially if any two of L, M, N coincide, but we'll ignore that.) So the nine points X_1, X_2, X_3, X_4, X_5, X_6, L, M, N are distinct.

We now examine two reducible cubics, each consisting of three lines. They are $\Gamma_1 = X_1X_2 \cup X_3X_4 \cup X_5X_6$, and $\Gamma_2 = X_2X_3 \cup X_4X_5 \cup X_6X_1$. They meet *precisely* in our nine points X_1, X_2, X_3, X_4, X_5, X_6, L, M, N (Fig. 10.2), and so we have nine associated points. The cubic $\Gamma_3 = \Lambda \cup LM$ contains eight of these points, namely X_1, X_2, X_3, X_4, X_5, X_6 (on Λ) and L, M (on $LM!$), so by Theorem 10.12, Γ_3 contains N also. Now N does not lie on Λ, so it must lie on LM, which is what we wanted. $\square$

Proposition 10.14 (Brianchon's theorem). *Let x_1, x_2, x_3, x_4, x_5, x_6 be six distinct tangents to a proper conic Λ (Fig. 9.21). Let $\ell = x_1x_2 \cdot x_4x_5$ (the join of the meets x_1x_2 and x_4x_5); let $m = x_2x_3 \cdot x_5x_6$, and let $n = x_3x_4 \cdot x_6x_1$ (similarly). Then ℓ, m, and n*

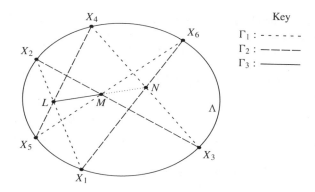

Fig. 10.2 Pascal's theorem, via nine associated points

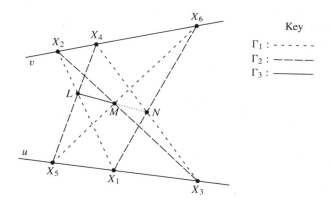

Fig. 10.3 Pappus' theorem

are concurrent. In words: if the sides of a hexagram touch a proper conic, then the joins of opposite vertices are concurrent.

Proof. This is the dual of Pascal's theorem. □

There is a version of Pascal's theorem for an improper conic, a line-pair:

Proposition 10.15 (Pappus' theorem). *Let u, v be distinct lines. Let X_1, X_3, X_5 be distinct points of u, not on v, and let X_2, X_4, X_6 be distinct points of v, not on u. Let $L = X_1X_2 \cdot X_4X_5$ (the meet of the joins X_1X_2 and X_4X_5); let $M = X_2X_3 \cdot X_5X_6$, and let $N = X_3X_4 \cdot X_6X_1$ (similarly). Then L, M, N are collinear. In words: if a hexagram has its vertices lying alternately on the two lines of a line-pair, then the meets of opposite sides are collinear* (Fig. 10.3).

Proof.   The proof is much the same as for Pascal's theorem, using the reducible conic $\Lambda = u \cup v$. It is easy to check that L, M, N are distinct and none of them lies on Λ, so we can define the cubics Γ_1, Γ_2, Γ_3 as before, and the same proof works.   □

We leave the reader to explore what happens if we allow the six points to be distributed between u and v in a different way, or if one of them lies on both u and v.

- Ex.10.7:   *State the dual of Pappus' theorem, and draw a suitable diagram.*

- Ex.10.8:   *Just as with Desargues' theorem (see Exercise 9.28), the Pappus diagram can be relabelled many ways. For example, take the hexagram $X_1 X_5 L M X_2 X_6$, all of whose sides are already in the diagram, and whose vertices lie alternately on $X_1 X_2$ and $X_5 X_6$. Then $X_1 X_5 \cdot M X_2 = X_3$, $X_5 L \cdot X_2 X_6 = X_4$, $L M \cdot X_6 X_1 = N$, and the line through these three points is already in the diagram. Thus the permutation $(X_2 X_5)(X_3 L)(X_4 M)$ of the nine vertices 'preserves' Pappus' theorem. So how many such permutations are there, and (harder) what can you say about the subgroup of S_9 so obtained?*

Suppose we are given two points, A and B, and two conics, Λ_1 and Λ_2, through A and B. In general, Λ_1 and Λ_2 will meet again in two more points, C and D say, and we shall call the line CD the *residual common chord* of Λ_1 and Λ_2 (with respect to A and B). Our next result is about *three* conics through two given points, and their residual common chords, taken in pairs:

Proposition 10.16.   *Let Λ_1, Λ_2, Λ_3 be three conics through the points A and B. Let Λ_2 and Λ_3 meet again in C_1 and D_1 (only), let Λ_3 and Λ_1 meet again in C_2 and D_2 (only), and let Λ_1 and Λ_2 meet again in C_3 and D_3 (only), and assume that the eight points A, B, C_1, D_1, C_2, D_2, C_3, D_3 are distinct. Then the residual common chords $C_1 D_1$, $C_2 D_2$ and $C_3 D_3$ are concurrent* (Fig. 10.4).

Proof.   We may as well assume that no two of the lines $C_1 D_1$, $C_2 D_2$, $C_3 D_3$ coincide, else the result is trivial. Next, the line AB does not contain C_i or D_i for any i, else AB is a common component of two of the conics, which would then have infinitely many common points. Likewise the line $C_i D_i$ does not pass through A or B, for any i, else $C_i D_i$ is a common component of two of the conics. Next, AB is not a component of Λ_1, else the other component must contain C_2, D_2, C_3, and D_3, so that $C_2 D_2$ and $C_3 D_3$ are the same line, which we have assumed is not the case. Similarly, AB is not a component of Λ_2 or Λ_3. Likewise, $C_i D_i$ is not a component of Λ_j for $i \neq j$, else the other component must be AB.

We now define three cubics: put $\Gamma_i = \Lambda_i \cup C_i D_i$, $i = 1, 2, 3$. If $C_1 D_1$ and $C_2 D_2$ meet at E, then Γ_1 and Γ_2 contain A, B, C_1, D_1, C_2, D_2, C_3, D_3, and E. By hypothesis, the first eight of these points are distinct. Then $E \neq A$, else $C_1 D_1$ contains A, which has been ruled out; and similarly $E \neq B$. If $E = C_1$, then the points C_1, C_2, D_2 are distinct, collinear, and lie on Λ_3, so that $C_2 D_2$ is a component of Λ_3, which has also been ruled out. Similarly $E \neq D_1$, C_2, D_2. We may as well assume $E \neq C_3$, D_3 also,

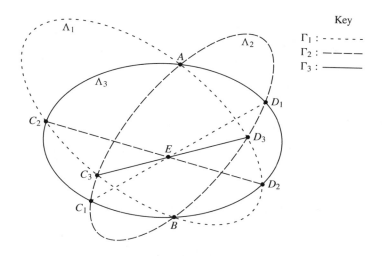

Fig. 10.4   Proposition 10.16

else the result is trivially true again. So we have nine distinct points on $\Gamma_1 \cap \Gamma_2$, and we must show there are no others.

So let $F \in \Gamma_1 \cap \Gamma_2$, that is,

$$F \in (\Lambda_1 \cup C_1 D_1) \cap (\Lambda_2 \cup C_2 D_2)$$

$$= (\Lambda_1 \cap \Lambda_2) \cup (\Lambda_1 \cap C_2 D_2) \cup (\Lambda_1 \cap C_1 D_1) \cup (C_1 D_1 \cap C_2 D_2).$$

If $F \in \Lambda_1 \cap \Lambda_2$, then $F = A$, B, C_3, or D_3. If $F \in \Lambda_1 \cap C_2 D_2$ then $F = C_2$ or D_2, for otherwise $C_2 D_2$ is a component of Λ_1, which has been ruled out. Likewise, if $F \in \Lambda_2 \cap C_1 D_1$, then $F = C_1$ or D_1. Finally, if $F \in C_1 D_1 \cap C_2 D_2$, then since $C_1 D_1 \neq C_2 D_2$, we must have $F = E$.

Thus we have nine associated points, and since Γ_3 contains eight of them, it contains the ninth, namely E, also. But E cannot lie on Λ_3, else $C_1 D_1$ is a component of Λ_3, which has been ruled out. So E lies on $C_3 D_3$, and our proof is complete.   □

Pascal's theorem and Pappus' theorem are special cases of Proposition 10.16, where two or three of the conics are line-pairs. For example, to deduce Pascal's theorem from Proposition 10.16, put $A = X_4$, $B = X_1$, $\Lambda_1 = X_1 X_2 \cup X_3 X_4$, $\Lambda_2 = X_4 X_5 \cup X_6 X_1$, and $\Lambda_3 = \Lambda$. Then it is easy to see that the residual common chords are $X_5 X_6$, $X_2 X_3$, and LN, so by Proposition 10.16 these meet at a point, namely M.

We shall now give several Euclidean manifestations of this theorem, by taking two of the points to be the circular points at infinity, or one of the lines to be the line at infinity, and interpreting the theorem in $\mathbb{R}^2$. The first we have met before, as Exercise 7.25:

Corollary 10.17.   *The radical axes of three circles, taken in pairs, are concurrent or all parallel (Fig. 10.5).*

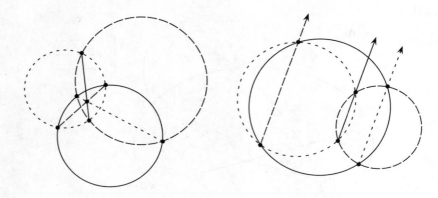

Fig. 10.5   Corollary 10.17, two cases

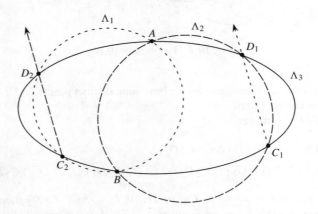

Fig. 10.6   Corollary 10.18

Proof.   Take A, B as the circular points at infinity, so that our three circles pass through A and B, and the residual common chords are the radical axes. (OK?) By Proposition 10.16 these meet at a point E; so either the lines are concurrent in $\mathbb{R}^2$ or else E is on the line at infinity, in which case the lines are all parallel.   □

Corollary 10.18.   *Let Λ_1, Λ_2 be circles meeting in two points A and B, and let Λ_3 be a conic through A and B. Let Λ_2 and Λ_3 meet again in C_1 and D_1 (only), and let Λ_3 and Λ_1 meet again in C_2 and D_2 (only). Then $C_1 D_1 \parallel C_2 D_2$ (Fig. 10.6).*

Proof. Λ_1 and Λ_2 meet again in the circular points at infinity, so their residual common chord (with respect to A and B) is the line at infinity. By Proposition 10.16, $C_1 D_1$ and $C_2 D_2$ meet on the line at infinity, so they are parallel.   □

Corollary 10.19.   *Let Λ_1 and Λ_2 be hyperbolas meeting in two points A and B, and let Λ_3 be a conic through A and B. Let Λ_2 and Λ_3 meet again in C_1 and D_1 (only), and*

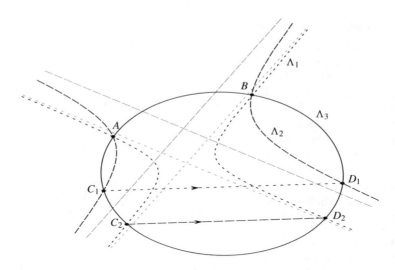

Fig. 10.7   Corollary 10.19

let Λ_3 *and* Λ_1 *meet again in* C_2 *and* D_2 *(only). If each asymptote of* Λ_1 *is parallel to one of the asymptotes of* Λ_2*, then also* $C_1D_1 \parallel C_2D_2$ *(Fig. 10.7).*

Proof. This time Λ_1 and Λ_2 meet in two *real* points at infinity; so once again their residual common chord (with respect to A and B) is the line at infinity, and the result follows as in the previous corollary. $\qquad\square$

We finish the applications of Proposition 10.16 with the solution of an amusing problem set by Coxeter in *The Mathematical Intelligencer*, 1986, problem 86-9, and which is considerably shorter than the published solution, in *ibid.*, vol. 9, no. 4, 1987, p. 35. Let Λ_1 and Λ_2 be two *real* conics, meeting (but not touching) at just *two* real points, A and B. Then as complex conics, Λ_1 and Λ_2 also meet in two complex points, the coordinates of one being the conjugates of the coordinates of the other. But this means that the line joining these points, the residual common chord, is self-conjugate, and hence real. Coxeter's problem: *construct* this line! Solution (using ruler only): draw a line through A and a line through B to form a line-pair Λ_3. The residual common chords of Λ_1 and Λ_3, and of Λ_2 and Λ_3, can now be put in, and they meet at some point of the required line. Repeat to obtain a second point, and join. (Cf. Exercise 7.26.)

- Ex.10.9:   *State and solve the dual of Coxeter's problem.*
- Ex.10.10:   *Let the lines* $\ell_1, \ell_2, \ell_3, \ell_4$ *form a complete quadrilateral in* $\mathbb{R}^2$*, so that no two of them are parallel and no three are concurrent. For* $1 \leq i \leq 4$*, let* Σ_i *be the circumcircle of the triangle formed by the three lines* ℓ_j *with* $j \neq i$*. Prove that* $\Sigma_1, \Sigma_2, \Sigma_3, \Sigma_4$ *have a common point (in* $\mathbb{R}^2$*).*

10.2.4   Singular and non-singular cubics

This section is just to give a flavour of how the theory of cubics develops. Many of the proofs are going to be very sketchy, or omitted. References for proper proofs and further reading are given at the end.

Cubics can have points with peculiar properties. First, *singularities*. Consider the parabola $y = x^2$. The line $y = mx$ meets this at $(0, 0)$, and to find the intersection multiplicity we substitute for y and solve $mx = x^2$, or $x(x - m) = 0$. So the intersection multiplicity of line and parabola at $(0, 0)$ is 1 if $m \neq 0$ and 2 if $m = 0$, when the line is a tangent and the equation $x(x - m) = 0$ becomes $x^2 = 0$ and so has a double root at $x = 0$. On the other hand, if we use the cubic $y^2 = x^3$ and try the same calculation, the equation we are trying to solve is now $m^2 x^2 = x^3$, or $x^2(x - m^2) = 0$, so that the intersection multiplicity is 2 when $m \neq 0$, when $x = 0$ is a double root; and 3 when $m = 0$, when the equation becomes $x^3 = 0$ and has a triple root at $x = 0$.

Definition 10.20.   *Let P be a point of the curve Γ. If the minimum intersection multiplicity of Γ with a variable line through P is r, then r is the **multiplicity** of P; P is a **simple point** of Γ if r = 1, and a **multiple point** or **singularity** of Γ if r > 1. It is a **double point** if r = 2, a **triple point** if r = 3, and so on. A curve is **singular** or **non-singular** according as it has or does not have a singularity.*

Examples.   Every point of a line is simple. A conic with a double point must be reducible, for the join of the double point to any other point on the conic meets it at least three times (adding the multiplicities), so this line is a component. Likewise, the only cubic with a triple point is three concurrent lines, and a cubic with two double points has the line joining them as a component, since $2 + 2 > 3$.

So an irreducible cubic can have at most one singularity, which must be a double point. Here is another example: $y^2 = x^2(x + a)$. This meets the line $y = mx$ where $m^2 x^2 = x^2(x + a)$, or $x^2(x + a - m^2) = 0$, so we have a double point at $(0, 0)$; but this time, if $a \neq 0$, there are two values of m, namely $m = \pm\sqrt{a}$, that give $x^3 = 0$, a triple intersection. The corresponding lines $y = \pm\sqrt{a}x$ are the *singular tangents* at the double point, and a double point with two singular tangents is called a *node* (Fig. 10.8). If we let $a \to 0$, then *both* the singular tangents become $y = 0$, the singular tangent

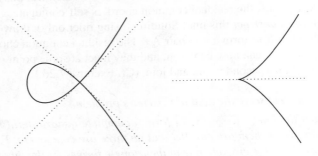

Fig. 10.8   Nodal and cuspidal cubics, with their singular tangents

to the limiting cubic, $y^2 = x^3$. This can therefore be thought of as a *repeated* singular tangent, and this type of double point is called a *cusp* (Fig. 10.8 again).

● Ex.10.11: *Plot the curves* $y^2 = x^2(x + 1)$ *and* $y^2 = x^2(x - 1)$, *paying especial attention to what happens near* $(0, 0)$. *Why are they so different?*

So irreducible cubics come in two kinds, singular (as above), and non-singular (examples soon); and the singular cubics are either nodal (two singular tangents) or cuspidal (a repeated singular tangent).

The other sort of peculiar point on a cubic is an *inflexion*, or *flex*. In calculus, we learn that the familiar cubic $y = x^3$ has a 'point of inflexion' at $(0, 0)$, where its first two derivatives vanish, and its *gradient* has a minimum. The line $y = mx$ meets $y = x^3$ where $x(x^2 - m) = 0$, so that $(0, 0)$ is a simple point of the curve; but when $m = 0$ the equation becomes $x^3 = 0$ and instead of a double intersection for the tangent, we have a triple intersection. A *flex* is a simple point of a curve for which the tangent has intersection multiplicity at least 3. So, for example, $y = x^4$ also has a flex at $(0, 0)$. This is slightly at odds with the language in elementary calculus, where $(0, 0)$ would be called a minimum, but *not* a 'point of inflexion' of this curve. For the tangent to a cubic, however, the intersection number can never exceed 3, so there is no problem.

We now sketch how to find the 'peculiar points' of a cubic curve. To see if the cubic $\Gamma : p(x, y) = 0$ has a singularity at $(0, 0)$, we write it as $p_0 + p_1 + p_2 + p_3 = 0$, where p_k is homogeneous of degree k. So p_0 is a constant, and Γ passes through $(0, 0)$ if $p_0 = 0$. In this case, $p_1 = 0$ is a linear equation, and is the equation of the tangent at $(0, 0)$, if this is a simple point. If p_1 is identically zero (and $p_0 = 0$), then we have a double point at $(0, 0)$, whose pair of singular tangents is given by the quadratic equation $p_2 = 0$. So we have a node if this gives distinct lines, and a cusp if it gives a repeated line. To find a singularity elsewhere, make a change of variables. For example, $p(x, y) = 0$ has a singularity at (a, b) iff $p(a + x, y + b) = 0$ has a singularity at $(0, 0)$. This can be done by a double Taylor expansion:

$$p(a + x, y + b) = p(a, b) + (xp_x(a, b) + yp_y(a, b)) + \text{higher terms},$$

where p_x, p_y stand for the partial derivatives $\partial p / \partial x$, $\partial p / \partial y$. The singularity (if any) can now be found by solving $p_x(a, b) = p_y(a, b) = p(a, b) = 0$ for (a, b). This only deals with finite singularities: there may be something lurking at infinity, on $z = 0$, and to deal with this we homogenize (or use a triple Taylor expansion), which means the curve is now $p(x, y, z) = 0$ and the singularity (if any) is at (a, b, c) where $p_x(a, b, c) = p_y(a, b, c) = p_z(a, b, c) = 0$. (This automatically entails $p(a, b, c) = 0$.)

● Ex.10.12: *Find the singularities (if any) of the cubic curves (i)* $y^2 = x(x - 1)^2$, *(ii)* $y = x^3$, *(iii)* $y^2 = x(x^2 - 1)$, *(iv)* $x^2 y = 1$, *(v)* $(x^2 - 1)y = 1$.

To find the flexes of Γ is more tricky. The triple Taylor expansion about $P = (a, b, c)$ gives

$$p(x, y, z) = p(a, b, c) + \text{linear terms} + \text{quadratic terms} + \text{cubic terms}.$$

Equating the linear terms to zero gives a line, the *polar line* of P, which in the case where P is a simple point of the curve is just the tangent at P. Equating the quadratic terms to zero gives a conic, the *polar conic* of P. This conic meets Γ in $2 \times 3 = 6$ points, in general, and these are the points where tangents to Γ from P touch Γ. This is why a non-singular cubic has class $= 6$. If P is on Γ, the polar conic of P, like the polar line, touches Γ at P, and so gobbles up two of the six intersections. So from a general point of a non-singular cubic Γ, there are four tangents to Γ (excluding the tangent at P itself).

If P is a singularity *or* a flex of Γ, then at P the polar conic degenerates into a line-pair. So to find where the flexes and singularities are, we look at those points (a, b, c) for which the polar conic is improper. Now the polar conic is

$$
\begin{pmatrix} x & y & z \end{pmatrix} \begin{pmatrix} p_{xx} & p_{xy} & p_{xz} \\ p_{yx} & p_{yy} & p_{yz} \\ p_{zx} & p_{zy} & p_{zz} \end{pmatrix} \begin{pmatrix} x \\ y \\ z \end{pmatrix} = 0,
$$

where $p_{xx} = \partial^2 p/\partial x^2$, $p_{xy} = \partial^2 p/\partial x \partial y$, etc., all evaluated at (a, b, c). The condition for this to be improper is

$$
\begin{vmatrix} p_{xx} & p_{xy} & p_{xz} \\ p_{yx} & p_{yy} & p_{yz} \\ p_{zx} & p_{zy} & p_{zz} \end{vmatrix} = 0,
$$

and this is a homogeneous cubic equation in a, b, c. Replace a by x, b by y, and c by z in this equation and we have another cubic curve, the *Hessian* of Γ, which meets Γ at its singularities and flexes. Since $3 \times 3 = 9$, Bezout's theorem tells us to expect a non-singular cubic to have nine flexes (but, it turns out, no more than three of them can be real), and a singular cubic to have fewer, because the singularity gobbles up some of the intersections of Γ with its Hessian. In fact, a nodal cubic has three flexes, and a cuspidal cubic has only one flex.

● Ex.10.13:   *Find the singularities and flexes of the cubic curves (i) $y^2 = x^3$, (ii) $y^2 = x^2(x - 1)$, (iii) $xy = x^3 + y^3$.*

We shall now make a tentative start on the classification of irreducible cubics. Let Γ be such a curve; we now know it has at least one flex, so choose the coordinates to put a flex at $(0, 1, 0)$. This means that the equation of Γ has no term in y^3. If we also insist that $z = 0$ is the tangent at this flex, then putting $z = 0$ in the equation of Γ must reduce it to $x^3 = 0$, which means that the equation of Γ has no terms in xy^2 or x^2y either. So its equation is of the form

$$
y^2z + 2yz(ax + bz) = \text{a homogeneous cubic in } x \text{ and } z.
$$

The term in y^2z must be present, else $(0, 1, 0)$ is a double point and not a flex; so we have already divided through by its coefficient. We now make a projective change of variable, replacing y by $y - ax - bz$ (and leaving x and z alone), which gives

$$
y^2z = \text{another homogeneous cubic in } x \text{ and } z.
$$

Since the curve is irreducible, the coefficient of x^3 is not zero, else $z = 0$ is a component. Replacing x by a suitable scalar multiple, we can fix it so that the coefficient of x^3 is 1. Dehomogenizing, we obtain

$$y^2 = (x - a)(x - b)(x - c),$$

say. Three cases arise. Firstly, if $a = b = c$, replacing x by $x + a$ gives the equation $y^2 = x^3$, a cuspidal cubic. Secondly, if two of a, b, c are equal, say $a = b \neq c$, then replacing x by $x + a$ gives the equation $y^2 = x^2(x - d)$, say; and replacing x by dx and y by $d^{3/2}y$ reduces this to $y^2 = x^2(x - 1)$, a nodal cubic. Thirdly, if a, b, c are distinct, then replacing x by $x + a$ gives $y^2 = x(x - d)(x - e)$ say, where $d \neq e$ and $de \neq 0$. Replacing x by dx and y by $d^{3/2}y$ reduces this to $y^2 = x(x - 1)(x - \lambda)$, say, where $\lambda \neq 0, 1$. This is a non-singular cubic.

Let us summarize. Every cuspidal cubic is projectively equivalent to $y^2 = x^3$, and so all cuspidal cubics are projectively equivalent. Every nodal cubic is projectively equivalent to $y^2 = x^2(x - 1)$, and so all nodal cubics are projectively equivalent. Every non-singular cubic is projectively equivalent to $y^2 = x(x - 1)(x - \lambda)$ for some $\lambda \neq 0, 1$. Certainly two such cubics that give the same λ are projectively equivalent; but what about cubics that give different values of λ?

Write Γ_λ for the cubic $y^2 = x(x - 1)(x - \lambda)$, where $\lambda \neq 0, 1$. Let us try some changes of coordinates. First, if we replace x by $1 - x$, then $x(x - 1)(x - \lambda)$ becomes $(1-x)(-x)(1-x-\lambda)$, or $-x(x-1)(x-(1-\lambda))$; so if we also replace y by iy, we change Γ_λ to $\Gamma_{1-\lambda}$. Next, replacing x by λx changes $x(x - 1)(x - \lambda)$ to $\lambda x(\lambda x - 1)(\lambda x - \lambda)$, or $\lambda^3 x(x - 1)(x - \lambda^{-1})$. If we also replace y by $\lambda^{3/2}y$, then we have changed Γ_λ to $\Gamma_{\lambda^{-1}}$. Repeating the process with the new values of λ, we see that Γ_μ is projectively equivalent to Γ_λ when $\mu = \lambda, 1 - \lambda, \lambda^{-1}, (1 - \lambda)^{-1}, 1 - \lambda^{-1}$, or $(1 - \lambda^{-1})^{-1}$, and we recognize these quantities from equation (9.2) as the possible cross-ratios of four points when we vary the order in which they are taken. Surely there must be something to do with a cross-ratio here, too?

The connection is via the statement made above, that from an arbitrary point P of a non-singular cubic Γ, four tangents to Γ can be drawn (excluding the tangent at P itself, unless P is a flex). These four lines, concurrent at P, have a cross-ratio, or (rather) they have in general *six* possible cross-ratios, according to the order in which they are taken. *Salmon's theorem* (which we shall not prove here) says that the set of six values obtained depends only on Γ, and not on the choice of P. Since cross-ratios are preserved by projective transformations, non-singular cubics that are projectively equivalent will give the same set of six values. To avoid having to handle six values all the time, one introduces the function

$$J(\lambda) = \frac{(\lambda^2 - \lambda + 1)^3}{\lambda^2(\lambda - 1)^2}$$

called the *modulus* or *J-invariant* of Γ_λ. It is an easy exercise to check that, if

$$\mu = \lambda,\ 1 - \lambda,\ \lambda^{-1},\ (1 - \lambda)^{-1},\ 1 - \lambda^{-1},\ (1 - \lambda^{-1})^{-1}, \tag{10.2}$$

then $J(\mu) = J(\lambda)$, and in fact the converse is true. To prove this, write

$$F(y) = (y^2 - y + 1)^3 - J(x)y^2(y - 1)^2,$$

a polynomial in y of degree 6, monic, with coefficients from the field $\mathbb{C}(x)$ of complex rational functions in x. By what has just been said, plus the remainder theorem, $F(y)$ must be divisible by

$$(y - x)(y - 1 + x)(y - x^{-1})(y - (1 - x)^{-1})(y - 1 + x^{-1})(y - (1 - x^{-1})^{-1}),$$

which is also a polynomial in y of degree 6, monic, with coefficients from the field $\mathbb{C}(x)$ of complex rational functions in x, and so must be equal to $F(y)$. Thus

$$J(y) - J(x) = \frac{(y - x)(y - 1 + x)(y - x^{-1})(y - (1 - x)^{-1})(y - 1 + x^{-1})(y - (1 - x^{-1})^{-1})}{y^2(y - 1)^2}$$

from which it is immediate that, for $\lambda, \mu \neq 0, 1$, we have $J(\mu) = J(\lambda)$ if and only if (10.2) holds. If using the remainder theorem over $\mathbb{C}(x)$ (instead of $\mathbb{C}$) makes you queasy, then your only recourse is to bring $J(y) - J(x)$ to a common denominator, and then factorize the numerator, which should lead you to the same conclusion, but more slowly.

Now $(0, 1, 0)$ is a flex of $\Gamma_\lambda : y^2z = x(x - z)(x - \lambda z)$, and the four tangents to Γ_λ from this point are $x = 0$, $z = 0$ (the tangent at the flex), $x = \lambda z$, and $x = z$. These cut $y = 0$ where $x = 0$, ∞, λ, and 1 respectively, and $\{0, \infty\,; \lambda, 1\} = \lambda$. So we have:

Theorem 10.21 (Classification of irreducible cubics). *All cuspidal cubics are projectively equivalent; all nodal cubics are projectively equivalent; and every non-singular cubic is projectively equivalent to Γ_λ for some $\lambda \neq 0, 1$, with Γ_λ and Γ_μ projectively equivalent if and only if $J(\lambda) = J(\mu)$.* □

We are now going to return to the theorem of nine associated points, in order to do group theory on a cubic. Let Γ be an irreducible cubic, and put $G =$ the set of all simple points of Γ. So G is Γ with at most one point removed, and in particular, if Γ is non-singular, then $G = \Gamma$.

First we define an operation $\circ$ on G which does *not* make it a group. For any A, $B \in G$, let the line AB meet Γ again in C, and define $A \circ B = C$ (Fig. 10.9). This is easy if A, B, C are distinct, but the other cases call for some comment. If $A = B$ then the line AB is just the tangent to Γ at A, and C is where this tangent meets Γ again. This point $C = A \circ A$ is then called the *tangential* of A. If A is a flex of Γ and $B = A$, then $C = A$ also, so that $A \circ A = A$: a flex of a cubic is a simple point that coincides with its tangential. The remaining special cases are when $A \neq B$ but AB is the tangent to Γ at A, in which case $C = A$ and $A \circ B = A$; and similarly if $A \neq B$ but AB is the tangent to Γ at B, in which case $C = B$ and $A \circ B = B$. Note that $C \in G$, in all cases.

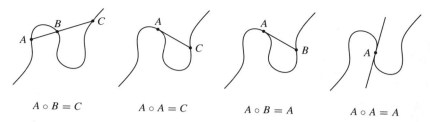

$A \circ B = C$ $A \circ A = C$ $A \circ B = A$ $A \circ A = A$

Fig. 10.9 The operation $\circ$ on a cubic

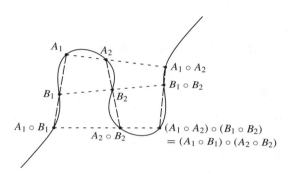

Fig. 10.10 Lemma 10.12(iii)

We collect some properties of this operation:

Lemma 10.22. *(i) For any A, $B \in G$, we have $A \circ B = B \circ A$. (ii) For any A, $B \in G$, we have $A \circ (A \circ B) = B$. (iii) For any A_1, A_2, B_1, $B_2 \in G$, we have $(A_1 \circ A_2) \circ (B_1 \circ B_2) = (A_1 \circ B_1) \circ (A_2 \circ B_2)$ (Fig. 10.10).*

Proof. (i) is trivial, and (ii) says that if $A \circ B = C$, then $A \circ C = B$. For distinct points, this is the obvious statement that if AB meets Γ again at C, then AC meets Γ again at B. The other cases are left to the reader.

(iii) Let $A_1 \circ A_2 = A_3$ and $B_1 \circ B_2 = B_3$, so that $(A_1 \circ A_2) \circ (B_1 \circ B_2) = A_3 \circ B_3$. Let $A_1 \circ B_1 = C_1$ and $A_2 \circ B_2 = C_2$; and finally let $C_1 \circ C_2 = C_3$. Thus $(A_1 \circ B_1) \circ (A_2 \circ B_2) = C_1 \circ C_2 = C_3$, so we want to show that $A_3 \circ B_3 = C_3$.

Now the cubic $A_1 A_2 \cup B_1 B_2 \cup C_1 C_2$ meets Γ in A_1, A_2, A_3, B_1, B_2, B_3, C_1, C_2, C_3. Assuming these nine point are distinct, then we have nine associated points. The cubic $A_1 B_1 \cup A_2 B_2 \cup A_3 B_3$ contains A_1, B_1, C_1, A_2, B_2, C_2, A_3, B_3, so it must contain C_3 also. Now C_3 cannot lie on $A_1 B_1$ or $A_2 B_2$, else the cubic contains four collinear points and is reducible; so C_3 lies on $A_3 B_3$, and thus $C_3 = A_3 \circ B_3$, as required.

The remaining cases are when two or more of our nine points coincide, and here we are just going to wave our hands and say *it can be shown that* the special cases are all limiting versions of the case above, and the result holds in those cases also, by continuity. (So go and ask an analyst.) ☐

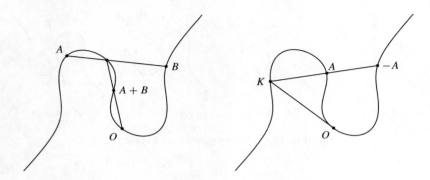

Fig. 10.11   The group operations on a cubic

- Ex.10.14:   *If $A \circ B = A \circ C$, then $B = C$.*

Now for our group:

Proposition 10.23.   *Pick any point $O \in G$ and then, for $A, B \in G$, define $A + B = O \circ (A \circ B)$ (Fig. 10.11). With this operation, G is an abelian group, with neutral element O.*

Proof.   Most of this is easy. Certainly $A + B = B + A$, since $A \circ B = B \circ A$. Next, $O + A = O \circ (O \circ A) = A$. Then, given A, put $B = A \circ K$, where $K = O \circ O$ is the tangential of O. We have $A + B = O \circ (A \circ (A \circ K)) = O \circ K = O \circ (O \circ O) = O$, and thus $B = -A$, the inverse of A (Fig. 10.11 again).

The only tricky part is the associative law: we must show that, for $A, B, C \in G$, we have $A + (B + C) = (A + B) + C$. This reads $O \circ (A \circ (B + C)) = O \circ ((A + B) \circ C)$, so it is enough to show that $A \circ (B + C) = (A + B) \circ C$. But

$$A \circ (B + C) = ((A \circ B) \circ B) \circ (O \circ (B \circ C))$$
$$= ((A \circ B) \circ O) \circ (B \circ (B \circ C))$$
$$= (A + B) \circ C$$

by Lemma 10.22, and we have finished. (See also Fig. 10.12.)   □

- Ex.10.15:   *Show that a different choice of neutral element gives an isomorphic group. Specifically, let $O' \in G$ and define $A * B = O' \circ (A \circ B)$, giving a second group structure on the set G. Then let $L = O \circ O'$, and for $A \in G$, define $A' = L \circ A$, so that $A \mapsto A'$ is a bijection $G \to G$. Show that $A' * B' = (A + B)'$, and deduce that the two groups G (with the operation $+$) and G (with the operation $*$) are isomorphic.*

- Ex.10.16:   *Let K be the tangential of O, as before, and let $A, B, C \in G$. Show that A, B, C are collinear if and only if $A + B + C = K$.*

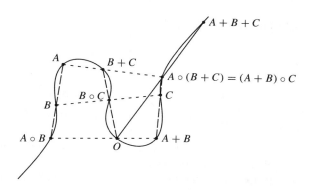

Fig. 10.12 The associative law

The result of the last exercise is more useful if we make a special choice of neutral element O: choose O to be a flex of Γ, so that $O = K$, and then A, B, C are collinear if and only if $A + B + C = O$. All sorts of corollaries flow from this. First, if A', B', C' are the tangentials of A, B, C respectively, then $A' = A \circ A$, so $2A = A + A = O \circ (A \circ A) = O \circ A' = -A'$, or $A' = -2A$, and similarly for B' and C'. Then if $A + B + C = O$, we have $-2A - 2B - 2C = O$, or $A' + B' + C' = O$, i.e., the tangentials of collinear points are collinear. Next, A is a flex iff it is equal to its tangential, that is, $A = -2A$, or $3A = O$. If $A + B + C = O$, then $3A + 3B + 3C = O$; if A and B are flexes then $3A = O$ and $3B = O$, so that $3C = O$ also, and C is a flex: the join of any two flexes passes through a third flex, as we have already seen in a few special cases, in Exercise 10.13.

Suppose a conic Λ meets Γ in the $2 \times 3 = 6$ points A_1, A_2, A_3, B_1, B_2, B_3, all in G. Let $C_i = A_i \circ B_i$ for $i = 1, 2, 3$, so that Γ meets the cubic $A_1 B_1 \cup A_2 B_2 \cup A_3 B_3$ in the points A_1, B_1, C_1, A_2, B_2, C_2, A_3, B_3, C_3. Assuming these are distinct, the nine associated points theorem says that the cubic $\Lambda \cup C_1 C_2$, which contains A_1, A_2, A_3, B_1, B_2, B_3, C_1, C_2, must contain C_3 also; this does not lie on Λ, so it is on $C_1 C_2$, that is, $C_3 = C_1 \circ C_2$. So we have

$$A_1 + B_1 + C_1 = O$$
$$A_2 + B_2 + C_2 = O$$
$$A_3 + B_3 + C_3 = O$$
$$C_1 + C_2 + C_3 = O$$

from which it follows that the six points on $\Lambda \cap \Gamma$ add up to zero:

$$A_1 + A_2 + A_3 + B_1 + B_2 + B_3 = O.$$

The converse is also true: given six points A_1, A_2, A_3, B_1, B_2, B_3 of G with sum zero, they lie on a conic. For let Λ be the conic through A_1, A_2, A_3, B_1, B_2, and define C_1,

C_2, C_3 as before. This time we have

$$A_1 + B_1 + C_1 = O$$
$$A_2 + B_2 + C_2 = O$$
$$A_3 + B_3 + C_3 = O$$
$$A_1 + A_2 + A_3 + B_1 + B_2 + B_3 = O$$

from which $C_1 + C_2 + C_3 = O$, and C_1, C_2, C_3 are collinear. Once again, Γ meets the cubic $A_1 B_1 \cup A_2 B_2 \cup A_3 B_3$ in the points $A_1, B_1, C_1, A_2, B_2, C_2, A_3, B_3, C_3$. Assuming these are distinct, the nine associated points theorem says that the cubic $\Lambda \cup C_1 C_2$ contains $A_1, A_2, A_3, B_1, B_2, C_1, C_2, C_3$, and so must contain B_3 also. Since B_3 cannot lie on $C_1 C_2$ (else this line is a component of Γ), it lies on Λ. So three points of G lie on a line if and only if their sum is zero, and six points of G lie on a conic if and only if their sum is zero. These are the cases $r = 1$ and 2 of the following theorem: $3r$ points of G lie on a curve of order r if and only if their sum is zero. From this, if a curve of order n meets G in $3n$ points (as Bezout's theorem tells us it must, in general), and if $3r$ of these points lie on a curve of order r, then these $3r$ points add up to zero, and so do all $3n$ points, and hence so do the $3(n - r)$ points *not* on the curve of order r. These $3(n - r)$ points therefore lie on a curve of order $n - r$. This in turn is the case $m = 3$ of yet another theorem: if a curve Γ of order m and a curve of order n meet in mn points (Bezout), all simple points of Γ, and if mr of these points lie on a curve of order r, then the remaining $m(n - r)$ points lie on a curve of order $n - r$.

It is easy to recognize the group G of the cubic Γ when Γ is singular. For the cuspidal cubic $\Gamma : yz^2 = x^3$, every simple point is $(t, t^3, 1)$ for some $t \in \mathbb{C}$. (The cusp is at $(0, 1, 0)$, reached by letting $t \to \infty$.) The line $ax + by + cz = 0$ meets Γ at $(t, t^3, 1)$ where $at + bt^3 + c = 0$. This is a cubic in t, having as roots the parameters t_1, t_2, t_3 of the points A_1, A_2, A_3 where the line meets Γ, and since the cubic has no term in t^2, we deduce that $t_1 + t_2 + t_3 = 0$. So there is a bijection $\mathbb{C} \to G$ given by $t \mapsto (t, t^3, 1)$, having the property that $t_1 + t_2 + t_3 = 0$ if and only if $A_1 + A_2 + A_3 = O$, and from this it follows that G is isomorphic to $\mathbb{C}$, or $\mathbb{R}^2$, under addition.

Now take $\Gamma : xyz = y^3 - x^3$, a nodal cubic. Here every simple point is $(t, t^2, t^3 - 1)$ for some $t \in \mathbb{C}^*$, the non-zero complex numbers. (The node is at $(0, 0, 1)$, reached by letting $t \to 0$ or $t \to \infty$.) The line $ax + by + cz = 0$ meets Γ where $at + bt^2 + c(t^3 - 1) = 0$. As long as the line avoids $(0, 0, 1)$, the node, we have $c \neq 0$, and then once again we have to solve a cubic in t for the parameters t_1, t_2, t_3 of the points A_1, A_2, A_3 where the line meets Γ. This time, the coefficient of t^3 is c and the constant term is $-c$, so we deduce that $t_1 t_2 t_3 = 1$ if and only if $A_1 + A_2 + A_3 = O$. Thus in this case G is isomorphic to the group $\mathbb{C}^*$ of non-zero complex numbers under multiplication.

Recall that there is a map $\mathbb{R} \to T$, the circle group, given by $y \mapsto e^{2\pi i y}$. This is a homomorphism from the additive group $\mathbb{R}$ to the multiplicative group T, whose kernel is $\mathbb{Z}$, the integers. In effect, the exponential map winds the real line $\mathbb{R}$ around and around the circle T, each element $e^{2\pi i y}$ of T corresponding to an entire coset $y + \mathbb{Z}$ of the subgroup $\mathbb{Z}$ of $\mathbb{R}$. (Group theorists then write $T \cong \mathbb{R}/\mathbb{Z}$, the *quotient* of $\mathbb{R}$ by $\mathbb{Z}$.)

We can extend this map to a map $\mathbb{R}^2 \to \mathbb{C}^*$, by $(x, y) \mapsto e^{2\pi(x+iy)}$ (or alternatively $\mathbb{C} \to \mathbb{C}^*$, $z \mapsto e^{2\pi z}$), where each such element corresponds to a coset $(x, y) + (0 \times \mathbb{Z})$ of the subgroup $0 \times \mathbb{Z}$ of $\mathbb{R}^2$. Here the points (x, y), $(x, y \pm 1)$, $(x, y \pm 2)$, $\ldots$, all have the same image, and instead of just winding the y-axis around a circle, we are rolling up the entire plane, like a roll of carpet, to form a cylinder. So for a nodal cubic, the group G is a cylinder, $(\mathbb{R} \times \mathbb{R})/(0 \times \mathbb{Z})$, or $\mathbb{R} \times (\mathbb{R}/\mathbb{Z})$, or $\mathbb{R} \times T$.

A non-singular cubic is an *elliptic curve*. Its points can be parametrized by $\mathbb{R} \times \mathbb{R}$ in such a way that the kernel of the map is not 0×0 (as for the cuspidal cubic), nor $0 \times \mathbb{Z}$ (as for the nodal cubic), but $\mathbb{Z} \times \mathbb{Z}$, though this does not necessarily lie inside $\mathbb{R} \times \mathbb{R}$ in the obvious way. The nodal cubic was parametrized by the function $f(z) = e^{2\pi z}$, which is *periodic* with period i, that is, $f(z+ni) = f(z)$, for all $z \in \mathbb{C}$ and all $n \in \mathbb{Z}$, and to find a parametrization of the non-singular cubic we need a *doubly periodic* or *elliptic* function f, which is a complex function having two independent periods, that is, two complex numbers ω_1 and ω_2, linearly independent over $\mathbb{R}$, such that $f(z + n_1\omega_1 + n_2\omega_2) = f(z)$, for all $z \in \mathbb{C}$ and all $n_1, n_2 \in \mathbb{Z}$. This gives rise to a map $\mathbb{C} \to G$ whose kernel is $\{n_1\omega_1 + n_2\omega_2 : n_1, n_2 \in \mathbb{Z}\}$, which is called a *lattice*, and is isomorphic to $\mathbb{Z} \times \mathbb{Z}$. So the group G is isomorphic to $(\mathbb{R} \times \mathbb{R})/(\mathbb{Z} \times \mathbb{Z}) = (\mathbb{R}/\mathbb{Z}) \times (\mathbb{R}/\mathbb{Z}) = T \times T = T^2$, a torus. Think of this as rolling the plane up one way, to make a cylinder, and then the other way, like coiling a hosepipe, to make a torus.

Let us use this theory to count the flexes on a cubic, once more. The only element of $\mathbb{R}$ whose order divides 3 is the zero element, whereas T has three such elements: 1, $e^{2\pi i/3}$ and $e^{-2\pi i/3}$. So the number of elements of order dividing 3 in (i) $\mathbb{R} \times \mathbb{R}$, (ii) $\mathbb{R} \times T$, (iii) $T \times T$, is (respectively) (i) $1 \times 1 = 1$, (ii) $1 \times 3 = 3$, (iii) $3 \times 3 = 9$. So a cuspidal cubic has one flex, a nodal cubic has three, and a non-singular cubic has nine.

Recall that every non-singular cubic can be put in the form $y^2 = x(x - \lambda)(x - \mu)$ with 0, λ, μ distinct; that if we let $\mu \to 0$ then the limit $y^2 = x^2(x - \lambda)$ is nodal; and if $\lambda \to 0$ as well then the limit $y^2 = x^3$ is cuspidal. Can we therefore apply some limiting process to a torus to obtain first a cylinder, and then a plane? A method is suggested in the following exercises.

- Ex.10.17: *Show that the torus obtained by rotating the circle $(x-a)^2+z^2 = b^2$, $y = 0$ about the z-axis has equation $(x^2+y^2+z^2+a^2-b^2)^2 - 4a^2(x^2+y^2) = 0$, or $(x^2 + y^2 + z^2 + rs)^2 - (r + s)^2(x^2 + y^2) = 0$, where $r = a - b$ and $s = a + b$.*

- Ex.10.18: *Show that the limit of this torus as $s \to \infty$ is a cylinder.*

- Ex.10.19: *Starting with the same torus, move the origin to $(a, 0, 0)$ and let $a \to \infty$, and show that the limit is another cylinder.*

- Ex.10.20: *Show that the limit of a cylinder of radius a, as $a \to \infty$, is a plane.*

It is possible to include the singularities in this. The cuspidal cubic is parametrized by $\mathbb{C} \cup \{\infty\}$, if we include the cusp itself, and this, via stereographic projection, is a sphere. This can also be seen by a three-dimensional inversion (with respect to a sphere): take the plane as $x - 1 = 0$ and invert with respect to the sphere $x^2 + y^2 + z^2 = 1$. This means replacing x by $x/(x^2 + y^2 + z^2)$, (and y, if present, by $y/(x^2 + y^2 + z^2)$), and likewise

z by $z/(x^2 + y^2 + z^2)$), giving $x - (x^2 + y^2 + z^2) = 0$, or $(x - \frac{1}{2})^2 + y^2 + z^2 = \frac{1}{4}$, a sphere. (Here $\infty \mapsto (0,0,0)$, of course.) This creates a small worry: what happened to the singularity? The cusp is a 'peculiar' point on the cubic, but its image $(0,0,0)$ on the sphere looks the same as any other point on the sphere. We'll return to this below.

The nodal cubic, complete with node, is (cylinder $\cup$ $\{\infty\}$), which, if we take the cylinder as $x^2 + y^2 - c^2 = 0$, inverts to $x^2 + y^2 - c^2(x^2 + y^2 + z^2)^2 = 0$. This we recognize as the limit of the torus $(x^2 + y^2 + z^2 + rs)^2 - (r+s)^2(x^2 + y^2) = 0$ when $s = 1/c$ and $r \to 0$. Here r was the radius of the hole through the middle of the torus: thinking of it as a tyre, r is the radius of the wheel to which the tyre is to be fitted. When $r = 0$, the hole has shrunk to nothing, and we have a surface we might describe as a 'pinched' torus: the hole has been pinched or constricted, and so closed up. This surface is the rotation of the circle $(x - a)^2 + z^2 = a^2$, $y = 0$ (which passes through the origin) about the z-axis, and the singularity corresponds to the 'pinch-point' $(0,0,0)$.

Clearly the pinched torus (for a nodal cubic) is the limit of a (proper) torus (for a non-singular cubic); how do we get a sphere (for a cuspidal cubic) as the limit of a pinched torus (for a nodal cubic)? Starting again with the cylinder $x^2 + y^2 - c^2 = 0$, move the origin to $(a, 0, 0)$ to get $(x + a)^2 + y^2 - c^2 = 0$. Inverting this gives $(x + a(x^2 + y^2 + z^2))^2 + y^2 - c^2(x^2 + y^2 + z^2)^2 = 0$, which is another pinched torus, except that if $a > c > 0$ then the pinch-point $(0,0,0)$ is not at the centre, but on the rim of the wheel to which we are fitting the tyre: the wheel has non-zero radius, but at one point on the rim the tyre has been constricted to a point. Write this equation as $x^2 + 2ax(x^2 + y^2 + z^2) + y^2 + (a^2 - c^2)(x^2 + y^2 + z^2)^2 = 0$, or $x^2 + y^2 + (r + s)x(x^2 + y^2 + z^2) + rs(x^2 + y^2 + z^2)^2 = 0$, where $r = a - c$ and $s = a + c$. Divide this through by s and then let $s \to \infty$ and we obtain $x(x^2 + y^2 + z^2) + r(x^2 + y^2 + z^2)^2 = 0$, or $(x^2 + y^2 + z^2)((x - d)^2 + y^2 + z^2 - d^2) = 0$, where $d = (2r)^{-1}$. The second bracket is a sphere, as we had hoped, and goes through $(0,0,0)$, which corresponds to the singularity of the cubic. The first bracket is a point-sphere, or sphere of radius zero, at $(0,0,0)$, so that the peculiar point of our cubic is still peculiar. This clears up our earlier worry: we have not just a sphere, but a sphere with an additional point-sphere to represent the cusp.

Further reading. Most of the missing proofs in this section can be found in [32], [28], or [13]. The modern (ring-theoretic) approach to algebraic geometry is in [15]; all the algebraic background is given at the beginning, but if this is your first encounter with most of it, you will find the rest of that book heavy going. Nonetheless, it is more digestible than most of the alternatives.

An excellent introduction to the modern way of doing things is given in [25]. This book is short, self-contained, entertaining, and has plenty of pictures, sadly lacking in too many modern geometry books. It also contains (at the end) a good account of the development of the subject during the twentieth century. Highly recommended.

A distressing aspect of modern mathematics is the extent to which mathematicians, having ascended to a new level of understanding in the subject, then pull the ladder up after them. Algebraic geometers are more guilty of this than most, and in consequence their subject has a fearsome reputation, even among mathematicians. The present text may not replace any ladders, but perhaps provides you with enough lumber to build one or two of your own.

ANSWERS TO EXERCISES

10.1: Let f be given by the real matrix $\mathbf{A} = (a_{ij})$, and suppose $f(I) = I$. Then

$$\begin{pmatrix} a_{11} & a_{12} & a_{13} \\ a_{21} & a_{22} & a_{23} \\ a_{31} & a_{32} & a_{33} \end{pmatrix} \begin{pmatrix} 1 \\ i \\ 0 \end{pmatrix} = \begin{pmatrix} \lambda \\ \lambda i \\ 0 \end{pmatrix}$$

for some $\lambda = ke^{i\theta}$. From this, $a_{11} + a_{12}i = k\cos\theta + ik\sin\theta$, and $a_{21} + a_{22}i = ik\cos\theta - k\sin\theta$, and $a_{31} + a_{32}i = 0$. So

$$\mathbf{A} = \begin{pmatrix} k\cos\theta & k\sin\theta & a_{13} \\ -k\sin\theta & k\cos\theta & a_{23} \\ 0 & 0 & a_{33} \end{pmatrix},$$

which is the matrix of a direct similarity. (OK?) If $f(I) = J$ the calculation is similar (sorry!) and yields an opposite similarity. The converse is easy.

10.2: Let the four lines AB, AC, AI, AJ be $y = m_k x + c_k$, for $k = 1, 2, 3, 4$. So $m_3 = i$ and $m_4 = -i$; let $m_1 = \tan\theta$ and $m_2 = \tan\varphi$. By Exercise 9.32 the required cross-ratio is

$$\frac{(i - \tan\theta)(-i - \tan\varphi)}{(i - \tan\varphi)(-i - \tan\theta)} = \frac{(i\cos\theta - \sin\theta)(-i\cos\varphi - \sin\varphi)}{(i\cos\varphi - \sin\varphi)(-i\cos\theta - \sin\theta)} = \frac{(ie^{i\theta})(-ie^{-i\varphi})}{(ie^{i\varphi})(-ie^{-i\theta})} = e^{2i(\theta - \varphi)}.$$

But of course $\theta - \varphi = \angle BAC$. For the last part, if $BA \perp AC$ then $\theta - \varphi = \pm\pi/2$ so that $e^{2i(\theta - \varphi)} = e^{\pm i\pi} = -1$, and conversely.

10.3: In $\mathbb{R}^3$, putting $z = \theta$ gives $y = z^2 = \theta^2$ and $x = yz = \theta^2\theta = \theta^3$, so we have all the common points in $\mathbb{R}^3$ already, and any further common points must be on the plane at infinity, $w = 0$. Substituting back, this gives $yz = 0$ and $z^2 = 0$, that is, $z = 0$. So the intersection consists of the line $z = 0$, $w = 0$ together with the twisted cubic $(x, y, z, w) = (\theta^3, \theta^2, \theta, 1)$, where $\theta \in \mathbb{R} \cup \{\infty\}$. (Here $\theta = \infty$ gives the point $(1, 0, 0, 0)$ (OK?), which is on $z = 0$, $w = 0$ anyway.)

10.4: No, it also factorizes as $(2kx)(k^{-1}y)$ or $(k^{-1}y)(2kx)$, for any non-zero constant k.

10.5: One way should be obvious: if p_1 and p_2 are homogeneous, so is p. For the converse, write each of p, p_1, p_2 as the sum of its homogeneous parts, that is, bracket together terms of the same degree. On forming the product $p_1 p_2$, notice that the product of the homogeneous parts of highest (lowest) degree of p_1 and p_2 is the homogeneous part of highest (lowest) degree of p. Thus if p_1 or p_2 has more than one homogeneous part, so does p.

10.6: If either Γ is proper and ℓ is a tangent; or if Γ is a line-pair and ℓ passes through the common point of this line-pair; or if Γ is a repeated line.

10.7: Given distinct points X and Y, let u_1, u_3, u_5 be distinct lines through X, not through Y, and let u_2, u_4, u_6 be distinct lines through Y, not through X. Let $\ell = u_1u_2 \cdot u_4u_5$ (the join of the meets u_1u_2 and u_4u_5); let $m = u_2u_3 \cdot u_5u_6$, and let $n = u_3u_4 \cdot u_6u_1$ (similarly). Then ℓ, m, and n are concurrent. If, on the diagram for Pappus' theorem, we put $X = X_5$, $Y = X_2$, $u_1 = X_5X_6$, $u_2 = v$, $u_3 = X_4X_5$, $u_4 = X_1X_2$, $u_5 = u$, and $u_6 = X_2X_3$, then we get $\ell = X_6X_1$, $m = X_3X_4$, and $n = LM$. The dual of Pappus' theorem says that ℓ, m, and n are concurrent; but ℓ and m meet at $X_6X_1 \cdot X_3X_4 = N$, so the dual of Pappus' theorem says that N lies on n, i.e., that L, M, and N are collinear. But this is Pappus' theorem itself; so Pappus' theorem and its dual are the same: it is *self-dual*.

10.8: There are nine points in the figure, and nine lines. Any of the nine points can be chosen as X_1. There are three lines through X_1, and six points on these lines other than X_1, namely L, X_2, N, X_6, X_3, X_5. Any of these six points can be chosen as X_2. Then X_3 has to be on a line through X_1 and on a line through X_2 (but not on X_1X_2), and there are two such points, namely X_3 and X_6. Either can be chosen as X_3. The rest of the labelling is now determined: X_6 is the 'other choice' for X_3, and L, X_5, N, M are the 'other points' on X_1X_2, X_1X_3, X_1X_6, X_2X_3 respectively. So the size of the 'relabelling group' $G \subset S_9$ is $9 \times 6 \times 2 = 108$. It is possible (though hard) to recognize G as isomorphic to a certain group of matrices, namely as a subgroup of $GL_3(\mathbb{Z}_3)$, the invertible 3×3 matrices over the field $\mathbb{Z}_3$ of integers modulo 3. (So $\mathbb{Z}_3 = \{0, 1, -1\}$, with $1 + 1 = -1$, or $1 + 1 + 1 = 0$.) Here is a sketch: work in the projective plane over $\mathbb{Z}_3$, that is, with homogeneous coordinates (x, y, z) with $x, y, z \in \mathbb{Z}_3$. This is a *finite* geometry, that is, it contains only a finite number of points (and lines). In fact, it has exactly 13 points and (dually!) 13 lines, with four points on each line and (dually!) four lines through each point. For example, the 'line at infinity' $z = 0$ contains the points $(1, 0, 0)$, $(0, 1, 0)$, $(1, 1, 0)$, $(1, -1, 0)$, and no others, and the point $(1, 0, 0)$ lies on the lines $y = 0$, $z = 0$, $y + z = 0$, $y - z = 0$, and no others. If we omit these four points and four lines, the remaining $13 - 4 = 9$ points and lines are arranged in a Pappus configuration! So we are looking for the projective transformations in our strange little geometry that fix the point $(1, 0, 0)$ and leave the line $z = 0$ invariant. The transformation given by $\mathbf{A} = (a_{ij}) \in GL_3(\mathbb{Z}_3)$ fixes $(1, 0, 0)$ iff $a_{21} = a_{31} = 0$, and it leaves $z = 0$ invariant iff $a_{31} = a_{32} = 0$. Since we must now have $a_{ii} \neq 0$, all i, we divide through by a_{33} (which does not affect the projective transformation), and finally obtain that G is isomorphic to the subgroup of all matrices

$$\begin{pmatrix} a & b & c \\ 0 & d & e \\ 0 & 0 & 1 \end{pmatrix}$$

in $GL_3(\mathbb{Z}_3)$. As a check, this means $a, d \in \{1, -1\}$ and $b, c, e \in \{0, 1, -1\}$, which is $2^2 \times 3^3 = 108$ choices, again.

10.9: Let Λ_1, Λ_2 be two real conics having just two real common tangents (each having distinct points of contact with Λ_1 and Λ_2). Then as complex conics, Λ_1 and Λ_2 also have two complex common tangents, the equation of one being the conjugate of the equation of the other. The point where these tangents meet is thus self-conjugate, and hence real. We have to construct this point. Solution: choose a point on each of the real tangents to form a point-pair, a reducible conic envelope Λ_3. Through these points, construct the other tangents to the two conics, which are the other common tangents of Λ_1 and Λ_3, and of Λ_2 and Λ_3. Mark where the two new tangents to Λ_1 meet, and where the two new tangents to Λ_2 meet, and join these two points. This line contains the required point. Repeat to obtain a second line, and mark where it meets the first line, which is the required point. (This construction is much longer than the original, because the dual of marking where a line meets a conic—which hardly counts as a construction—is the construction of the tangents from a point to a conic. See Exercise 9.36 for a method for this step of the construction.)

10.10: Let A_{ij} be the meet of ℓ_i and ℓ_j, for $i, j = 1, 2, 3, 4$ and $i < j$. These are the six distinct vertices of our quadrilateral. The circles Σ_1 and Σ_2 meet at A_{34}, but they do not touch at A_{34}, otherwise $\ell_1 \parallel \ell_2$, which is forbidden. (Please check this.) So Σ_1 and Σ_2 meet in a second point, B, with $B \neq A_{34}$. Then $B \neq A_{ij}$ for all other i, j, since a circle cannot contain three collinear points. (OK?) Now define the cubics Γ_i by $\Gamma_i = \Sigma_i \cup \ell_i$, for $1 \leq i \leq 4$. Then $\Gamma_1 \cap \Gamma_2 = (\Sigma_1 \cup \ell_1) \cap (\Sigma_2 \cup \ell_2) = (\Sigma_1 \cap \Sigma_2) \cup (\Sigma_1 \cap \ell_2) \cup (\Sigma_2 \cap \ell_1) \cup (\ell_1 \cap \ell_2) = \{A_{34}, B, I, J\} \cup \{A_{23}, A_{24}\} \cup \{A_{13}, A_{14}\} \cup \{A_{12}\}$, where I and J are the circular points at infinity. So the points A_{ij}, together with I, J, and B, are nine associated points. All the A_{ij} and I, J lie on

Γ_3, and hence so does B. Since B does not lie on ℓ_3 (OK?), it lies on Σ_3. Similarly for Σ_4. (The reader might like also to find an elementary proof of this theorem, just by chasing angles around in $\mathbb{R}^2$.)

10.11: These nodal cubics correspond to $a = 1$ and $a = -1$, respectively. In the first case, the singular tangents $y = \pm x$ are real; this type of node is called a *crunode*. In the second case the singular tangents are $y = \pm ix$ and are imaginary, but meet at the real point $(0, 0)$, which is an isolated point of the real cubic. It is called an *acnode*: the curve has acne! (The distinction is only made for real curves; as complex curves, they are projectively equivalent and there is no distinction.)

10.12: (i) Node at $(1, 0)$, singular tangents $y = \pm(x - 1)$. (ii) Cusp at infinity, at $(0, 1, 0)$. Singular tangent $z = 0$. (iii) Non-singular. (iv) Cusp at infinity, at $(0, 1, 0)$. Singular tangent $x = 0$ (the asymptote). (v) Node at infinity, at $(0, 1, 0)$. Singular tangents $x = \pm z$, or (dehomogenizing) $x = \pm 1$, the asymptotes.

10.13: (i) Homogenizing, $x^3 - y^2z = 0$. The Hessian is $xy^2 = 0$, and solving together gives $(0, 0, 1)$, a cusp, and $(0, 1, 0)$, a flex. (ii) Homogenizing, $x^3 - x^2z - y^2z = 0$. The Hessian is $3xy^2 - x^2z - y^2z = 0$, and the two curves meet at $(0, 0, 1)$, a node, and $(0, 1, 0)$, $(4\sqrt{3}, \pm4, 3\sqrt{3})$, three flexes. (Notice that the three flexes are collinear.) (iii) Homogenizing, $x^3 + y^3 - xyz = 0$. The Hessian is $3x^3 + 3y^3 + xyz = 0$, and the two curves meet at $(0, 0, 1)$, a node, and $(1, -1, 0)$, $(e^{2\pi i/3}, -1, 0)$, $(e^{4\pi i/3}, -1, 0)$, three flexes. All the flexes are at infinity, and two are imaginary; but again they are collinear.

10.14: $B = A \circ (A \circ B) = A \circ (A \circ C) = C$.

10.15: $A' * B' = O' \circ ((L \circ A) \circ (L \circ B)) = O' \circ ((L \circ L) \circ (A \circ B)) = (L \circ O) \circ ((L \circ L) \circ (A \circ B)) = (L \circ (L \circ L)) \circ (O \circ (A \circ B)) = L \circ (A + B) = (A + B)'$.

10.16: We have $A + B + C = O \circ (A \circ (O \circ (B \circ C)))$. If A, B, C are collinear, then $B \circ C = A$, so $A + B + C = O \circ (A \circ (O \circ A)) = O \circ O = K$. If $A + B + C = K$, then $O \circ (A \circ (O \circ (B \circ C))) = O \circ O$, so $A \circ (O \circ (B \circ C)) = O$, whence $O \circ (B \circ C) = O \circ A$, and $B \circ C = A$.

10.17: Use cylindrical polar coordinates (ρ, φ, z), where $x = \rho \cos \varphi$ and $y = \rho \sin \varphi$. Then the plane $\varphi =$ constant meets the torus in the two circles $(\rho \pm a)^2 + z^2 - b^2 = 0$, or in $((\rho - a)^2 + z^2 - b^2)((\rho + a)^2 + z^2 - b^2) = 0$. This gives $(\rho^2 + z^2 + a^2 - b^2 - 2a\rho)(\rho^2 + z^2 + a^2 - b^2 + 2a\rho) = 0$, or $(\rho^2 + z^2 + a^2 - b^2) - 4a^2\rho^2 = 0$. Substitute $\rho^2 = x^2 + y^2$.

10.18: We have, on dividing through by s^2,

$$\left(\frac{x^2 + y^2 + z^2}{s} + r\right)^2 - \left(\frac{r}{s} + 1\right)^2 (x^2 + y^2) = 0,$$

and on letting $s \to \infty$, this becomes $r^2 - (x^2 + y^2) = 0$, a cylinder.

10.19: We have $((x + a)^2 + y^2 + z^2 + a^2 - b^2)^2 - 4a^2((x + a)^2 + y^2) = 0$, or $((x + a)^2 + y^2 + z^2 + a^2 - b^2 - 2a(x + a))((x + a)^2 + y^2 + z^2 + a^2 - b^2 + 2a(x + a)) - 4a^2y^2 = 0$, or $(x^2 + y^2 + z^2 - b^2)((x + 2a)^2 + y^2 + z^2 - b^2) - 4a^2y^2 = 0$, or $(x^2 + z^2 - b^2)((x + 2a)^2 + y^2 + z^2 - b^2) - ((x + 2a)^2 + y^2 + z^2 - b^2 - 4a^2)y^2 = 0$, or $(x^2 + z^2 - b^2)((x + 2a)^2 + y^2 + z^2 - b^2) - (x^2 + 4ax + y^2 + z^2 - b^2)y^2 = 0$. Dividing through by $4a^2$ and then letting $a \to \infty$ gives $x^2 + z^2 - b^2 = 0$, another cylinder.

10.20: Take the cylinder $x^2 + y^2 - a^2 = 0$, and move the origin to $(a, 0, 0)$ to get $(x + a)^2 + y^2 - a^2 = 0$, or $x^2 + 2ax + y^2 = 0$. Now divide through by $2a$ and let $a \to \infty$, giving $x = 0$, a plane.

Bibliography

[1] Allenby, R. B. J. T. *Rings, Fields and Groups*. Arnold, 1983.

[2] Birkhoff, G. D. A set of postulates for plane geometry (based on scale and protractor). *Annals of Mathematics*, Vol. 33, 1932.

[3] Birkhoff, G. D. and MacLane, S. *A Survey of Modern Algebra (revised edn)*. Macmillan, 1953.

[4] Birkill, J. C. *A First Course in Mathematical Analysis*. Cambridge, 1962.

[5] Budden, F. J. and Wormell, C. P. *Mathematics through Geometry*. Pergamon, 1964.

[6] Cohn, P. M. *Algebra, Volume 1* (2nd edn). Wiley, 1982.

[7] Cohn, P. M. *Algebra, Volume 2* (2nd edn). Wiley, 1982.

[8] Cohn, P. M. *Algebra, Volume 3* (2nd edn). Wiley, 1982.

[9] Coxeter, H. S. M. *Introduction to Geometry*. Wiley, 1961.

[10] Coxeter, H. S. M. *Regular Polytopes*. Dover, 1973.

[11] Cundy, H. M. and Rollett, A. P. *Mathematical Models* (3rd edn). Tarquin, 1981.

[12] Durell, C. V. *Projective Geometry*. Macmillan, 1926.

[13] Du Val, P. *Elliptic Functions and Elliptic Curves*. Cambridge, 1973.

[14] Evelyn, C. J. A., Money-Coutts, G. B. and Tyrrell, J. A. *The Seven Circles Theorem and Other New Theorems*. Stacey, 1974.

[15] Fulton, W. *Algebraic Curves*. Benjamin, 1969.

[16] Gauss, C. F. *Disquisitiones Arithmeticae* (translated by A. A. Clarke). Yale, 1966.

[17] Heath, T. L. *A Manual of Greek Mathematics*. Dover, 1963.

[18] Hilbert, D. *The Foundations of Geometry*. Open Court, 1902.

[19] Ledermann, W. *Introduction to Group Theory*. Oliver and Boyd, 1973.

[20] Lockwood, E. H. *A Book of Curves*. Cambridge, 1961.

[21] Milne, A. A. *The House at Pooh Corner*. Methuen, 1928.

[22] Newman, J. R. *The World of Mathematics*. Allen & Unwin, 1960.

[23] Pedoe, D. *A Course of Geometry for Schools and Colleges*. Cambridge, 1970.

[24] Porteous, I. R. *Toplogical Geometry*. Van Nostrand, 1969.

[25] Reid, M. *Undergraduate Algebraic Geometry*. Cambridge, 1988.

[26] Roe, J. *Elementary Geometry*. Oxford, 1993.

[27] Russell, B. *Mathematics and the Metaphysicians*. Reprinted in [22].

[28] Seidenberg, A. *Elements of the Theory of Algebraic Curves*. Addison-Wesley, 1968.

[29] Semple, J. G. and Kneebone, G. T. *Algebraic Projective Geometry*. Oxford, 1952.
[30] Stewart, I. N. *Galois Theory* (2nd edn). Chapman & Hall, 1989.
[31] Smith, D. E. and Latham, M. L., translators. *The Geometry of René Descartes*. Dover, 1954. (Part of this also appears in [22].)
[32] Walker, R. J. *Algebraic Curves*. Princeton, 1950.

Notation

Index

AAA, 66
abelian group, 85
acnode, 299
acute angle, 20
acute-angled triangle, 20
additive group, 85
adjacent angles, 17
affine
   equivalence, 206
   geometry, 204
   regular polygon, 207
   transformation, 172, 205
algebraic geometry, 271
allied angles, 17
alternate angles, 17
alternate segment theorem, 184
alternating group, 131
altitude, 21, 54
analytic geometry, 6
anchor-ring, *see* torus
angle, 37, 39
   acute, 20
   auxiliary, 169
   bisector, 15, 49, 58
   at centre, 181
   circular measure, 39
   external, 17
   obtuse, 20
   radians, 39
   reflex, 20, 37
   right, 20
   in semicircle, 1, 18, 181
   between tangent and chord, 185
   trisecting, 18–20
angles, 229
   adjacent, 17
   allied, 17
   alternate, 17
   complementary, 17
   corresponding, 17

   interior, 17
   opposite, 182
   in same segment, 180
   supplementary, 17
   vertically opposite, 17
angular region, 52
anti-prism, 141
apex
   of cone, 166
   of pyramid, 140
Apollonius' circle, 200, 224
applied mathematicians, 192
applied mathematics, 68, 172
Archimedes, 62
area
   of ellipse, 212
   signed, 104
   spherical excess formula, 261
   of triangle, 55, 62
areal coordinates, 69
ASA, 56
associative, 85
asymptote, 170
auxiliary
   angle, 169
   circle, 169
axiom, 2
   parallel, 3, 222
axis
   major, 168
   minor, 169
   of an opposite isometry, 97
   order r (r-axis), 146
   principal, 168
   radical, 187, 235, 283
   of a reflection, 33
   semi-, 170
   semi-major, 169
   semi-minor, 169
   x-, y-, 6, 25